An Introduction to
the Mathematics of Biology:
with Computer Algebra Models

Edward K. Yeargers
School of Biology
Georgia Institute of Technology

Ronald W. Shonkwiler
School of Mathematics
Georgia Institute of Technology

James V. Herod
School of Mathematics
Georgia Institute of Technology

BIRKHÄUSER
BOSTON • BASEL • BERLIN

Ronald W. Shonkwiler
James V. Herod
School of Mathematics
Georgia Institute of Technology
Atlanta, GA 30332-0160

Edward K. Yeargers
School of Biology
Georgia Institute of Technology
Atlanta, GA 30332-0160

Library of Congress Cataloging In-Publication Data

Yeargers, Edward K.
    An introduction to the mathematics of biology: with computer
algebra models / Edward K. Yeargers, Ronald W. Shonkwiler, James
V. Herod.
        p.   cm.
    Includes bibliographical references and index.
    ISBN 0-8176-3809-1 (alk. paper). -- ISBN 3-7643-3809-1 (alk.
paper)
      1. Biomathematics.   I. Shonkwiler, Ronald K., 1942- .
II. Herod, J. V., 1937- .   III. Title.
QH323.5.Y435   1996
574'.01'51--dc20   96-1385
                                          CIP

Printed on acid-free paper
© Birkhäuser Boston 1996

Birkhäuser

ISBN 0-8176-3809-1
ISBN 3-7643-3809-1

Designed and typeset by Merry Obrecht Sawdey, ShadeTree Designs, Minneapolis, MN.
Cover design by Joseph Sherman, Dutton and Sherman Design, Hamden, CT, USA.
Printed and bound by The Maple Press Company, York, PA, USA.
Printed in the U.S.A.

9 8 7 6 5 4 3 2 1

Contents

Preface

Biology is a source of fascination for most scientists, whether their training is in the life sciences or not. In particular, there is a special satisfaction in discovering an understanding of biology in the context of another science like mathematics. Fortunately there are plenty of interesting (and fun) problems in biology, and virtually all scientific disciplines have become the richer for it. For example, two major journals, *Mathematical Biosciences* and *Journal of Mathematical Biology*, have tripled in size since their inceptions 20–25 years ago.

The various sciences have a great deal to give to one another, but there are still too many fences separating them. In writing this book we have adopted the philosophy that mathematical biology is not merely the intrusion of one science into another, but has a unity of its own, in which both the biology and the mathematics should be equal and complete, and should flow smoothly into and out of one another. We have taught mathematical biology with this philosophy in mind and have seen profound changes in the outlooks of our science and engineering students: The attitude of "Oh no, another pendulum on a spring problem!," or "Yet one more LCD circuit!" completely disappeared in the face of applications of mathematics in biology. There is a timeliness in calculating a protocol for administering a drug. Likewise, the significance of bones being "sinks" for lead accumulation while bone meal is being sold as a dietary calcium supplement adds new meaning to mathematics as a *life science*. The dynamics of a compartmentalized system are classical; applications to biology can be novel. Exponential and logistic population growths are standard studies; the delay in the increase of AIDS cases behind the increase in the HIV-positive population is provocative.

With these ideas in mind we decided that our book would have to possess several important features. For example, it would have to be *understandable to students of either biology or mathematics*, the latter referring to any science students who normally take more than a year of calculus, i.e., majors in mathematics, physics, chemistry, and engineering.

No prior study of biology would be necessary. Mathematics students rarely take biology as part of their degree programs, but our experience has been that very rapid progress is possible once a foundation has been laid. Thus, the *coverage of biology would be extensive,* considerably more than actually needed to put the mathematics of the book into context. This would permit mathematics students to have much greater latitude in subsequent studies, especially in the "what-if" applications of a computer algebra system. It would also help to satisfy the intense intellectual interest that mathematics students have in the life sciences, as has been manifested in our classes.

One year's study of calculus would be required. This would make the material accessible to most biology majors and is the least mathematics preparation we felt to be practical for the study of the subject matter. We use more advanced material, such as partial derivatives, but we explain them fully. Our biology students have had no problems with this approach.

Part of every section would have applications of a computer algebra system. This hands-on approach provides a rich source of information through the use of "what-if" input and thus allows students to grasp important biological and mathematical concepts in a way that is not possible otherwise. To facilitate this we have posted various *Maple* programs to the World Wide Web in care of the Birkhäuser web page (http://www.birkhauser.com/books/isbn/0-8176-3809-1). In particular, changes in syntax as a result of new releases of Maple will be posted on these Web pages. On the other hand, we realize that computer algebra systems may not be available to everyone, and each lesson is complete without their use.

Most importantly, *the biology and mathematics would be integrated.* Each chapter deals with a major topic, such as lead poisoning, and we begin by presenting a thorough foundation of fundamental biology. This leads into a discussion of a related mathematical concept and its elucidation with the computer algebra system. Thus, for each major topic, the biology and the mathematics are combined into an integrated whole.

To summarize, we hope that mathematics students will look at this book as a way to learn enough biology to make good models and that biology students will see it as an opportunity to understand the dynamics of a biological system. For both these students and their engineering classmates, perhaps this book can present a new perspective for a life's work.

Acknowledgements

We are grateful to our wives for their patience while we were spending so much time on writing this book. Ms. Annette Rohrs did a wonderful job of typing the manuscript while we pulled her three different ways. In particular, we are indebted to the late Dr. Robert Pierrotti, head of the Georgia Tech Center for Education Integrating Science, Mathematics and Computing. He brought us together and supported us in several ways. The birth of this book would have been far more complicated without his help.

Chapter 1

Biology, Mathematics, and a Mathematical Biology Laboratory

Section 1.1

The Natural Linkage Between Mathematics and Biology

Mathematics and biology have a synergistic relationship. Biology produces interesting problems, mathematics provides models to understand them, and biology returns to test the mathematical models. Recent advances in computer algebra systems have facilitated the manipulation of complicated mathematical systems. This has made it possible for scientists to focus on understanding mathematical biology, rather than on the formalities of obtaining solutions to equations.

What is the function of mathematical biology?

Our answer to this question, and the guiding philosophy of this book, is simple: The function of mathematical biology is to exploit the natural relationship between biology and mathematics. The linkage between the two sciences is embodied in the reciprocal contributions that they make to each other: Biology generates complex problems, and mathematics can provide ways to understand them. In turn, mathematical models suggest new lines of inquiry that can only be tested on real biological systems.

We believe that an understanding of the relationship between two subjects must be preceded by a thorough understanding of the subjects themselves. Indeed, the excitement of mathematical biology begins with the discovery of an interesting and uniquely biological problem. The excitement grows when we realize the mathematical tools at our disposal can profitably be applied to the problem. The interplay between mathematical tools and biological problems constitutes mathematical biology.

The time is right for integrating mathematics and biology.

Biology is a rapidly expanding science; research advances in the life sciences leave virtually no aspects of our public and private lives untouched. Newspapers bombard us with information about *in vitro* fertilization, industrial pollution, radiation effects, AIDS, genetic manipulation, and forensics.

Quite apart from the news pouring onto us from the outside world, we have an innate interest in biology. We have a natural curiosity about ourselves. Every day we ask ourselves a non-stop series of questions: What happens to our bodies as we get older? Where does our food go? How do poisons work? Why do I look like my mother? What does it mean to "think"? Why are HIV infections spreading so rapidly in certain population groups?

Professional biologists have traditionally made their living by trying to answer these kinds of questions. But scientists with other kinds of training have also seen ways that they could enter the fray. As a result, chemists, physicists, engineers, and mathematicians have all made important contributions to the life sciences. These contributions often have been of a sort that required specialized training or a novel insight that only specialized training could generate.

In this book we present some mathematical approaches to understanding biological systems. This approach has the hazard that an in-depth analysis could quickly lead to unmanageably complex numerical and symbolic calculations. However, technical advances in the computer hardware and software industries have put powerful computational tools into the hands of anyone who is interested. Computer algebra systems allow scientists to bypass some of the details of solving mathematical problems. This then allows them to spend more time on the interpretation of biological phenomena, as revealed by the mathematical analysis.[1]

Section 1.2

The Use of Models in Biology

Scientists must represent real systems by models. Real systems are too complicated, and, besides, observation may change the real system. A good model should be simple and should exhibit the behaviors of the real system that interest us. Further, it should suggest experimental tests of itself that are so revealing that we must eventually discard the model in favor of a better one. We therefore measure scientific progress by the production of better and better models, not by whether we find some absolute truth.

---

[1]References 1–4 at the end of this chapter are recent articles that describe the importance of mathematical biology.

A model is a representation of a real system.

The driving force behind the creation of models is this admission: Truth is elusive, but we can gradually approximate it by creating better and better representations.

There are at least two reasons why the truth is so elusive in real systems. The first reason is obvious: The universe is extremely complicated. People have tried unsuccessfully to understand it for millennia, running up countless blind alleys and only occasionally finding enlightenment. Claims of great success abound, usually followed by their demise. Physicists in the late nineteenth century advised their students that Maxwell's equations had summed up everything important about physics, and that further research was useless. Einstein then developed the theory of general relativity, which contained Maxwell's equations as a mere subcategory. The unified field theory ("The Theory of Everything") will contain Einstein's theory as a subcategory. Where will it end?

The second reason for the elusivity of the truth is a bit more complicated: It is that we tend to change reality when we examine any system too closely. This concept, which originates in quantum mechanics, suggests that the disturbances that inevitably accompany all observations will change the thing being observed. Thus, "truth" will be changed by the very act of looking for it.[2] At the energy scale of atoms and molecules the disturbances induced by the observer are especially severe. This has the effect of rendering it impossible to observe a single such particle without completely changing some of the particle's fundamental properties. There are macroscopic analogs to this effect. For example, what is the "true" color of the paper in this book? The answer depends on the color of the light used to illuminate the paper, white light being merely a convenience; most other colors would also do. Thus, you could be said to have chosen the color of the paper by your choice of observation method.

Do these considerations make a search for ultimate explanations hopeless? The answer is "No, because what is really important is the progress of the search, rather than some ultimate explanation that is probably unattainable anyway."

Science is a rational, continuing search for better models.

Once we accept the facts that a perfect understanding of very complex systems is out of reach and that the notion of "ultimate explanations" is merely a dream, we will have freed ourselves to make scientific progress. We are then able to take a reductionist approach, fragmenting big systems into small ones that are individually amenable to understanding. When enough small parts are understood, we can take a holistic approach, trying to understand the relationships among the parts, thus reassembling the entire system.

In this book we reduce complicated biological systems to relatively simple mathematical models, usually of one to several equations. We then solve the

---

[2]This situation is demonstrated by the following exchange: Question: How would you decide which of two gemstones is a real ruby and which is a cheap imitation? Answer: Tap each sharply with a hammer. The one that shatters used to be the real ruby.

equations for variables of interest and ask if the functional dependencies of those variables predict salient features of the real system.

There are several things we expect from a good model of a real system.

a.   The model must exhibit properties that are similar to those of the real system, and those properties must be the ones in which we are interested.[3] A six-inch replica of a 747 airliner may have the exact fluid dynamical properties of the real plane, but would be useless in determining the comfort of the seats of a real 747.

b.   It must self-destruct. A good model must suggest tests of itself and predict their outcomes. Eventually a good model will suggest a very clever experiment whose outcome will not be what the model predicted. The model must then be discarded in favor of a new one.

The search for better and better models thus involves the continuous testing and replacement of existing models. This search must have a rational foundation, being based on phenomena that can be directly observed. A model that cannot be tested by the direct collection of data, and that therefore must be accepted on the basis of faith, has no place in science.

Many kinds of models are important in understanding biology phenonema.

Models are especially useful in biology. The most immediate reason is that living systems are much too complicated to be truly understood as whole entities. Thus, to design a useful model we must strip away irrelevant, confounding behaviors, leaving only those that directly interest us. We must walk a fine line here: In our zeal to simplify, we may strip away important features of the living system, and the other extreme, a too-complicated model, is intractable and useless.

Models in biology span a wide spectrum. The next box lists some that are commonly used.

Why is there so much biological information in this book?

It is possible to write a mathematical biology book that contains only a page or two of biological information at the beginning of each chapter. We see that format as the source of two problems: First, it is intellectually limiting. A student cannot apply the powerful tools of mathematics to biological problems he or she does not understand. This limitation can be removed by a thorough discussion of the underlying biological systems, which can suggest further applications of mathematics. Thus, a strong grounding in biology helps students to move further into mathematical biology.

---

[3]One characteristic of the real system that we definitely do *not* want is its response to the observation process, as described earlier. In keeping with the concept of a model as an idealization, we want the model to represent the real system in a "native state," divorced from the observer.

MODEL	WHAT THE MODEL REPRESENTS
$aa \times Aa$	Gene behavior in a genetic cross
$\dfrac{dA}{dt} = -kA$	Rate of elimination of a drug from the blood
$\boxed{\text{R}} \rightarrow \boxed{\text{C}} \rightarrow \boxed{\text{E}}$	Reflex arc involving a stimulus $\underline{\text{R}}$eceptor, the $\underline{\text{C}}$entral nervous system and an $\underline{\text{E}}$ffector muscle
a camera	The eye of a vertebrate or of an octopus

Second, giving short shrift to biology reinforces the misconception that each of the various sciences sits in a vacuum. In fact, it has been our experience that many students of mathematics, physics and engineering have a genuine interest in biology, but little opportunity to study it. Taking our biological discussions well beyond the barest facts can help these students to understand the richness of biology, and thereby to encourage interdisciplinary thinking.

Section 1.3

What Can Be Derived from a Model and How Is It Analyzed?

A model is more than the sum of its parts. Its success lies in its ability to discover new results, results that transcend the individual facts built into it. One result of a model can be the observation that seemingly dissimilar processes are in fact related. In an abstract form, the mathematical equations of the process might be identical to those of other phenomena. In this case the two disciplines reinforce each other: A conclusion difficult to see in one might be an easy consequence in the other.

To analyze the mathematical equations that arise in this book, we draw on the fundamentals of matrix calculations, counting principles for permutations and combinations, the calculus, and fundamentals of differential equations. However, we will make extensive use of the power of symbolic computational software—a computer algebra system. All of the displayed calculations and graphs in this text are done using *Maple*. The *Maple* syntax immediately accompanies the calculation or graph and is indicated by a special font.

Deriving consequences: the other side of modeling

After a model has been formulated and the mathematical problems have been defined, the problems must be solved. In symbolic form, a problem takes on a life of its own, no longer necessarily tied to its physical origins. In symbolic form, the system may even apply to other, totally unexpected, phenomena. What do the seven bridges at Königsberg have to do with discoveries about DNA? The mathematician Euler formed an abstract model of the bridges and their adjoining land masses and founded the principles of Eulerian graphs on this model. Today, Eulerian graphs are used, among other ways, to investigate the ancestry of living things by calculating the probability of matches of DNA base pair sequences (see Kandel [5]). The differential equations describing spring-mass systems and engineering vibrations are identical to those governing electrical circuits with capacitors, inductors, and resistors. And again these very same equations pertain to the interplay between glucose and insulin in humans. The abstract and symbolic treatment of these systems through mathematics allows the transfer of intuition between them. Through mathematics, discoveries in any one of these areas can lead to a breakthrough in the others. But mathematics and applications are mutually reinforcing: the abstraction can uncover truths about the application, suggest questions to ask and experiments to try; the application can foster mathematical intuition and form the basis of the results from which mathematical theorems are distilled.

In symbolic form, a biological problem lends itself to powerful mathematical processing techniques, such as differentiation or integration, and is governed by mathematical assertions known as theorems. Theorems furnish the conclusions that may be drawn about a model so long as their hypotheses are fulfilled. Assumptions built into a model are there to allow its equations to be posed and its conclusions to be mathematically sound. The validity of a model is closely associated with its assumptions, but experimentation is the final arbiter of its worth. The assumption underlying the exponential growth model, namely $\frac{dy}{dt} = ky$ (see Section 2.4 and Chapter 3), is unlikely to be fulfilled precisely in any case, yet exponential growth is widely observed for biological populations. However, exponential growth ultimately predicts unlimited population size, which never materializes precisely due to a breakdown in the modeling assumption. A model is robust if it is widely applicable. In every case, the assumptions of a model *must* be spelled out and thoroughly understood. The validity of a model's conclusions must be experimentally confirmed. Limits of applicability, robustness, and regions of failure need to be determined by carefully designed experiments.

Some biological systems involve only a small number of entities or are greatly influenced by a few of them, maybe even one. Consider the possible DNA sequences 100 base pairs long. Among the possibilities, one or two might be critical to life. (It is known that tRNA molecules can have as few as 73 nucleotide residues (Lehninger [6]).) Or consider the survival prospects of a skein of Canadian geese blown off migratory course, toward the Hawaiian islands. Their survival analysis must keep track of detailed events for each goose and possibly even details of their individual genetic makeups, for the loss of a single goose or the birth of defective goslings could spell extinction for the small colony. (The Ney-Ney, indigenous to Hawaii, is thought to be related to Canadian geese.) This is the mathematics of discrete systems, i.e., the mathematics of a finite number of states. The main tools we will need here are knowledge of matrices and their arithmetic, counting principles for permutations and combinations, and some basics of probability calculations.

Other biological systems or processes involve thousands, even millions, of entities, and the fate of a few of them has little influence on the entire system. Examples are the diffusion process of oxygen molecules or the reproduction of a bacterial colony. In these systems, individual analysis gives way to group averages. An average survival rate of 25% among goslings of a large gaggle of Canadian geese still ensures exponential growth of the gaggle in the absence of other effects; but this survival probability sustained by exactly four offspring of an isolated gaggle might not result in exponential growth at all but, rather, in total loss instead. When there are large numbers involved, the mathematics of the continuum may be brought to bear, principally calculus and differential equations. This greatly simplifies the analysis. The techniques are powerful and mature and a great many are available.

Computer algebra systems make the mathematics accessible.

Students in biology and allied fields such as immunology, epidemiology, or pharmacology need to know how to quantify concepts and to make models, yet these students typically have only one year of undergraduate study in mathematics. This one year may be very general and not involve any examples from biology. When the need arises, they are likely to accept the models and results of others, perhaps without deep understanding.

On the other side of campus, students in mathematics read in the popular technical press of biological phenomena, and wish they could see how to use their flair for mathematics to get them into biology. The examples they typically see in mathematics classes have their roots in physics. Applications of mathematics to biology seem far away.

How can this dilemma be resolved? Should the biology students be asked to take a second year of mathematics in order to be ready to use the power of differential equations for modeling? And what of discrete models, probabilistic models, or statistics? One might envision some collection of biology students having almost a minor in mathematics in order to be prepared to read, understand, and create biological models.

Must the mathematics students take a course in botany, and then zoology, before they can make a model for the level to which the small vertebra population must be immunized in a geographic region in order to reduce the size of the population of ticks carrying Lyme Disease? Such a model is suggested by Kantor [7].

There is an alternative. Computer algebra systems create a new paradigm for designing, analyzing, and drawing conclusions from models in science and engineering. The technology in the computer algebra systems allows concepts to be paramount while computations and details become less important. With such a symbolic computational engine it is possible to read about models that are being actively explored in the current literature, and to analyze these new models on a home computer, a lap-top, or an office workstation.

The theorems from which our conclusions are derived often result from carefully tracking evolving system behavior over many iterations in discrete systems or infinite time in continuous ones. Where possible, the mathematical equations are solved and the solutions exhibited. Predictions of the model are made under a range of starting conditions and possibly unusual parameter regimes. These are the bases of "what if?" experiments. For example, given a satisfactory model for a fishery, *what if* one imposes various levels of harvesting? To answer this and related questions, the computer algebra system can carry out the technical computations: calculate roots, symbolically differentiate, integrate, and solve differential equations, perform matrix arithmetic, track system evolution, and graphically display results.

In this book we will use *Maple*. It should be emphasized that any *Maple* capability we use here is likely to be available on other symbolic processing systems as well. *Maple* code will be indented, set in smaller type and led off with the *Maple* prompt. *Maple* code will usually appear immediately preceeding a graphical figure or displayed results. That code (possibly along with lines from a preceeding display for lengthy developments) produces the figure or the display.

References and Suggested Further Reading

1. **The future of mathematical biology:** "Mathematics and biology: The interface, challenges and opportunities," A National Science Foundation workshop held at Lawrence Berkeley Lab, Department of Energy, 1992.
2. **The importance of mathematics in the life sciences:** Louis J. Gross, "Quantitative training for life-science students," *Bioscience*, vol. 44, No. 2, February, p. 59, 1994.
3. **Modeling in biology:** W. Daniel Hillis, "Why physicists like models and why biologists should," *Current Biology*, vol. 3, no. 2, pp. 79–81, 1993.
4. **Applications of mathematical biology:** Frank Hoppensteadt, "Getting Started in Mathematical Biology," *Notices of the AMS*, vol. 42, no. 9, p. 969, 1995.
5. **DNA sequences:** D. Kandel, Y. Matias, R. Unger, P. Winkler, "Shuffling biological sequences," AT&T Bell Lab preprint, 600 Mountain Ave., Murray Hill, NJ 07974, 1995.
6. **Biochemistry of nucleic acids:** Albert Lehninger, *Biochemistry*, p. 935, Worth Publishers, New York, 1975.
7. **Conquering Lyme disease:** Fred S. Kantor, "Disarming Lyme disease," *Scientific American* **271**, pp. 34–39, September 1994.

Chapter 2

Some Mathematical Tools

Introduction to this chapter

This book is about biological modeling—the construction of mathematical abstractions intended to characterize biological phenomena and the derivation of predictions from these abstractions under real or hypothesized conditions. A model must capture the essence of an event or process but at the same time not be so complicated that it is intractable or dilutes the event's most important features. In this regard, the field of differential equations is the most widely invoked branch of mathematics across the broad spectrum of biological modeling. Future values of the variables that describe a process depend on their rates of growth or decay. These in turn depend on present, or past, values of these same variables through simple linear or power relationships. These are the ingredients of a differential equation. We discuss linear and power laws between variables and their derivatives in Section 2.1 and differential equations in Section 2.4.

Once formulated, a model contains parameters which must be specialized to the particular instance of the process being modeled. This requires gathering and treating experimental data. It requires determining values of the parameters of a model so as to agree with, or fit, the data. The universal technique for this is the method of least squares, the subject of Sections 2.2 and 2.3. Even though experimental data is subject to small random variations, or *noise*, and imprecision, least squares is designed to deal with this problem.

Describing noisy data and other manifestations of variation is the province of statistics. Distributions of values can be graphically portrayed as histograms or distilled to a single number, the average or mean. The most widely occurring distribution in the natural world is the normal or Gaussian distribution. These topics are taken up in Section 2.5.

Finally, to a greater extent in biological phenomena than in other fields of science and engineering, random processes play a significant role in shaping the course of events. This is true at all scales from diffusion at the atomic level, to random combinations of genes, to the behavior of whole organisms. Being in the wrong place at the wrong time can mean being a victim (or finding a meal). In Section 2.6 we discuss the basics of probabilities.

Fortunately, while an understanding of these mathematical tools is required for this book, deep knowledge of mathematical techniques is not. This is a consequence of the fruition of symbolic mathematical software such as *Maple*. We will use the power of this software to execute the calculations, know the special functions, simplify the algebra, solve the differential equations and generally perform the technical work. Above all, *Maple* will display the results. Therefore, the curious are free to let their imaginations roam and focus on perfecting and exercising the models themselves.

Section 2.1

Linear Dependence

The simplest, non-constant, relationship between two variables is a linear one. The simplest linear relationship is one of proportionality: if one of the variables doubles or triples or halves in value the other does likewise. Proportionality between variables x and y is expressed as $y = kx$ for some constant k. Proportionality can apply to derivatives of variables as well as to variables themselves since they are just rates of change. Historically, one of the major impacts of calculus is the improved ability to model by the use of derivatives in just this way.

Relationships among variables can be graphically visualized.

When studying almost any phenomenon, among the first observations to be made about it are its changing attributes. A tropical storm gains in wind speed as it develops; the intensity of sound decreases with distance from its source; living things increase in weight in their early period of life. The measurable quantities associated with a given phenomenon are referred to as *constants, variables,* or *parameters.* Constants are unchanging quantities such as the mathematical constant $\pi = 3.14159\ldots$ or the physical constant named after Boltzmann: $k = 1.38 \times 10^{-16}$ ergs per degree. Variables are quantitative attributes of a phenomenon that can change in value, like the wind speed of a tropical storm, or the intensity of sound, or the weight of an organism.

Parameters are quantities that are constant for a particular instance of a phenomenon but can be different in another instance. For example, the strength of hair fibers is greater for thicker fibers, and the same holds for spider web filaments, but the latter has a much higher strength per unit cross-section.[1] Strength per unit cross-section is a property of material that tends to be constant for a given type of material but varies over different materials and therefore is a parameter in modeling filament strength.

---

[1] The strength of a material per unit cross-section is known as *Young's modulus.*

Often two variables of a phenomenon are *linearly related,* that is, a graphical representation of their relationship is a straight line. Temperatures as measured on the Fahrenheit scale, F, and on the Celsius scale, C, are related this way; see Figure 2.1.1. Knowing that the temperatures $C = 0$ and $C = 100$ correspond to $F = 32$ and $F = 212$ respectively allows one to derive their linear relationship, namely

$$F = \frac{9}{5}C + 32. \tag{2.1.1}$$

In this, both C and F have *power* or *degree* one, that is, their exponent is 1. (Being understood, the 1 is not explicitly written.) When two variables are algebraically related and all terms in the equation are of degree one (or constant), then the graph of the equation will be a straight line. The multiplier, or *coefficient,* $9/5$ of C in equation (2.1.1) is the *slope* of the straight line or the *constant of proportionality* between the variables. The constant term 32 in the equation is the *intercept* of the straight line or *translational term* of the equation. These parameters are shown graphically in Figure 2.1.1.

We can isolate the constant of proportionality by appropriate translation. Absolute zero on the Celsius scale is $-273.15C$, which is usually expressed in de-

> plot([C,9/5*C+32,C=0..100],-10..100,-30..220,tickmarks=[5,0]):

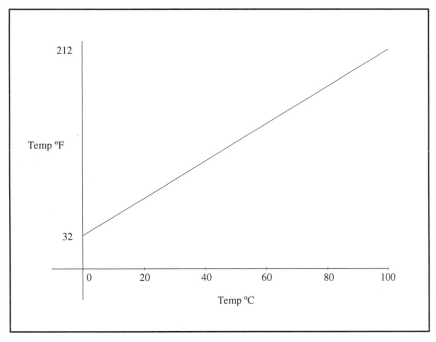

Figure 2.1.1 Temperature conversion

grees *Kelvin*, *K*. Translation from degrees *K* to degrees *C* involves subtracting the fixed amount 273.15,

$$C = K - 273.15. \tag{2.1.2}$$

From equation (2.1.1) we calculate absolute zero on the Fahrenheit scale as

$$F = \frac{9}{5}(-273.15) + 32 = -459.67,$$

or about -460 degrees *Rankin*, *R*, that is

$$F = R - 459.67. \tag{2.1.3}$$

Hence, substituting equations (2.1.2) and (2.1.3) into equation (2.1.1), we find *R* is related to *K* by

$$R = \frac{9}{5}K.$$

Thus *R* is proportional to *K* and both are zero at the same time, so there is no translational term.

It is often observed that the relationship between two variables is one of proportionality, in which the constant is not yet known. Thus if variables x and y are linearly related (and both are zero at the same time), we write

$$y = kx,$$

with the constant of proportionality k to be subsequently determined (see Section 2.2 on least squares).

Power laws can be converted to linear form.

The area of a circle does not vary linearly with the radius but rather quadratically, $A = \pi r^2$; the power, or degree, of r is two. Heat radiates in proportion to the fourth power of absolute temperature, gravitational force varies in proportion to the inverse square power of distance, and diffusivity varies with the one-third power of density (see Chapter 6). These are examples in which the relationship between variables is by a power law with the power different from one. There are many more.

In general, a power law is of the form

$$y = Ax^k \tag{2.1.4}$$

for some constants A and k. Due to the particular ease of graphing linear relationships, it would be advantageous if this equation could be put into linear form. This can be done by taking the logarithm of both sides of the equation. Two popular bases for logarithms are 10 and $e = 2.718281828\ldots$; the former is often denoted by *log* while the latter by *ln*. Either will work:

$$\log y = k \log x + \log A, \qquad (2.1.5)$$

and the relationship between $\log y$ and $\log x$ is linear. Plotting pairs of (x, y) data values on special *log-log paper* will result in a straight line with slope k. Of course, on a log-log plot there is no point corresponding to $x = 0$ or $y = 0$. However, if $A = 1$ then $\log y$ is proportional to $\log x$ and the graph goes through the point $(1, 1)$. In general, A appears on the graph as the value of y when $x = 1$.

Another frequently encountered relationship between variables is the *exponential* one given by

$$y = Ca^x. \qquad (2.1.6)$$

Note that the variable x is now in the exponent. Exponential functions grow (or decay) much faster than polynomial functions; that is, if $a > 1$, then, as an easy consequence of L'Hôpital's rule, for any power k:

```
> assume(a>1); assume(k>0);
> limit(x^k/a^x,x=infinity);
```

$$\lim_{x \to \infty} \frac{x^k}{a^x} = 0. \qquad (2.1.7)$$

Figure 2.1.2 demonstrates this with $k = 3$ and $a = 2$. We have drawn graphs of $y = x^3$, $y = 2^x$, and $y = 100 \cdot x^3/2^x$. The graphs of the first two cross twice, the last time about $x \approx 10$:

```
> solve(x^3=2^x,x); evalf(");
```

$$1.3734, \qquad\qquad 9.939.$$

Taking logarithms of equation (2.1.6) with base e gives

$$\ln y = x \ln a + \ln C. \qquad (2.1.8)$$

Here it is $\ln y$ that is proportional to x. A *semi-log plot* of exponentially related variables, as in equation (2.1.8), produces a straight line whose slope is $\ln a$.

```
> plot({[x,x^3,x=0..12],[x,2^x,x=0..12],[x,100*x^3/2^x,x=0..14]},
    x=0..14,y=0..4000);
```

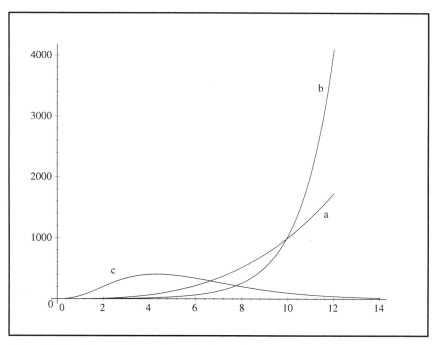

Figure 2.1.2 Exponential vs polynomial rate of growth.
Graphs of (a) x^3, (b) 2^x, and (c) $100x^3/2^x$.

By defining $r = \ln a$ and exponentiating both sides of equation (2.1.8), we get

$$y = Ce^{rx} \quad \text{where} \quad r = \ln a. \tag{2.1.9}$$

This is an alternate form of relationship equation (2.1.6) and shows that an exponential relationship can be expressed in base e if desired.

Proportionality can pertain to derivatives, too.

A natural and simplifying assumption about the growth of a population is that *the number of offspring at any given time is proportional to the number of adults at that time* (see Chapter 3). There is exactly a linear relationship between the number of offspring and the number of adults. Let $y(t)$ (or just y in brief) denote the number of adults at time t. In any given small interval of time Δt, the number of offspring in that time represents the change in the population Δy. The ratio $\Delta y / \Delta t$ is the average rate of growth of the population over the time period Δt.

The derivative dy/dt is the instantaneous rate of growth at time t, or just the rate of growth at time t, instantaneous being understood. Making the questionable, but simplifying, assumption that new offspring are immediately adults, leads to a mathematical expression of the italicized statement above:

$$\frac{dy}{dt} = ky$$

for some constant of proportionality k. That is, the derivative or rate of growth is proportional to the number present.

This particular differential equation is easily solved by integration,

$$\frac{dy}{y} = k\,dt \quad \text{or} \quad \ln y = kt + \ln A$$

with constant of integration $\ln A$. Exponentiating both sides gives

$$y = Ae^{kt}.$$

This situation is typical and we will encounter similar ones throughout the book.

Exercises

1. Proportionality constants associated with changes in units are often used in making conversions after measurements have been made. Convert from the specified units to the indicated units.

   a. Convert: x inches to centimeters, y pounds per gallon to kilograms per liter, z miles per hour to kilometers per hour.
   Maple has the ability to change units (type: ?convert). Here is syntax for the above.

   ```
   > convert(x*inches,metric);
   > convert(y*pounds/gallon,metric,US);
   > convert(z*miles/hour,metric);
   ```

   b. Sketch three graphs similar to Figure 2.1.1 to show the changes in units indicated above. Syntax similar to that which generated Figure 2.1.1 can be used here.

2. In this exercise, we compare graphs of power law relations with standard graphs, log graphs, and log-log graphs.

   a. Sketch the graphs of πr^2 and $\frac{4}{3}\pi r^3$ on the same graph. Then sketch both as log-log plots.

   ```
   > plot({Pi*r^2,4/3*Pi*r^3},r=0..1);
   > plots[loglogplot]({Pi*r^2,4/3*Pi*r^3},r=0.1..1);
   ```

b. Sketch the graphs of $3x^5$ and $5x^3$ on the same graph. Then sketch both log plots.

```
> plot({3*5^x,5*3^x},x=0..1);
> plots[logplot]({3*5^x,5*3^x},x=0..1);
```

3. This exercise examines limits of quotients of polynomials and exponentials. Sketch the graphs of $3x^2 + 5x + 7$ and 2^x on the same axis. Also, sketch the graph of the quotient $(3x^2 + 5x + 7)/2^x$. Evaluate the limit of this quotient.

```
> plot({3*x^2+5*x+7,2^x},x=0..7);
> plot((3*x^2+5*x+7)/2^x,x=0..10,y=0..10);
> limit((3*x^2+5*x+7)/2^x,x=infinity);
```

4. This exercise solves differential equations such as those seen in Section 2.1. Give the solution and plot the graph of the solution for each of these differential equations.

$$\frac{dy}{dt} = 3y(t), \qquad y(0) = 2,$$

$$\frac{dy}{dt} = 2y(t), \qquad y(0) = 3,$$

$$\frac{dy}{dt} = 2y(t), \qquad y(0) = -3,$$

$$\frac{dy}{dt} = -2y(t), \qquad y(0) = 3.$$

Here is syntax that will do the first problem and will un-do the definition of y to prepare for the remaining problems.

```
> eq:=diff(y(t),t) = 3*y(t); dsolve({eq,y(0)=2},y(t));
> y:=unapply(rhs("),t); plot(y(t),t=0..1);
> y:='y';
```

Section 2.2

Linear Regression, the Method of Least Squares

In this section we introduce the method of least squares for fitting straight lines to experimental data. By transformation, the method can be made to work for data related by power laws and exponential laws as well as for linearly related data.

The method is illustrated with two examples.

The method of least squares calculates a linear fit to experimental data.

Imagine performing the following simple experiment: Record the temperature of a bath as shown on two different thermometers, one calibrated in Fahrenheit and the other in Celsius, as the bath is heated. We plot the temperature F against the temperature C. Surprisingly, if there are three or more data points observed to high precision, they will not fall on a single straight line because the mathematical line established by two of the points will dictate infinitely many digits of precision for the others—no measuring device is capable of infinite precision. This is one source of error; there are others. Thus experimental data, even data for linearly related variables, are not expected to fall perfectly along a straight line.

How, then, can we conclude experimentally that two variables are linearly related, and, if they are, how can the slope and intercept of the correspondence be determined? The answer to the latter question is by the method of least squares fit and is the subject of this section; the answer to the first involves theoretical considerations and the collective judgment of scientists familiar with the phenomenon.

Assume the variables x and y are suspected to be linearly related and we have 3 experimental points for them, for example C and F in the example above. For the 3 data points (x_1, y_1), (x_2, y_2), and (x_3, y_3) shown in Figure 2.2.1, consider a possible straight line fit, $\ell(x)$. Let e_1, e_2, and e_3 be the errors

$$e_i = y_i - \ell(x_i), \qquad i = 1 \ldots 3,$$

defined as the difference between the data value y_i and the linear value $\ell(x_i)$ for each point. Note that we assume all x-data values are exact and that the errors are in the y-values only. This is reasonable because x is the independent variable; the x-values are the ones determined by the experimenter.

We want to choose a line ℓ that minimizes all of the errors at the same time; thus a first attempt might be to minimize the sum $e_1 + e_2 + e_3$. The difficulty with this idea is that these errors can cancel because they are signed values. Their sum could even be zero. But squaring each error eliminates this problem. And we choose the line ℓ so as to minimize

$$E = \sum_{i=1}^{3} e_i^2 = \sum_{i=1}^{3} [y_i - \ell(x_i)]^2,$$

that is, the least of the squared errors.

A line is determined by two parameters, slope m and intercept b, $\ell(x) = mx + b$. Therefore the mathematical problem becomes to find m and b to minimize

$$E(m, b) = \sum_{i=1}^{n} [y_i - (mx_i + b)]^2, \tag{2.2.1}$$

for n equal to the number of data points, 3 in this example. We emphasize that this

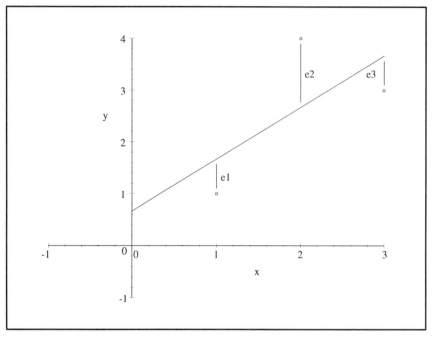

Figure 2.2.1   The differences $e_i = y_i - \ell(x_i)$

error E is a function of m and b (not x and y, as the x_i and y_i are given numbers). Solving such a minimization problem is standard practice: Set the derivatives of E with respect to its variables m and b equal to zero, and solve for m and b.[2]

$$0 = \frac{\partial E}{\partial m} = -2\sum_{i=1}^{n}[y_i - (mx_i + b)]x_i$$

$$0 = \frac{\partial E}{\partial b} = -2\sum_{i=1}^{n}[y_i - (mx_i + b)].$$

These equations simplify to

$$0 = \sum_{i=1}^{n} x_iy_i - m\sum_{i=1}^{n} x_i^2 - b\sum_{i=1}^{n} x_i$$

$$0 = \sum_{i=1}^{n} y_i - m\sum_{i=1}^{n} x_i - nb,$$

(2.2.2)

---

[2] Since E is a function of two independent variables m and b, it can vary with m while b is held constant or vice versa. To calculate its derivatives, we do just that: Pretend b is a constant, and differentiate with respect to m as usual. This is called the *partial derivative* with respect to m and is written $\partial E/\partial m$ in deference to the variables held fixed. Similarly, hold m constant and differentiate with respect to b to get $\partial E/\partial b$. At a minimum point of E, both derivatives must be zero since E will be momentarily stationary with respect to each variable.

which may be easily solved, for instance by Cramer's Rule. The least squares solution is

$$m = \frac{n \sum_{i=1}^{n} x_i y_i - (\sum_{i=1}^{n} x_i)(\sum_{i=1}^{n} y_i)}{n \sum_{i=1}^{n} x_i^2 - (\sum_{i=1}^{n} x_i)^2}$$

(2.2.3)

$$b = \frac{(\sum_{i=1}^{n} x_i^2)(\sum_{i=1}^{n} y_i) - (\sum_{i=1}^{n} x_i)(\sum_{i=1}^{n} x_i y_i)}{n \sum_{i=1}^{n} x_i^2 - (\sum_{i=1}^{n} x_i)^2}.$$

The expression for b simplifies to[3]

$$b = \bar{y} - m\bar{x} \quad \text{where} \quad \bar{y} = \frac{1}{n}\sum_{i=1}^{n} y_i \text{ and } \bar{x} = \frac{1}{n}\sum_{i=1}^{n} x_i.$$

We will illustrate the least squares method with two examples.

Example 2.2.1

Juvenile height vs. age is only approximately linear.

In Table 2.2.1 we show age and averaged height data for children.

Table 2.2.1 Average height versus age for children

height (cm)	75	92	108	121	130	142	155
age	1	3	5	7	9	11	13

SOURCE: *The Merck Manual of Diagnosis and Therapy*, 15th ed., David N. Holvey, editor, Merck Sharp & Dohme Research Laboratories, Rahway, NJ, 1987.

With $n = 7$, *age* and *height* interpreted as x and y respectively in equation (2.2.1), and using the data of the table, parameters m and b can be evaluated from equations (2.2.3):

```
> ht:=[75,92,108,121,130,142,155]; age:=[1,3,5,7,9,11,13];
> sumy:=sum(ht[n],n=1..7); sumx:=sum(age[n],n=1..7);
> sumx2:=sum(age[n]^2,n=1..7); sumxy:=sum(age[n]*ht[n],n=1..7);
> m:=evalf((7*sumxy-sumx*sumy)/(7*sumx2-sumx^2));
> b:=evalf((sumx2*sumy-sumx*sumxy)/(7*sumx2-sumx^2));
```

$$m = 6.46, \quad \text{and} \quad b = 72.3.$$

These data are plotted in Figure 2.2.2 along with the least squares fit for an assumed linear relationship $ht = m \cdot age + b$ between height and age. This process

[3]Starting from $\bar{y} - m\bar{x}$ with m from equation (2.2.3), make a common denominator, cancel the terms $-(\sum x_i)^2 \bar{y} + \bar{x} \sum x_i \sum y_i$, and the expression for b emerges.

has been developed as a routine called fit[leastsquare]. To introduce this shortcut, clear *m* and *b*.

```
> m:='m'; b:='b';
> ht:=[75,92,108,121,130,142,155]; age:=[1,3,5,7,9,11,13];
    pts:=[seq([age[i],ht[i]],i=1..7)]:
> with(plots): with(stats):
> Data:=plot(pts,style=POINT,symbol=CIRCLE):
> fit[leastsquare[[x,y],y=m*x+b]]([age,ht]); m:=op(1,op(1,rhs(" ")));
    b:=op(2,rhs(" "));
> Fit:=plot(m*x+b,x=0..14):
```

This demonstrates the mechanics of the least squares method. But it must be kept in mind that the method is merely statistical; it can demonstrate that data are consistent or not with a linear assumption, but it can not prove linearity. In this example, a linear fit to the data is reasonably good, but no rationale for a linear relationship has been provided.

```
> display({Data,Fit});
```

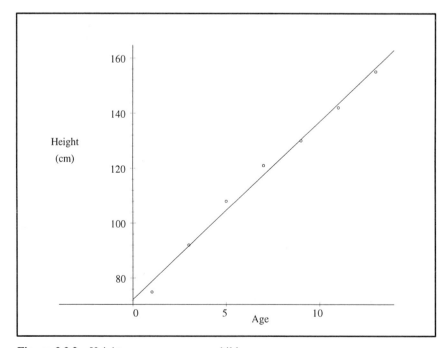

Figure 2.2.2   Height versus age among children

Example 2.2.2

The number of AIDS cases increase cubically.

As we saw in the first part of this section, when the data are obviously not linear, we can try to fit a power law of the form $y = Ax^k$. Consider the following data as reported in the HIV/AIDS Surveillance Report published by the U.S. Department of Health and Human Services concerning the reported cases of AIDS by half-years shown in Table 2.2.2. The third column is the sum of all the cases reported up to that time, i.e., the cumulative AIDS cases (CAC).

A graph of the AIDS data is shown in Figure 2.2.4. The circle symbols of the figure give the CAC data versus year; the solid curve is the least squares fit, which we discuss next. In this figure, CAC is measured in thousands and t is decades from 1980, that is, $t = (\text{year} - 1980)/10$.

We begin by first reading in the data.

```
> AIDS:=([97, 206, 406, 700, 1289, 1654, 2576, 3392, 4922,
      6343, 8359, 9968, 12990, 14397, 16604, 17124, 19585,
      19707, 21392, 20846, 23690, 24610, 26228, 22768, 4903]);
> CAC:=[seq(sum(AIDS[j]/1000.0, j=1..i),i=1..24)];
> Time:=[seq(1981+(i-1)/2,i=1..24)]:
```

Table 2.2.2 Total and Reported Cases of AIDS in the U.S.

Year	Reported Cases of AIDS	Cumulative AIDS Cases (thousands)
1981	97	.097
1981.5	206	.303
1982	406	.709
1982.5	700	1.409
1983	1289	2.698
1983.5	1654	4.352
1984	2576	6.928
1984.5	3392	10.320
1985	4922	15.242
1985.5	6343	21.585
1986	8359	29.944
1986.5	9968	39.912
1987	12990	52.902
1987.5	14397	67.299
1988	16604	83.903
1988.5	17124	101.027
1989	19585	12.0612
1989.5	19707	140.319
1990	21392	161.711
1990.5	20846	181.557
1991	23690	206.247
1991.5	24610	230.857
1992	26228	257.085
1992.5	22768	279.853

> display({LnFit,LnData});

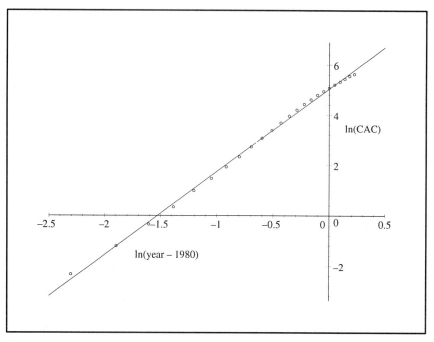

Figure 2.2.3 Log-log plot of cumulative AIDS cases and its fit

To produce the fit we proceed as before using equation (2.2.1) but this time performing least squares on $y = \ln(\text{CAC})$ versus $x = \ln t$,

$$\ln(\text{CAC}) = k * \ln t + \ln A. \qquad (2.2.4)$$

Here we rescale time to be decades after 1980 and calculate the logarithm of the data.

```
> LnCAC:=map(ln,CAC);
> Lntime:=map(ln,[seq((i+1)/2/10,i=1..24)]);
```

It remains to calculate the coefficients.

```
> with(stats):
> fit[leastsquare[[x,y],y=k*x+LnA]]([Lntime,LnCAC]);
> k:=op(1,op(1,rhs(" "))); LnA:=(op(2,rhs(" "))); A:=exp(LnA);
```

$$k = 3.29, \quad \text{and} \quad \ln A = 5.04 \quad A = 155.$$

> plots[display](Fit,Data);

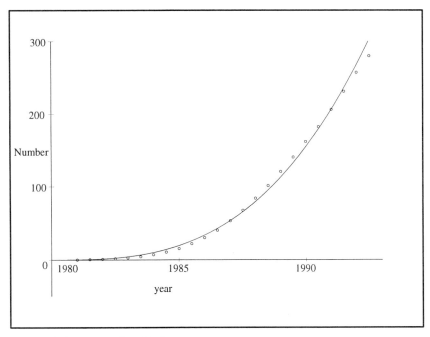

Figure 2.2.4   Cumulative AIDS cases

We draw the graph of ln(CAC) versus ln(time) to emphasize that their relationship is nearly a straight line. The log-log plot of best fit is shown in Figure 2.2.3 and is drawn as follows:

```
> Lndata:=plot([seq([Lntime[i],LnCAC[i]],i=1..24)]
    style=POINT,symbol=CIRCLE):
  Lnfit:=plot(k*x+ln(A),x=-2.5..0.5):
  plots[display]({Lndata,Lnfit});
```

The curve of best fit is, from equation (2.2.4),

$$CAC = 155\, t^{3.29}.$$

But we want an integer exponent, hence the comparative graph to the data will be taken as

```
> n:=trunc(k);
```

$$n = 3$$

$$CAC = 155 t^3 = 155 \left(\frac{year - 1980}{10} \right)^3.$$

Figure 2.2.4 is drawn as an overlay of the data and this fit.

```
> pts:=[seq([Time[i], CAC[i]], i=1..24];
> Fit:=plot(A*((t-1980)/10)^n,t=1980..1993):
> Data:=plot(pts,style=POINT,symbol=CIRCLE):
```

Again, we see the fit is good. Turning from the mechanical problem of fitting the data to the scientific problem of explaining the fit, why should a cubic fit so well?

In the studies of populations and infectious diseases, it is common to ask at what rate an infected population is growing. Quite often, populations grow exponentially in their early stages, that is according to equation (2.1.8). We will investigate this idea in Chapters 3 and 4.

In the first decade after the appearance of AIDS and the associated HIV, an analysis of the data for the total number of reported cases of AIDS led to the announcement that the population was growing cubically as a function of time. This was a relief of sorts because the growth was not exponential as expected, since exponential growth is much faster than polynomial growth, see equation (2.1.6).

Colgate, et al. [2] construct a model for HIV infection which leads to the result that the growth rate should be cubic in the early stages. A central idea in the model is the recognition that the disease spreads at different rates in different "risk groups," and that there is a statistically predictable rate at which the disease crosses risk groups.

In the exercises we attempt an exponential fit to this data.

Exercises

1. Ideal weights for medium built males are listed from Reference [3].

Table 2.2.3 Ideal weights for medium built males

Height (in)	Weight (lb)
62	128
63	131
64	135
65	139
66	142
67	146
68	150
69	154
70	158
71	162
72	167
73	172

a.   Show that a linear fit for this data is

$$wt = 4.04 \cdot ht + 124.14.$$

b.   In many geometric solids, volume changes with the cube of the height.
     Give a cubic fit for this data.

c.   Using the techniques of Example 2.2.2, find n and A such that

$$wt = A \cdot (ht - 60)^n.$$

The following code can be used for (b). A modification of one line can
be used for (a). For (c), modify the code for Example 2.2.2.

```
> ht:=[62,63,64,65,66,67,68,69,70,71,72,73,74];
> wt:= [128,131,135,139,142,146,150,154,158,162,167,172, 177];
> with(stats): fit[leastsquare[[x,y],y=a*x^3+b*x^2 +c*x+d]]([ht,wt]);
> y:=unapply(rhs(" "),x);
> pts:=[seq([ht[i],wt[i]],i=1..13)];
> J:=plot(pts,style=POINT,symbol=CROSS): K:=plot(y(x),x=62..74):
> with(plots): display({J,K});
> errorLinear:=sum('(4.04*ht[i]-124.14- wt[i])^2','i'=1..13);
> errorcubic:=sum('(y(ht[i])- wt[i])^2','i'=1..13);
> evalf(" );
```

2.   Changes in the human life span are illustrated graphically on page 110 in the
     October 1994 issue of *Scientific American*. The data appear below in three
     rows: The first row indicates the age category. The next two rows indicate
     the percentage of people who survived to that age in the United States in the
     years 1900 and 1960. The last row is the percentage of people who survived
     to that age in ancient Rome. Get a least squares fit for these data sets. Syntax
     that provides such a fit is given for the 1960 data.

```
> age60:=[0,10,20,30,40,50,60,80,100]:
> percent60:=[100,98.5,98,96.5,95,92.5,79,34,4]:
> with(stats):
> fit[leastsquare[[x,y],y=a*x^4+b* x^3+c*x^2+d*x+e]]([age60,percent60]);
> yfit60:=unapply(rhs(" "),x):
> pts60:=[seq([age60[i],percent60[i]],i=1..9)]:
> J6:=plot(pts60,style=POINT,symbol=CROSS):
> K6:=plot(yfit60(x),x=0..100):
> with(plots): display({J6,K6});
```

Table 2.2.4   Survival Rates for Recent U.S. and Ancient Rome

age	0	10	20	30	40	50	60	80	100
1900	100	82	78	75	74	60	43	19	3
1960	100	98.5	98	96.5	95	92.5	79	34	4
Rome	90	73	50	40	30	22	15	5	0.5

3. We have found a fit for the cumulative U.S. AIDS data as a cubic polynomial. We saw that, in a sense, a cubic polynomial is the appropriate choice. On first looking at the data as shown in Figure 2.2.3, one might guess that the growth is exponential growth. Find an exponential fit for the data. Such a fit would use equation (2.1.8). *Maple* code to perform the calculations is only slightly different from that for the cubic fit.

```
> AIDS:=([97, 206, 406, 700, 1289, 1654, 2576, 3392, 4922, 6343, 8359, 9968,
    12990, 14397, 16604, 17124, 19585, 19707, 21392, 20846, 23690, 24610,
    26228, 22768, 4903]);
> CAC:=[seq(sum(AIDS[j]/1000.0,j=1..i),i=1..24)];
> Time:=[seq(1981+(i-1)/2,i=1..24)]:
> pts:=[seq([Time[i],CAC[i]],i=1..24)]:
> LnCAC:=map(ln,CAC);
> Times:=[seq((i+1)/2/10,i=1..24)];
> with(stats):
> fit[leastsquare[[x,y],y=m*x+b]]([Times,LnCAC]);
> k:=op(1,op(1,rhs(" "))); A:=op(2,rhs(" "));
> y:=t->exp(A)*exp(k*t);
> J:=plot(y((t-1980)/10),t=1980..1992):
> K:=plot(pts,style=POINT,symbol=CIRCLE):
> plots[display]({J,K});
```

4. Table 2.2.5 presents unpublished data that was gathered by Dr. Melinda Millard-Stafford at the Exercise Science Laboratory in the Department of Health and Performance Sciences at Georgia Tech. It relates the circumference of the forearm with grip strength. The first two columns are for a group of college women and the second two columns are for college men. Find regression lines for both sets of data.

```
> CW:=[24.2,22.9,27.,21.5,23.5,22.4,23.8,25.5, 24.5,25.5,22.,24.5];
> GSW:=[38.5,26.,34.,25.5,37.,30.,34.,43.5,30.5, 36.,29.,32];
> with(stats): fit[leastsquare[[x,y],y=m*x+b]]([CW,GSW]);
> pts:=[seq([CW[i],GSW[i]],i=1..12)];
> J:=plot(pts,style=POINT,symbol=CROSS):
> K:=plot(2.107*x-17.447,x=21..28):
> CM:=[28.5,24.5,26.5,28.25,28.2,29.5,24.5,26.9,28.2,25.6,28.1, 27.8,29.5,29.5,29];
> GSM:=[45.8,47.5,50.8,51.5,55.0,51.,47.5,45.,56.0,49.5,57.5,51.,59.5, 58.,68.25];
> fit[leastsquare[[x,y],y=m*x+b]]([CM,GSM]);
> pts:=[seq([CM[i],GSM[i]],i=1..15)];
```

Table 2.2.5 Forearm and Grip Strength in Males and Females

Females		Males	
Circumference (cm)	Grip (kg)	Circumference (cm)	Grip (kg)
24.2	38.5	28.5	45.8
22.9	26.0	24.5	47.5
27.0	34.0	26.5	50.8
21.5	25.5	28.25	51.5
23.5	37.0	28.2	55.0
22.4	30.0	29.5	51.0
23.8	34.0	24.5	47.5
25.5	43.5	26.9	45.0
24.5	30.5	28.2	56.0
25.5	36.0	25.6	49.5
22.0	29.0	28.1	57.5
24.5	32.0	27.8	51.0
		29.5	59.5
		29.5	58.0
		29.0	68.25

```
> L:=plot(pts,style=POINT,symbol=CIRCLE):
> M:=plot(2.153*x-6.567,x=24..30):
> with(plots): display({J,K,L,M});
```

Section 2.3

Multiple Regression

The least squares method extends to experimental models with arbitrarily many parameters, however, the model must be linear in the parameters. The mathematical problem of their calculation can be cast in matrix form, and, as such, the parameters emerge as the solution of a linear system.

The method is again illustrated with two examples.

Least squares can be extended to more than two parameters.

In the previous section we learned how to perform linear regression, or least squares, on two parameters, to get the slope m and intercept b of a straight line fit to data. We also saw that the method applies to other "models" for the data than just the linear model. By a *model* here we mean a mathematical formula of a given form involving unknown parameters. Thus the *exponential model* for (x, y) data is

$$y = Ae^{rx}.$$

And to apply linear regression we transform it to the form, see equation (2.1.5),

$$\ln y = rx + \ln A.$$

Here the transformed data are $Y = \ln y$ and $X = x$, while the transformed parameters are $M = r$ and $B = \ln A$. The key requirement of a regression model is that it be linear in the parameters.

Regression Principle. The method of least squares can be adapted to calculate the parameters of a model if there is some transformation of the model that is linear in the transformed parameters.

Consider the Michaelis–Menten equation for the initial reaction rate v_0 of the enzyme-catalyzed reaction of a substrate having a concentration denoted by $[S]$,

$$v_0 = \frac{v\text{max}[S]}{K_m + [S]};$$

the parameters are $v\text{max}$ and K_m. By reciprocating both sides of this equation we get the *Lineweaver–Burk equation*

$$\frac{1}{v_0} = \frac{K_m}{v\text{max}} \frac{1}{[S]} + \frac{1}{v\text{max}}.$$

Now the transformed model is linear in its parameters $M = K_m/v\text{max}$ and $B = 1/v\text{max}$ and the transformed data are $Y = 1/v_0$ and $X = 1/[S]$. After determining the slope M and intercept B of a *double reciprocal plot* of $1/v_0$ vs $1/[S]$ by least squares, then calculate $v\text{max} = 1/B$ and $K_m = M/B$.

So far we have only looked at two parameter models, but the principles apply to models of any number of parameters. For example, the Merck Manual, 14th Edition, published by the Merck Sharp & Dohme Research Laboratories, in 1982, gives a relationship between the outer surface area of a human as a function of height and weight as follows:

$$\text{surface area} = c \cdot wt^a \cdot ht^b,$$

with parameters a, b, and c (a and b have been determined to be 0.425 and 0.725 respectively). A transformed model, linear in parameters, for this is

$$\ln(\text{surface area}) = a\ln(wt) + b\ln(ht) + \ln c.$$

The transformed data are triples of values (X_1, X_2, Y) where $X_1 = \ln(wt)$, $X_2 = \ln(ht)$, and $Y = \ln(\text{surface area})$.

We now extend the method of least squares to linear models of r generalized independent variables $X_1, \ldots, X_r$ and one generalized dependent or response variable Y,

$$Y = a_1 X_1 + a_2 X_2 + \ldots + a_r X_r.$$

Note that we can recover the two variable case of Section 2.2 by taking $r = 2$ and $X_2 = 1$. Assume there are n data points $(X_{1,i}, \ldots, X_{r,i}, Y_i)$, $i = 1, \ldots, n$. As before, let e_i denote the error between the experimental value Y_i and the predicted value,

$$e_i = Y_i - (a_1 X_{1,i} + \ldots + a_r X_{r,i}), \quad i = 1, \ldots, n.$$

And as before, we choose parameter values $a_1, \ldots, a_r$ to minimize the squared error,

$$E(a_1, \ldots, a_r) = \sum_{i=1}^{n} e_i^2$$
$$= \sum_{i=1}^{n} [Y_i - (a_1 X_{1,i} + \ldots + a_r X_{r,i})]^2.$$

To minimize E differentiate it with respect to each parameter a_j and set the derivative to zero,

$$0 = \frac{\partial E}{\partial a_j} = -2 \sum_{i=1}^{n} X_{j,i} [Y_i - (a_1 X_{1,i} + \ldots + a_r X_{r,i})], \quad j = 1, \ldots, r.$$

The resulting linear system for the unknowns $a_1, \ldots, a_r$ can be rearranged to the following form (compare with equations (2.2.2))

$$
\begin{array}{ccccccc}
a_1 \sum_i^n X_{1,i} X_{1,i} & + & \ldots & + & a_r \sum_i^n X_{1,i} X_{r,i} & = & \sum_i^n X_{1,i} Y_i \\
\vdots & & & & \vdots & & \vdots \\
a_1 \sum_i^n X_{r,i} X_{1,i} & + & \ldots & + & a_r \sum_i^n X_{r,i} X_{r,i} & = & \sum_i^n X_{r,i} Y_i.
\end{array}
\tag{2.3.1}
$$

It is possible to write this system in a very compact way using matrix notation. Let M^T be the matrix of data values of the independent variables

$$M^T = \begin{bmatrix} X_{1,1} & X_{1,2} & \cdots & X_{1,n} \\ X_{2,1} & X_{2,2} & \cdots & X_{2,n} \\ \vdots & \vdots & \cdots & \vdots \\ X_{r,1} & X_{r,2} & \cdots & X_{r,n} \end{bmatrix}.$$

The ith row of this matrix is the vector of data values of X_i. Represent the data values of the dependent variable Y as a column vector and denote the whole column as $\mathbf{Y}$,

$$\mathbf{Y} = \begin{bmatrix} Y_1 \\ Y_2 \\ \vdots \\ Y_n \end{bmatrix}.$$

Denoting by M the transpose of M^T, the system of equations (2.3.1) can be written in matrix form as

$$M^T M \mathbf{a} = M^T \mathbf{Y} \tag{2.3.2}$$

where $\mathbf{a}$ is the column vector of regression parameters.

Example 2.3.1

Can body mass and skin fold predict body fat?

Sparling et al. [4] investigate the possibility of predicting body fat from height, weight, and skin fold measurements for black women. Percentage body fat can be estimated by two methods: hydrostatic weighing and bio-electric impedance analysis. As in standard practice, height and weight enter the prediction as the fixed combination of weight divided by height squared to form a factor called *body-mass-index*,

$$\text{body-mass-index} = \frac{\text{weight}}{\text{height}^2}.$$

The assumed relationship is taken as

$$\text{percent-body-fat} = a * \text{body-mass-index} + b * \text{skin-fold} + c$$

for some constants a, b, and c.

Table 2.3.1 gives a subset of data from Sparling et al. [4] that we will use in this example to find the constants. Weight and height measurements were made in pounds and inches respectively; body mass index is to be in kilograms per square meter, so the conversions 39.37 inches $= 1$ meter and 2.2046 pounds $= 1$ kilogram have been done to calculate the body mass index column of the table.

We compute the third column of Table 2.3.1 from the first two:

```
> ht:=[63,65,61.7,65.2,66.2,65.2,70.0,63.9,63.2,68.7,68,66];
  wt:=[109.3,115.6,112.4,129.6,116.7,114.0,152.2,115.6,121.3,
  167.7,160.9,149.9];
> convert([seq(wt[i]*lbs/(ht[i]/12*feet)^2,i=1..12)],metric);
```

Table 2.3.1   Height, Weight, Skin Fold, and % Body Fat for Black Women

height (in)	weight (lbs)	body mass (kg/m²)	skin fold	% body fat
63.0	109.3	19.36	86.0	19.3
65.0	115.6	19.24	94.5	22.2
61.7	112.4	20.76	105.3	24.3
65.2	129.6	21.43	91.5	17.1
66.2	116.7	18.72	75.2	19.6
65.2	114.0	18.85	93.2	23.9
70.0	152.2	21.84	156.0	29.5
63.9	115.6	19.90	75.1	24.1
63.2	121.3	21.35	119.8	26.2
68.7	167.7	24.98	169.3	33.7
68.0	160.9	24.46	170.0	36.2
66.0	149.9	24.19	148.2	31.0

To apply equation (2.3.2), we take X_1 to be body-mass-index, X_2 to be skin-fold, and $X_3 = 1$ identically. From the table,

$$M^T = \begin{bmatrix} 19.36 & 19.24 & 20.76 & 21.43 & 18.72 & \ldots & 24.19 \\ 86.0 & 94.5 & 105.3 & 91.5 & 75.2 & \ldots & 148.2 \\ 1 & 1 & 1 & 1 & 1 & \ldots & 1 \end{bmatrix},$$

and the response vector is

$$\mathbf{Y}^T = \begin{bmatrix} 19.3 & 22.2 & 24.3 & 17.1 & 19.6 & \ldots & 31.0 \end{bmatrix}.$$

Solving the system of equations (2.3.2) gives the values of the parameters. We solve equations (2.3.2) in Exercise 2. Here we use the "fit[leastsquares]" routine.

```
> BMI:=[19.36,19.24, 20.76, 21.43, 18.72, 18.85, 21.84, 19.90, 21.35,
     24.98,24.46, 24.19];
   SF:=[86.0, 94.5,105.3, 91.5, 75.2, 93.2,156.0, 75.1, 119.8, 69.3, 170.0, 148.2];
   PBF:=[19.3, 22.2, 24.3, 17.1, 19.6, 23.9, 29.5, 24.1, 26.2, 33.7, 36.2,31.0];
> with(stats):
   fit[leastsquare[[bdymass,sfld,c]]]([BMI,SF,PBF]);
> bdft:=unapply(rhs(" "),(bdymass,sfld));
```

$$a = .00656 \quad b = .1507 \quad c = 8.074.$$

Thus, we find that

percent-body-fat $\approx .00656 * $ body-mass-index $+ .1507 * $ skin-fold $+ 8.074.$

$$(2.3.3)$$

To test the calculations, here is a data sample not used in the calculation. The subject is 64.5 inches tall, weighed 135 pounds, and has skin-fold that measures

159.9 millimeters. Her body fat percentage is 30.8 as compared to the predicted value of 32.3.

```
> convert(135*lbs/((64.5/12*ft)^2),metric);
> bdft(22.815,159.9);
```

$$\text{bdft} = 32.3$$

Example 2.3.2

Can thigh circumference and leg strength predict vertical jumping ability?

Unpublished data gathered by Dr. Millard-Stafford in the Exercise Science Laboratory at Georgia Tech relates men's ability to jump vertically to the circumference of the thigh and leg strength as measured by leg press. The correlation was to find a, b, and c so that

$$\text{jump height} = a * (\text{thigh circum}) + b * (\text{bench press}) + c.$$

Hence the generalized variable X_1 is thigh circumference, X_2 is bench press and $X_3 = 1$.

Data from a sample of college-age men is shown in Table 2.3.2. From the table,

Table 2.3.2 Leg Size, Strength, and
Jumping Ability for Men

Thigh Average Circumference (cm)	Leg Press (lbs)	Vertical Jump (in.)
58.5	220	19.5
50.0	150	18.0
59.5	165	22.0
58.0	270	19.0
60.5	200	21.0
57.5	250	22.0
49.3	210	29.5
53.6	130	18.0
58.3	220	20.0
51.0	165	20.0
54.2	190	25.0
54.0	165	17.0
59.5	280	26.5
57.5	190	23.0
56.25	200	29.0

$$M^T = \begin{bmatrix} 58.5 & 50 & 59.5 & 58 & \cdots & 56.25 \\ 220 & 150 & 165 & 270 & \cdots & 200 \\ 1 & 1 & 1 & 1 & \cdots & 1 \end{bmatrix}$$

and

$$\mathbf{Y}^T = \begin{bmatrix} 19.5 & 18 & 22 & 19 & \cdots & 29 \end{bmatrix}.$$

Solutions for equation (2.3.2) using the data from Table 2.3.2 are found approximately:

```
> thigh:=[58.5, 50, 59.5, 58, 60.5, 57.5, 49.3, 53.6, 58.3, 51, 54.2, 54, 59.5,
     57.5, 56.25];
> press:=[220,150,165,270,200,250,210,130,220,165,190,165,280,190,200];
> jump:=[19.5,18,22,19,21,22,29.5,18,20,20,25,17,26.5,23,29];
> with(stats):
  fit[leastsquare[[x,y,z], z=a*x+b*y+c, {a,b,c}]]([thigh,press,jump]);
```

$$a = -.29, \quad b = .044, \quad c = 29.5.$$

And hence multilinear regression predicts that the height a male can jump is given by the formula

$$\text{jump height} \approx -.29 * (\text{thigh circum}) + .044 * (\text{bench press}) + 29.5. \quad (2.3.4)$$

Surprisingly, the coefficient of the thigh circumference term is negative, which suggests that thick thighs hinder vertical jumping ability.

Exercises

1. This exercise will review some of the arithmetic for matrices and vectors.

    ```
    > with (linalg):
    > A:=matrix([[a,b],[c,d],[e,f]]);
      c:=vector([c1,c2]);
    ```

    Multiplication of the matrix A and the vector c produces a vector.

    ```
    > evalm(A&*c);
    ```

    An interchange of rows and columns of A produces the *transpose* of A. Two matrices can be multiplied.

    ```
    > transpose(A) &* A;
    ```

2. Compute the solution for Example 2.3.1 using matrix structure and *Maple*.
    The following syntax will accomplish this.

```
> with(linalg):
> M:=matrix([[19.36, 86, 1], [19.24, 94.5, 1], [20.76, 105.3, 1], [21.43, 91.5, 1],
    [18.72, 75.2, 1],[18.85, 93.2, 1], [21.84, 156.0, 1],[19.9, 75.1, 1],
    [21.35, 119.8, 1], [24.98, 169.3, 1], [24.46, 170., 1], [24.19, 148.2, 1]]);
> transpose(M);
> A:=evalm(transpose(M) &* M);
> z:=vector([19.3, 22.2, 24.3, 17.1, 19.6, 23.9, 29.5, 24.1, 26.2, 33.7,
    36.2, 31.0]);
> y:=evalm(transpose(M)&*z);
> evalm(inverse(A)&*y);
```

3. a. In this exercise, we get a linear regression fit for some hypothetical data
    relating age, percentage body fat, and maximum heart rate. Maximum heart
    rate is determined by having an individual exercise until near complete ex-
    haustion.

Table 2.3.3   Data for Age, % Body Fat, and
Maximum Heart Rate

Age (Years)	% Body Fat	Maximum Heart Rate
30	21.3	186
38	24.1	183
41	26.7	172
38	25.3	177
29	18.5	191
39	25.2	175
46	25.6	175
41	20.4	176
42	27.3	171
24	15.8	201

The syntax that follows will produce a linear regression fit for the data. This
syntax will also produce a plot of the regression plane. Observe that it shows
a steep decline in maximum heart rate as a function of age and a lesser decline
with increased percentage body fat.

b. As an example of the use of this regression formula, compare the predicted
maximum heart rate for two persons at age 40 where one has maintained
a 15% body fat and the other has gained weight to a 25% body fat. Also,
compare two people with 20% body fat where one is age 40 and the other is
age 50.

```
> age:=[30,38,41,38,29,39,46,41,42,24];
  BF:= [21.3,24.1,26.7,25.3,18.5,25.2,25.6,20.4,27.3,15.8];
  hr:=[186,183,172,177,191,175,175,176,171,201];
> with(stats):
```

```
fit[leastsquare[[a,b,c]]]([age,BF,hr]);
> h:=unapply(rhs(" "),(a,b));
> plot3d(h(a,b),a=30..60,b=10..20,axes=NORMAL);
> h(40,15); h(40,25); h(40,20); h(50,20);
```

4. The following is further data relating leg size, strength, and the ability to
    jump. The data were gathered for college women.

Table 2.3.4   Leg Size, Strength, and
Jumping Ability for Women

Thigh Circumference (cm)	Leg Press (lb)	Vertical Jump (in.)
52.0	140	13.0
54.2	110	8.5
64.5	150	13.0
52.3	120	13.0
54.5	130	13.0
58.0	120	13.0
48.0	95	8.5
58.4	180	19.0
58.5	125	14.0
60.0	125	18.5
49.2	95	16.5
55.5	115	10.5

Find a least squares data fit for the data, which are from unpublished work
by Dr. Millard-Stafford in the Health and Performance Science Department
at Georgia Tech.

Section 2.4

Modeling with Differential Equations

Understanding a natural process quantitatively almost always leads to a differential equation model. Consequently a great deal of effort has gone into the study
of differential equations. The theory of linear differential equations, in particular,
is well-known, and not without reason, as this type occurs widely.

   Besides their exact solution in terms of functions, numerical and asymptotic
solutions are also possible when exact solutions are not available.

In differential equations, as with organisms, there is need of a nomenclature.

In Section 2.1 we proposed a simple differential equation for mimicking the
growth of a biological population, namely

$$\frac{dy}{dt} = ky. \tag{2.4.1}$$

A *differential equation* refers to any equation involving derivatives. Other examples are

$$\frac{d^2y}{dt^2} - 4\frac{dy}{dt} + 4y = e^{-t}, \tag{2.4.2}$$

and

$$\frac{dy}{dt} = y - \frac{y^2}{2 + \sin t}, \tag{2.4.3}$$

and many others. If only first order derivatives appear in a differential equation, then it is called a *first order* equation. Both equations (2.4.1) and (2.4.3) are first order, but equation (2.4.2) is a second order equation. Every first order differential equation can be written in the form

$$\frac{dy}{dt} = f(t, y) \tag{2.4.4}$$

for some function f of two variables. Thus $f(t, y) = ky$ in the first equation above and $f(t, y) = y - (y^2/(2 + \sin t))$ in the third.

A *solution* of a differential equation means a function $y = y(t)$ which satisfies the equation for all values of t (over some specified range of t values). Thus $y = Ae^{kt}$ is a solution of equation (2.4.1) because then $dy/dt = kAe^{kt}$, and substituting into equation (2.4.1) gives

$$kAe^{kt} = k(Ae^{kt}),$$

true for all t. Note that A is a parameter of the solution and can be any value, so it is called an *arbitrary constant*. Recalling Section 2.1, A arose as the constant of integration in the solution of equation (2.4.1). In general the solution of a first order differential equation will incorporate such a parameter. This is because a first order differential equation is making a statement about the slope of its solution rather than the solution itself.

To fix the value of the inevitable arbitrary constant arising in the solution of a differential equation, a point in the plane through which the solution must pass is also specified, for example at $t = 0$. A differential equation along with such a side condition is called an *initial value problem*,

$$\frac{dy}{dt} = f(t, y) \quad \text{and} \quad y(0) = y_0. \tag{2.4.5}$$

It is not required to specify the point for which $t = 0$, just conventional. The *domain of definition*, or simply domain, of the differential equation is the set of points (t, y) for which the right-hand side of equation (2.4.4) is defined. Often this is the entire t, y plane.

Initial value problems can be solved analytically.

Exact solutions are known for many differential equations, cf. Kamke [5]. For the most part, solutions derive from a handful of principles. Although we will not study solution techniques here to any extent, we make two exceptions: We discuss methods for linear systems below and the separation of variables next.

Actually we have already seen separable variables at work in Section 2.1: The idea is to modify the differential equation algebraically in such a way that all instances of the independent variable are on one side of the equation and all those of the dependent variable are on the other. Then the solution results as the integral of the two sides. For example, consider

$$\frac{dy}{dt} = ay - by^2.$$

Dividing by the terms on the right hand side and multiplying by dt separates the variables, leaving only the integration:

$$\int \frac{dy}{y(a - by)} = \int dt.$$

Instead of delving into solution methods further, our focus in this text is on deciding what solutions mean and which equations should comprise a model in the first place. Happily, some of the solution techniques, such as separation of variables, are sufficiently mechanical that computers can handle the job, leaving us to higher level tasks. Here, then, are solutions to equations (2.4.2) and (2.4.3):

```
> dsolve(diff(y(t),t,t)-4*diff(y(t),t)+4*y(t) =exp(-t),y(t));
```

$$y(t) = \frac{1}{9} + C_1 e^{2t} + C_2 t e^{2t},$$

and

```
> dsolve(diff(y(t),t)=y(t)-y(t)^2/(2+sin(t)), y(t));
```

$$\frac{1}{y(t)} = e^{-t} \int \frac{e^t}{2 + \sin(t)}\, dt + e^{-t} C_1.$$

```
> with(DEtools):
> dfieldplot(diff(y(t),t)-y(t)+y(t)^2/(2+sin(t))=0,y(t), t=0..5,y=-1..5);
```

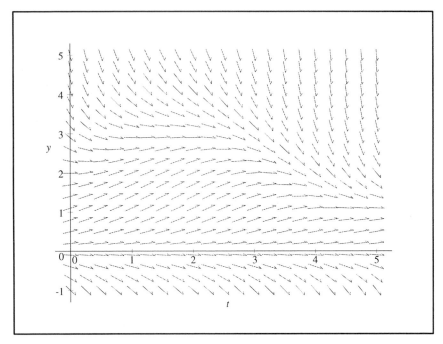

Figure 2.4.1   Direction field for equation (2.4.3)

Initial value problems can be solved numerically.

As mentioned above, equation (2.4.4) specifies the slope of the solution required by the differential equation at every point (t, y) in the domain. This may be visualized by plotting a short line segment having that slope at each point. This has been done in Figure 2.4.1 for equation (2.4.3). Such a plot is called a *direction field*. Solutions to the equation must follow the field and cannot cross slopes. With such a direction field it is possible to sketch solutions manually. Just start at the initial point $(0, y(0))$ and follow the direction field. Keep in mind that a figure such as Figure 2.4.1 is only a representation of the true direction field, that is to say, it shows only a small subset of the slope segments.

The mathematician Euler realized three centuries ago that the direction field could be used to approximate solutions of an initial value problem in a numerically precise way. Since Euler's time, techniques have improved—*Runge–Kutta methods* are used today, but the spirit of *Euler's method* is common to most of them; namely, a small step Δt to the right and Δy up is taken, where

$$\Delta y = f(t_i, y_i) \cdot \Delta t.$$

```
> with(DEtools):
> DEplot1(diff(y(t),t)-y(t)+y(t)^2/(2+sin(t))=0,y(t), t=0..5, {[0,1], [0,3], [0,5]});
```

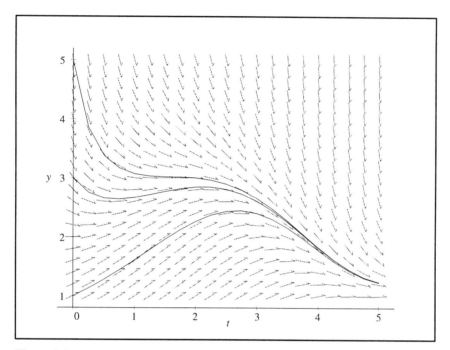

Figure 2.4.2 Solutions and direction field for equation (2.4.3)

The idea is that $\Delta y/\Delta t$ approximates dy/dt. These increments are stepped off one after another

$$y_{i+1} = y_i + \Delta y, \quad t_{i+1} = t_i + \Delta t, \quad i = 0, 1, 2, \ldots,$$

with starting values $y_0 = y(0)$ and $t_0 = 0$. Figure 2.4.2 shows some numerical solutions of equation (2.4.3).

Linear differential equations are among the simplest kind.

A differential equation that can be put into the form

$$a_n(t)\frac{d^n y}{dt^n} + \ldots + a_2(t)\frac{d^2 y}{dt^2} + a_1(t)\frac{dy}{dt} + a_0(t)y = r(t) \qquad (2.4.6)$$

is *linear*. The coefficients $a_i(t)$, $i = 0\ldots n$, can be functions of t, as can the *right-hand-side* $r(t)$. Equations (2.4.1) and (2.4.2) are linear, but equation (2.4.3) is not. When there are multiplications among the derivatives or the dependent variable y, such as y^2, the differential equation will not be linear. If $y_1(t)$ and $y_2(t)$

are both solutions to a linear differential equation with right-hand-side 0, then so is $Ay_1(t) + By_2(t)$ for any constants A and B. Consider the first order linear differential equation

$$\frac{dy}{dt} = my + R(t),\tag{2.4.7}$$

where we have taken $m = -a_0/a_1$ and $R(t) = r(t)/a_1$ in (2.4.6). Its solution is

$$y = Ae^{g(t)} + \Phi(t) \qquad \text{where} \quad g(t) = \int m\,dt.\tag{2.4.8}$$

In this A is the arbitrary constant and Φ is given below. To see this, first assume R is 0, and write the differential equation as

$$\frac{dy}{y} = m\,dt.$$

Now integrate both sides, letting $g(t) = \int m\,dt$ and C be the constant of integration,

$$\ln y = g(t) + C \qquad \text{or} \qquad y = Ae^{g(t)}$$

where $A = e^C$. By direct substitution it can be seen that

$$\Phi = e^{g(t)}\int e^{-g(t)}R(t)\,dt\tag{2.4.9}$$

is a solution.[4] But it has no arbitrary constant, so add the two to get equation (2.4.8). This is allowed by linearity. If m is a constant, then $\int m\,dt = mt$.

To see that finding this solution is mechanial enough that a computer can handle the job, one has only to explore.

```
> dsolve(diff(y(t),t)=m(t)*y(t)+R(t),y(t));
> dsolve(diff(y(t),t)=m*y(t)+R(t),y(t));
```

Systems of differential equations generalize their scalar counterparts.

Quite often modeling projects involve many more variables than two. Consequently it may require several differential equations to describe the phenomenon adequately. Consider the following model for small deviations about steady state

---

[4]A clever idea is to try a solution of the form $y = v(t)e^{g(t)}$ with $v(t)$ unknown. Substitute this into equation (2.4.7) to get $v'e^{g(t)} = R(t)$ since the term $vg'e^{g(t)} = vme^{g(t)}$ drops out. Now solve for v.

levels of a glucose/insulin system; g denotes the concentration of glucose and i the same for insulin,

$$\frac{dg}{dt} = -\alpha g - \beta i + p(t)$$
$$\frac{di}{dt} = \gamma g - \delta i. \tag{2.4.10}$$

As discussed in Section 2.1, the second equation expresses a proportionality relationship, namely, the rate of secretion of insulin increases in proportion to the concentration of glucose but decreases in proportion to the concentration of insulin. (Modeling coefficients are assumed to be positive unless stated otherwise.) The first equation makes a similar statement about the rate of removal of glucose except there is an additional term, $p(t)$, which is meant to account for ingestion of glucose. Because glucose and insulin levels are interrelated, each equation involves both variables. The equations define a system; the differential equations have to be solved simultaneously.

A system of differential equations can be written in vector form by defining a vector, say $\mathbf{Y}$, whose components are the dependent variables of the system. In vector notation equation (2.4.10) becomes

$$\frac{d\mathbf{Y}}{dt} = M\mathbf{Y} + \mathbf{P} \tag{2.4.11}$$

where the matrix M and vector $\mathbf{P}$ are

$$M = \begin{bmatrix} -\alpha & -\beta \\ \gamma & -\delta \end{bmatrix}, \qquad \mathbf{P} = \begin{bmatrix} p(t) \\ 0 \end{bmatrix}.$$

Since the system (2.4.10) is linear, its vector expression takes on the simple matrix form of equation (2.4.11). Furthermore, this matrix system can be solved in the same way as the scalar differential equation of equation (2.4.7). We have

$$\mathbf{Y} = e^{Mt}\mathbf{Y}_0 + e^{Mt} \int_0^t e^{-Ms}\mathbf{P}(s)\,ds. \tag{2.4.12}$$

Just as the exponential of the scalar product mt is

$$e^{mt} = 1 + mt + \frac{m^2 t^2}{2!} + \frac{m^3 t^3}{3!} + \cdots, \tag{2.4.13}$$

so the exponential of the matrix product Mt is

$$e^{Mt} = I + Mt + \frac{M^2 t^2}{2!} + \frac{M^3 t^3}{3!} + \cdots. \tag{2.4.14}$$

```
> GIdeq:= diff(g(t),t)=-g(t)-i(t), diff(i(t),t)=-i(t)+g(t);
> sol:=dsolve({GIdeq, g(0)=1, i(0)=0},{g(t),i(t)}):
> g:= unapply(subs(sol,g(t)),t); i:= unapply(subs(sol,i(t)),t);
> plot({g(t),i(t)},t=0..5);
```

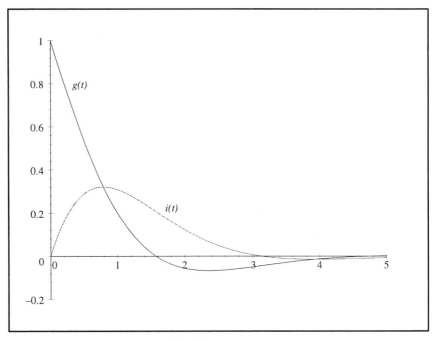

Figure 2.4.3 Plot of solutions $g(t)$, $i(t)$ of equation (2.4.10)

Since many properties of the exponential function stem from its power series expansion equation (2.4.13), the matrix exponential enjoys the same properties. In particular the property that makes for the same form of the solution,

$$\frac{d}{dt}e^{Mt}\mathbf{V(t)} = e^{Mt}\frac{d}{dt}\mathbf{V(t)} + e^{Mt}M\mathbf{V(t)}.$$

As in the case of a scalar differential equation, the system solutions can be plotted against t to help understand how the variables behave. For example, we could plot $g(t)$ and $i(t)$ using equation (2.4.12) (see Figure 2.4.3). But for a system there is an alternative; we can suppress t and plot $i(t)$ against $g(t)$. This is done, conceptually, by making a table of values of t and calculating the corresponding values of g and i. But we only plot (i, g) pairs. The coordinate plane of i and g is called the *phase plane* and the graph is called a *phase portrait* of the solution (see Figure 2.4.4).

```
> restart:
> with(DEtools):
> GIdeq:= diff(g(t),t)=-g(t)-i(t), diff(i(t),t)=-i(t)+g(t);
> inits:={[0,1,0],[0,2,0],[0,3,0],[0,4,0]};
> phaseportrait([GIdeq],[g,i],0..4,inits,stepsize=.1,g=-1..4,i=-1..1.5);
```

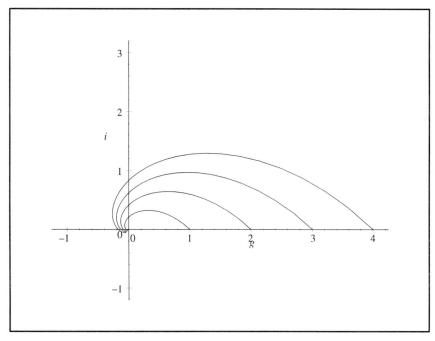

Figure 2.4.4 Phase portrait for equation (2.4.10)

Asymptotics predict the ultimate course of the model.

Often in science and engineering, we are interested in forecasting the future behavior of an observed process, $y(t)$. As t increases there are several possibilities,
among them: y can tend to a finite limit y_∞, known as an *asymptotic limit*,

$$\lim_{t \to \infty} y(t) = y_\infty;$$

y can tend to plus or minus infinity

$$\lim_{t \to \infty} y(t) = \pm\infty;$$

y can oscillate periodically, y can oscillate unboundedly,

$$\lim_{t \to \infty} |y(t)| = \infty;$$

or y can oscillate chaotically. If y is part of a system, its fate will be linked to that of the other variables; in this case we inquire about the vector solution $\mathbf{Y}$.

In the simplest case, $\mathbf{Y}$ has asymptotic limits. If the system is *autonomous*, meaning t appears nowhere in the system (except of course in the form d/dt), then to find the asymptotic limits, set all the derivatives of the system to zero. Solutions of the resulting algebraic system are called *critical points* or *stationary points*[5]. In the glucose/insulin example suppose the glucose ingestion term, $p(t)$, were constant at p, then setting the derivatives to zero leads to the algebraic system

$$0 = -\alpha g - \beta i + p$$
$$0 = \gamma g - \delta i.$$
(2.4.15)

> solve({-alpha *g-beta *i+p=0,gamma *g-delta*i=0},{g,i});

Its one critical point is $g = -\delta p/(\gamma\beta + \alpha\delta)$, $i = \gamma p/(\gamma\beta + \alpha\delta)$. If this point is taken as the initial point of the system, then for all time g will be $\delta p/(\gamma\beta + \alpha\delta)$ and i will be $\gamma p/(\gamma\beta + \alpha\delta)$.

It is not necessarily the case that a stationary point is also an asymptotic limit. Exponential growth, $\frac{dy}{dt} = y$, is an example since $y = 0$ is a stationary point, but, if $y(0) \neq 0$, then $y \to \infty$ as $t \to \infty$. On the other hand, when it can be shown that the solution of a system tends to an asymptotic limit, a giant step has been taken in understanding the system. For example, exponential decay, $\frac{dy}{dt} = -y$, has the asymptotic limit 0 for any starting point $y(0)$, for if $y > 0$ then $\frac{dy}{dt}$ is negative so y will decrease. Similarly, if $y < 0$ then $\frac{dy}{dt} > 0$ so y will increase. Either way, 0 is the asymptotic limit.

A complication here is that the existence or the value of the asymptotic limit can often depend on the starting point $\mathbf{Y}(0)$. Given that there is an asymptotic limit, $\mathbf{Y}_\infty$, the set of all starting points for which the solution tends to $\mathbf{Y}_\infty$ is called its *basin of attraction*, $\mathbf{B}_{\mathbf{Y}_\infty}$,

$$\mathbf{B}_{\mathbf{Y}_\infty} = \left\{ \mathbf{Y}_0 : \lim_{t\to\infty} \mathbf{Y}(t) = \mathbf{Y}_\infty \text{ when } \mathbf{Y}(0) = \mathbf{Y}_0 \right\}.$$

If the basin of attraction of a system is essentially the entire domain of definition, the asymptotic limit is said to be *global*. By way of example, the differential equation $\frac{dy}{dt} = -y(1 - y)$ has asymptotic limit $y = 0$ for solutions starting from $-\infty < y_0 < 1$; but when the starting point is beyond 1, solutions tend to infinity.

Periodicity is a more complicated asymptotic behavior. Further, just as in the asymptotic limit case, the solution can start out periodic, or can asymptotically tend to periodicity. An example of the former is $\frac{dy}{dt} = \cos t$, while the latter behavior is demonstrated by $\frac{dy}{dt} = -y + \cos t$. This second differential equation is solved by equation (2.4.8), $y = Ae^{-t} + \frac{1}{2}(\cos t + \sin t)$; A depends on the initial

---

[5]These are also called equilibrium points by some authors.

condition but the whole term tends to zero. A well-known periodic system is the one due to Lotka and Volterra modeling predator prey interaction. We study this in Chapter 4.

Exercises

1. Here are four differential equations with the same initial conditions:

$$\frac{d^2y}{dt^2} + 6y(t) = 0, \qquad y(0) = 1, \quad y'(0) = 0,$$

$$\frac{d^2y}{dt^2} - 6y(t) = 0, \qquad y(0) = 1, \quad y'(0) = 0,$$

$$\frac{d^2y}{dt^2} + 2\frac{dy}{dt} + 6y(t) = 0, \qquad y(0) = 1, \quad y'(0) = 0,$$

$$\frac{d^2y}{dt^2} - 2\frac{dy}{dt} + 6y(t) = 0, \qquad y(0) = 1, \quad y'(0) = 0.$$

While these four differential equations have a similar appearance, they have radically different behavior. Sketch the graphs of all four equations with the same initial values. Here is syntax that will draw the graphs.

```
> dsolve({diff(y(t),t,t)+6*y(t)=0,y(0) = 1, D(y)(0)= 0},y(t));
  y1:=unapply(rhs(" ),t);
> dsolve({diff(y(t),t,t)-6*y(t)=0,y(0) = 1, D(y)(0)= 0},y(t));
  y2:=unapply(rhs(" ),t);
> dsolve({diff(y(t),t,t)+2*diff(y(t),t)+6*y(t)=0,y( 0) = 1,
    D(y)(0)= 0},y(t));
  y3:=unapply(rhs(" ),t);
> dsolve({diff(y(t),t,t)- 2*diff(y(t),t)+6*y(t)=0,y(0) = 1,
    D(y)(0)= 0},y(t));
  y4:=unapply(rhs(" ),t);
> plot({y1(t),y2(t),y3(t),y4(t)},t=0..4, y=- 5..5);
```

2. We illustrate four ways to visualize solutions to a single first order differential equation in order to emphasize that different perspectives provide different insights. We use the same equation in all four visualizations:

$$\frac{d^2y}{dt^2} + y(t)/5 = \cos(t).$$

a.   Find and graph an analytic solution that starts at $y(0) = 0$.

```
> dsolve({diff(y(t),t,t)+y(t)/5 =cos(t), y(0) = 0,D(y)(0)=0},y(t));
> y:=unapply(rhs("),t);
> plot(y(t),t=0..4*Pi);
> restart:
```

b.   Give a direction field for the equation.

```
> with(DEtools):
> dfieldplot(diff(y(t),t)+y(t)/5=sin(t),y(t), t=0..4*Pi,y=-1..5);
> restart:
```

c.   Give several trajectories overlayed in the direction field.

```
> with(DEtools):
> DEplot1(diff(y(t),t)+y(t)/5=sin(t),y(t), t=0..4*Pi,{[0,1], [0,3], [0,5]});
```

d.   Give an animation to show the effect of the coefficient of $y(t)$ changing.

```
> with(plots):
> for n from 1 to 8 do
  a:=n/10:
  dsolve({diff(y(t),t)+a*y(t)/5=sin(t), y(0) = 1},y(t)):
  y:=unapply(rhs("),t):
  P.n:=plot([t,y(t),t=0..10*Pi],t=0..10*Pi):
  y:='y':
  od:
> display([seq(P.n,n=1..8)],insequence=true);
```

3.   Find the critical points for each of the following equations. Plot a few trajec-
     tories to confirm the locations of the basins of attractions.
     a. $\frac{dy}{dt} = -y(t)(1 - y(t))$.

```
> solve(y*(1-y)=0,y);
> with(DEtools):
> DEplot1(diff(y(t),t)=-y(t)*(1-y(t)),y(t), t=0..5,
    {[0,-1], [0,-1/2],[0,1/2]},y=-1..2);
```

b. $x' = 4x(t) - x^2(t) - x(t)y(t)$
     $y' = 5y(t) - 2y^2(t) - x(t)y(t)$

```
> solve({4*x-x^2-x*y=0,5*y-2*y^2-y*x=0},{x,y});
> with(DEtools):
> eqns:=[4*x-x^2-x*y, 5*y-2*y^2-y*x];
> inits:={[0,1,1],[0,1,4],[0,4,1],[0,4,4]};
> DEplot2(eqns,[x,y],t=0..4,inits,x=-1..5,y=-1..5);
```

4. The solution for $Z' = AZ(t)$, $Z(0) = C$, with A a constant square matrix and C a vector is $\exp(At)C$. Compute this exponential in the case

$$A = \begin{pmatrix} -1 & -1 \\ 1 & -1 \end{pmatrix}.$$

Evaluate $\exp(At)C$ where C is the vector

$$C = \begin{pmatrix} 1 \\ 0 \end{pmatrix}.$$

```
> with(linalg):
> A:=matrix([[-1,-1],[1,-1]]);
> exponential(A,t);
> evalm(" &* [1,0]);
```

Section 2.5

Matrix Analysis

The easiest kind of matrix to understand and calculate with is a diagonal matrix J, that is, one whose ik^{th} term is zero, $j_{ik} = 0$, unless $i = k$. The product of two diagonal matrices is again diagonal. The diagonal terms of the product are just the products of the diagonal terms of the factors. This pattern extends to all powers, J^r, as well. As a consequence, the exponential of a diagonal matrix is just the exponential of the diagonal terms.

It might seem that diagonal matrices are rare, but the truth is quite the contrary. For most problems involving a matrix, say, A, there is a change of basis matrix P so that PAP^{-1} is diagonal. We exploit this simplification to make predictions about the asymptotic behavior of solutions.

Eigenvalues predict the asymptotic behavior of matrix models.

Every $n \times n$ matrix A has associated with it a unique set of n complex numbers, $\lambda_1, \lambda_2, \ldots, \lambda_n$, called *eigenvalues*. Repetitions are possible, so the eigenvalues for A might not be distinct, but even with repetitions, they are always exactly n in number. In turn, each eigenvalue λ has associated with it a non-unique vector $\mathbf{e}$ called an *eigenvector*. An eigenvalue, eigenvector pair $\lambda, \mathbf{e}$ is defined by the matrix equation

$$A\mathbf{e} = \lambda\mathbf{e}. \tag{2.5.1}$$

An eigenvector for λ such as $\mathbf{e}$ is not unique because for every number a, the vector $\mathbf{e}' = a\mathbf{e}$ is also an eigenvector, as is easily seen from equation (2.5.1).

Example 2.5.1

The matrix

$$A = \begin{bmatrix} 1 & 3 \\ 0 & -2 \end{bmatrix}$$

has eigenvalues $\lambda_1 = 1$ and $\lambda_2 = -2$ with corresponding eigenvectors $\mathbf{e}_1 = \begin{pmatrix} 1 \\ 0 \end{pmatrix}$ and $\mathbf{e}_2 = \begin{pmatrix} 1 \\ -1 \end{pmatrix}$. Before invoking the computer on this one (see Exercise 1 in this section), work through it by hand.

Eigenvalues and eigenvectors play a central role in every mathematical model embracing matrices.

This statement cannot be overemphasized. The reason is largely a consequence of the following theorem.

Theorem 1

Let the $n \times n$ matrix A have n distinct eigenvalues; then there exists a non-singular matrix P such that the matrix

$$J = PAP^{-1} \tag{2.5.2}$$

is the diagonal matrix of eigenvalues of A,

$$J = \begin{bmatrix} \lambda_1 & 0 & \cdots & 0 \\ 0 & \lambda_2 & \cdots & 0 \\ \vdots & \vdots & \cdots & \vdots \\ 0 & 0 & \cdots & \lambda_n \end{bmatrix}.$$

The columns of P are the eigenvectors of A taken in the same order as the list of eigenvalues.

If the eigenvalues are not distinct, then we are not guaranteed that there will be a completely diagonal form; it can happen that there is not one. But even if not, there is an almost-diagonal form, called *Jordan Canonical form* (or just Jordan form) which has a pattern of 1's above the main diagonal. By calculating the Jordan form of a matrix, we get the diagonal form if the matrix has one. We will not need to discuss the Jordan form here, except to say that the computer algebra system can compute it.

The matrix product of this theorem, PAP^{-1}, is a change of basis modification of A; in other words, by using the eigenvectors as the reference system, the matrix A becomes the diagonal matrix J. Note that if $J = PAP^{-1}$ then the kth power of J and A are related as the k-fold product of PAP^{-1},

$$
\begin{aligned}
J^k &= (PAP^{-1})(PAP^{-1})\ldots(PAP^{-1}) \\
&= PA^k P^{-1}
\end{aligned}
\tag{2.5.2}
$$

since the interior multiplications cancel.

Diagonal matrices are especially easy to work with; for example, to raise J to a power, J^k, becomes raising the diagonal entries to that power:

$$
J^k =
\begin{bmatrix}
\lambda_1^k & 0 & \cdots & 0 \\
0 & \lambda_2^k & \cdots & 0 \\
\vdots & \vdots & \cdots & \vdots \\
0 & 0 & \cdots & \lambda_n^k
\end{bmatrix}.
$$

As a result, the exponential of J is just the exponential of the diagonal entries. From equation (2.4.13)

$$
e^{Jt} = I + Jt + \frac{J^2 t^2}{2!} + \frac{J^3 t^3}{3!} + \ldots
$$

$$
=
\begin{bmatrix}
(1 + \lambda_1 t + \frac{\lambda_1^2 t^2}{2!} \ldots) & 0 & \cdots & 0 \\
0 & (1 + \lambda_2 t \ldots) & & 0 \\
\vdots & \vdots & \cdots & \vdots \\
0 & 0 & \cdots & (1 + \lambda_n t \ldots)
\end{bmatrix}
\tag{2.5.3}
$$

$$
=
\begin{bmatrix}
e^{\lambda_1 t} & 0 & \cdots & 0 \\
0 & e^{\lambda_2 t} & \cdots & 0 \\
\vdots & \vdots & \cdots & \vdots \\
0 & 0 & \cdots & e^{\lambda_n t}
\end{bmatrix}.
$$

We illustrate the way in which these results are used.

The age-structure of a population evolves as dictated by a matrix L, such as the following (see Chapter 5):

$$
L =
\begin{bmatrix}
0 & 0 & 0 & 0 & 0.08 & 0.28 & 0.42 \\
.657 & 0 & 0 & 0 & 0 & 0 & 0 \\
0 & .930 & 0 & 0 & 0 & 0 & 0 \\
0 & 0 & .930 & 0 & 0 & 0 & 0 \\
0 & 0 & 0 & .930 & 0 & 0 & 0 \\
0 & 0 & 0 & 0 & .935 & 0 & 0 \\
0 & 0 & 0 & 0 & 0 & .935 & 0
\end{bmatrix}
$$

After k generations, the pertinent matrix is the kth power of L. From Theorem 1, there exists a matrix P such that $J = PLP^{-1}$, and, according to equation (2.5.2),

$$L^k = P^{-1} J^k P.$$

Letting λ_1 be the largest eigenvalue of L in absolute value, it is easy to see that

$$\frac{1}{\lambda_1^k} J^k = \begin{bmatrix} 1 & 0 & \cdots & 0 \\ 0 & \left(\frac{\lambda_2}{\lambda_1}\right)^k & \cdots & 0 \\ \vdots & \vdots & \cdots & \vdots \\ 0 & 0 & \cdots & \left(\frac{\lambda_n}{\lambda_1}\right)^k \end{bmatrix}$$

$$\longrightarrow \begin{bmatrix} 1 & 0 & \cdots & 0 \\ 0 & 0 & \cdots & 0 \\ \vdots & \vdots & \cdots & \vdots \\ 0 & 0 & \cdots & 0 \end{bmatrix} \quad \text{as } k \to \infty.$$

In other words, for large k, L^k is approximately λ_1^k times a fairly simple fixed matrix related to its eigenvectors, that is it grows or decays like λ_1^k.

In another example, consider the matrix form of the linear differential equation of Section 2.4, equation (2.5.11). From above, the matrix exponential, e^{Mt}, can be written as

$$e^{Mt} = P^{-1} e^{Jt} P$$

where e^{Jt} consists of exponential functions of the eigenvalues. If all those eigenvalues are negative, then no matter what P is, every solution will tend to 0 as $t \to \infty$. But if one or more eigenvalues are positive, then at least one component of a solution will tend to infinity.

In Chapter 7 we will consider compartment models. A *compartment* matrix C is defined as one whose terms c_{ij} satisfy the following conditions:

1. All diagonal terms c_{ii} are negative or zero;
2. All other terms are positive or zero;
3. All column sums $\sum_i c_{ij}$ are negative or zero.

Under these conditions, it can be shown that the eigenvalues of C have negative or zero real parts, so the asymptotic result above applies.

The fact that the eigenvalues have negative real parts under the conditions of a compartment matrix derives from *Gerschgorin's Circle Theorem*. (Note that since eigenvalues can be complex numbers, the circles are in the complex plane.)

Theorem 2

If A is a matrix and S is the following union of circles in the complex plane,

$$S = \bigcup_m \{complex\ z \ : \ |a_{mm} - z| \leq \sum_{j \neq m} |a_{jm}|\},$$

then every eigenvalue of A lies in S.

Notice that the mth circle above has center a_{mm} and radius equal to the sum of the absolute values of the other terms of the mth column.

Exercises

1. For both the following matrices, find the eigenvalues and eigenvectors. Then find the Jordan form. Note that the Jordan structure for the two is different.

$$A_1 = \begin{pmatrix} 0 & 0 & -2 \\ 1 & 2 & 1 \\ 1 & 0 & 3 \end{pmatrix} \quad \text{and} \quad A_2 = \begin{pmatrix} 3 & 1 & -1 \\ -1 & 2 & 1 \\ 2 & 1 & 0 \end{pmatrix}$$

Here is the syntax for working the problem for A_1. Define the matrix:

```
> with(linalg):
    A:=matrix([[0,0,-2],[1,2,1],[1,0,3]]);
```

   a. Find the eigenvalues and eigenvectors for A.

```
> eigenvects(A);
```

   b. Find the Jordan form and verify that the associated matrix P has the property that

$$PAP^{-1} = J$$

```
> J:=jordan(A,'P');
    evalm(P);
> evalm(P &* A &* inverse(P));
```

2. In a compartment matrix, one or more of the column sums may be zero. In this case, one eigenvalue can be zero and solutions for the differential equations

$$Z' = CZ(t)$$

may have a limit different from zero.

If all the column sums are negative in a compartment matrix, the eigenvalues will have negative real parts. All solutions for the differential equations

$$Z' = CZ(t)$$

will have limit zero in this case.

The following matrices contrast these ideas.

$$C_1 = \begin{pmatrix} -1 & 1 & 0 \\ 1 & -1 & 0 \\ 0 & 0 & -1 \end{pmatrix} \quad \text{and} \quad \begin{pmatrix} -1 & 0 & 1/2 \\ 1/2 & -1 & 0 \\ 0 & 1/2 & -1 \end{pmatrix}.$$

Let C be the matrix defined below.

```
> with(linalg):
  C:=matrix([[-1,1,0],[1,-1,0],[0,0,-1]]);
```

a. Find the eigenvalues and eigenvectors for C.

```
> eigenvects(C);
```

b. Graph each component of Z with $Z(0) = [1, 1, 1]$.

```
> exptC:=exponential(C,t);
> U:=evalm( exptC &* [1,0,1]);
> u:=unapply(U[1],t);
  v:=unapply(U[2],t);
  w:=unapply(U[3],t);
> plot(u(t),v(t),w(t),t=0..10);
```

Section 2.6

Statistical Data

Variation touches almost everything. A distribution is the fraction of observations having a particular value as a function of possible values. For example, the distribution of word lengths of the previous sentence is 3 of length 1, 4 of length 2, 1 of length 3, and so on (all divided by 17, the number of words in the sentence). The graph of a distribution with the observations grouped or made discrete to some resolution is a histogram. Distributions are approximately described by their mean or average value and the degree to which the observations deviate from the mean, their standard deviation. A widely occurring distribution is the normal or Gaussian distribution. This bell-shaped distribution is completely determined by its mean and standard deviation.

Histograms portray statistical data.

Given that the natural world is rife with variables, it is not surprising to find that variation is widespread. Trees have different heights, ocean temperatures change from place to place and from top to bottom, the individuals of a population have different ages, and so on. Natural selection thrives on variation. Variation is often due to chance events, thus the height of a tree depends on its genetic makeup, the soil it grows in, rainfall, and sunlight, among other things. Describing variation is a science all its own.

Since pictures are worth many words, we start with histograms. Corresponding to the phenomenon under study, any variation observed occurs within a specific range of possibilities, a *sample space*. This range of possibilities is then partitioned or divided up into a number of subranges or classes. A *histogram* is a graph of the fraction of observations falling within the various subranges plotted against those subranges.

Consider the recent age distribution data for the U.S. population, shown in Table 2.6.1. The possible range of ages, 0 to infinity, is partitioned into subranges or intervals of every 5 years from birth to age 80; a last interval, 80+, has been added for completeness. The table lists the percentage of the total population falling within the given interval; each percentage is also refined by sex. The cumulative percentage is also given, that is the sum of the percentages up to and including the given interval. A histogram is a graph of these data; on each partition interval is placed a rectangle or bar whose width is that of the interval and whose height is the corresponding percentage (see Figure 2.6.1).

The resolution of a histogram is determined by the choice of subranges; smaller and more numerous intervals mean better resolution and more accurate determination of the distribution, larger and fewer intervals entail less data storage and processing.

Table 2.6.1 Age Distribution for the U.S. Population

Age	% Female	% Male	% Population	Cumulative
0–4	3.6	3.6	7.2	7.2
5–9	3.9	3.7	7.6	14.8
10–14	4.1	3.9	8.0	22.8
15–19	4.7	4.3	9.0	31.8
20–24	5.0	4.2	9.2	41.0
25–29	4.3	4.0	8.3	49.3
30–34	4.0	3.5	7.5	56.8
35–39	3.6	2.9	6.5	63.3
40–44	2.7	2.2	4.9	68.2
45–49	2.8	2.0	4.8	73.0
50–54	3.0	2.2	5.2	78.2
55–59	3.1	2.1	5.2	83.4
60–64	2.8	1.9	4.7	88.1
65–69	2.3	1.8	4.1	92.2
70–74	2.0	1.4	3.4	95.6
75–79	1.7	0.8	2.5	98.1
80+	1.6	0.3	1.9	100

```
> mcent:=[3.6, 3.7, 3.9, 4.3, 4.2, 4.0, 3.5, 2.9, 2.2, 2.0, 2.2,
     2.1, 1.9, 1.8,1.4, 0.8, 0.3]:
  fcent:=[3.6, 3.9, 4.1, 4.7, 5.0, 4.3, 4.0, 3.6, 2.7, 2.8, 3.0,
     3.1, 2.8, 2.3, 2.0, 1.7, 1.6]:
  tot:=[seq(mcent[i]+fcent[i],i=1..17)]:
> ranges:=[0..5, 5..10, 10..15, 15..20, 20..25, 25..30, 30..35, 35..40, 40..45,
     45..50, 50..55, 55..60, 60..65, 65..70, 70..75, 75..80, 80..85]:
> with(stats): with(plots):
> mpop:=[seq(Weight(ranges[i],5*mcent[i]),i=1..17)]:
  fpop:=[seq(Weight(ranges[i],5*fcent[i]),i=1..17)]:
  pop:=[seq(Weight(ranges[i],5*tot[i]),i=1..17)]:
> statplots[histogram](pop);
```

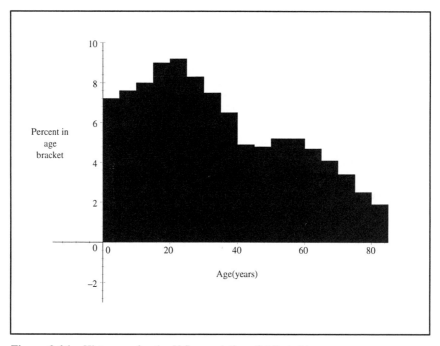

Figure 2.6.1   Histogram for the U.S. population distributed by age

The cumulative values are plotted in Figure 2.6.2. Since the percentage values have a resolution of 5 years, a decision has to be made about where the increments should appear in the cumulative plot. For example, 7.2% of the population is in the first age interval counting those who have not yet reached their 5th birthday. Should this increment be placed at age 0, at age 5, or maybe at age 2.5 in the cumulative graph?

We have chosen to do something different, namely to indicate this information as a line segment that is 0 at age 0 and 7.2 at age 5. In like fashion, we indicate in the cumulative graph the second bar of the histogram of height 7.6% as a line segment joining the points 7.2 at age 5 with $14.8(= 7.2 + 7.6)$ at age 10. Contin-

```
> age:=[2.5, 7.5, 12.5, 17.5, 22.5, 27.5, 32.5, 37.5, 42.5, 47.5, 52.5, 57.5,
    62.5, 67.5, 72.5, 77.5, 82.5];
> cummale:=[seq(sum('mcent[i]','i'=1..n),n=1..17)]:
  cumfale:=[seq(sum('fcent[i]','i'=1..n),n=1..17)]:
  cumtot:=[seq(sum('tot[i]','i'=1..n),n=1..17)]:
> ptsm:=[seq([age[i],cummale[i]],i=1..17)];
  ptsf:=[seq([age[i],cumfale[i]],i=1..17)]:
  ptsT:=[seq([age[i],cumtot[i]],i=1..17)]:
> plot(ptsm,ptsf,ptsT);
```

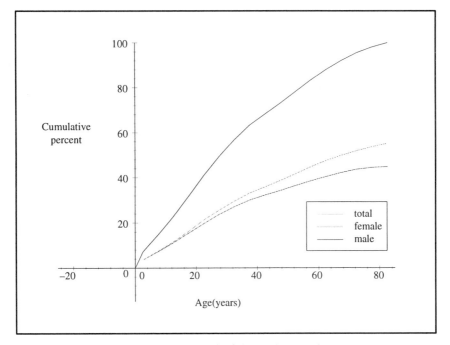

Figure 2.6.2 Cumulative populations (% of the total vs. age)

uing this idea for the balance of the data produces the figure. Our rationale here is the assumption that the people within any age group are approximately evenly distributed by age in this group. A graph that consists of joined line segments is called a *polygonal graph* or a *linear spline*.

This graph of accumulated percentages is called the *cumulative distribution function* or *cdf* for short. No matter what decision is made about placing the cumulative percentages, the *cdf* satisfies these properties: (1) it starts at 0, (2) it never decreases, and (3) it eventually reaches 1 (or in percentage, 100%).

The mean and median approximately locate the center of the distribution.

Sometimes it is convenient to summarize the information in a histogram. Of course no single number or pair of numbers can convey all the information; such a

summary is therefore a compromise but nevertheless a useful one. First, some information about where the data lie is given by the *mean* or *average*; it is frequently
denoted μ. Given the n values $x_1, x_2, \ldots, x_n$, their mean is

$$\mu = (x_1 + x_2 + \ldots + x_n)/n$$

$$\mu = \frac{1}{n} \sum_{i=1}^{n} x_i. \tag{2.6.1}$$

Another popular notation for this quotient is $\bar{x}$. It is necessarily true that some
values x_i are smaller than the mean and some are larger. In fact, one understands
the mean to be in the center of the x values in a sense made precise by equation (2.6.1). Of course, given $\bar{x}$ and n, the sum of the x's is easily computed,

$$\sum_{i=1}^{n} x_i = n\bar{x}.$$

Computing the mean of a *histogram* goes somewhat differently. Suppose the
total number of people referred to in Table 2.6.1 to be 100 million. (It no doubt
corresponds to many more than that, but it will be more convenient to calculate
percentages using 100 million, and we will see that, in the end, this choice is
irrelevant.) Then the 7.2% in the first group translates into 7.2 million people.
We don't know their individual ages, but as above, if they were evenly distributed
over ages 0 to 4.999..., then counting all 7.2 million as 2.5 gives the same result.
Hence in equation (2.6.1) these people contribute a value of 2.5 for 7.2 million
such people, or

$$\text{contribution of "0 to 5" group} = 2.5 \cdot 7.2 = \frac{0+5}{2} \cdot 7.2$$

in millions. Similarly, the second group contributes

$$\text{contribution of "5 to 10" group} = 7.5 \cdot 7.8 = \frac{5+10}{2} \cdot 7.8.$$

Continuing in this way we get, with $n = 100$ million,

```
> Sum('age[n]*tot[n]','n'=1..17)=sum('age[n]*tot[n]','n'=1..17);
```

$$\sum_{i=1}^{n} x_i = 2.5 \cdot 7.2 + 7.5 \cdot 7.6 + 12.5 \cdot 8.0 + \ldots + 82.5 \cdot 1.9 = 3431.0.$$

Divide the result by 100 (million) for the answer. But dividing by 100 million
means a quotient such as 7.2 million/100 million is just the fraction .072 (or
7.2%). In other words, we don't need to know the total population size; instead

we just use the fractions, such as .072, as multipliers or weights for their corresponding interval. Completing the calculation, then,

$$\bar{x} = 2.5 \cdot 0.072 + 7.5 \cdot 0.076 + \ldots + 82.5 \cdot 0.019 = 34.31. \quad (2.6.2)$$

Equation (2.6.2) illustrates a general principle for calculating the mean, which applies to equation (2.6.1) as well,

$$\mu = \sum_{\substack{\text{over possible} \\ \text{values } x}} x \cdot \text{fraction of values equal to } x. \quad (2.6.3)$$

In equation (2.6.2) the possible x's are 2.5, 7.5, and so on, while the fractions are .072, .076, and so on. In equation (2.6.1) the possible x's are x_1, x_2, and so on, while the fraction of values that are x_1 is just 1 out of n, that is $1/n$, and similarly for the other x_i's.

The *median* is an alternative to the mean for characterizing the center of a distribution. The median, $\hat{x}$, of a set of values $x_1, x_2, \ldots, x_n$ is such that one half the values are less than or equal to $\hat{x}$ and one half are greater than or equal to it. If n is odd, $\hat{x}$ will be one of the x's. If n is even, then $\hat{x}$ should be taken as the average of the middle two x values. For example, the median of the values 1, 3, 6, 7, and 15 is $\hat{x} = 6$, while the median of 1, 3, 6, and 7 is $(3 + 6)/2 = 4.5$.

The median is sometimes preferable to the mean because it is a more typical value. For example, for the values 3, 3, 3, 3, and 1000, the mean is 506 while the median is 3.

In the population data, the median age for men and women is between 29 and 30. This can be seen from an examination of the last column of Table 2.6.1. Contrast this median age with the average age, thus for men:

```
> Sum('mcent[n]*age[n]','n'=1..17)/Sum('mcent[n]','n'=1..17)
  = sum('mcent[n]*age[n]','n'=1..17)/sum('mcent[n]','n'=1..17);
```

$$\text{average age for men} = \frac{\sum_{n=1}^{17} [\text{percentage men at age } n] \cdot [\text{ age } [n]]}{\text{total percentage of men}}$$

$$= 32.17$$

In a similar manner, the average age for women in this data set is about 35.5, and the average age for the total population is about 33.8. The averages for these three sets of data—male population age distribution, female population age distribution, and total population age distribution—can be found with simple one-line commands and agree with our paper-and-pen calculations.

```
> with(describe): mean(pop); median(pop);
```

Variance and standard deviation measure dispersion.

As mentioned above, a single number will not be able to capture all the information in a histogram. The data set 60, 60, 60, 60 has a mean of 60 as does the data set 30, 0, 120, 90. If these data referred to possible speeds in miles per hour for a trip across Nevada by bus for two different bus companies, then we might prefer our chances with the first company. The *variance* of a data set measures how widely the data is dispersed from the mean; for n values $x_1, x_2, \ldots, x_n$, their variance v, or sometimes σ^2, is defined as

$$v = \frac{1}{n} \sum_{i=1}^{n} (x_i - \bar{x})^2 \tag{2.6.4}$$

where $\bar{x}$ is the mean as before.[6] Thus the speed variance for the first bus company is 0 and that for the second bus company is

$$\frac{1}{4}\left[(30 - 60)^2 + (0 - 60)^2 + (120 - 60)^2 + (90 - 60)^2\right] = 2,250.$$

As before, a more general equation for variance, one suitable for histograms, for example, is the following:

$$v = \sum_{\substack{\text{over possible} \\ \text{values } x}} (x - \bar{x})^2 \cdot \text{fraction of values equal to } x. \tag{2.6.5}$$

A problem with variance is that it corresponds to squared data values making it hard to interpret its meaning in terms of the original data. If the data have units, like miles per hour, then variance is in the square of those units. Closely related to variance is *standard deviation*, denoted σ. Standard deviation is defined as the square root of variance,

$$\text{standard deviation} = \sqrt{\text{variance}}.$$

Standard deviation is a measure of the dispersion of data on the same scale as the data itself. The standard deviation of bus speeds for the second company is 47.4 miles per hour. This is not saying that the average (unsigned) deviation of the data from the mean is 47.4 (for that would be $\frac{1}{n}\sum_1^n |x_i - \bar{x}| = 45$), but this is approximately what the standard deviation measures. For the bus companies, we make these calculations:

---

[6]For data representing a sample drawn from some distribution, $\bar{x}$ is only an estimate of the distribution's mean. For that reason, this definition of variance is a biased estimator of the distribution's variance. Divide by $n - 1$ in place of n for an unbiased estimator. Our definition is, however, the maximum likelihood estimator of the variance for normal distributions. Furthermore, this definition is consistent with the definition of variance for probability distributions, see Section 2.7, and for that reason we prefer it.

```
> bus1:=[60,60,60,60]; bus2:=[30,0,120,90];
> range(bus1), range(bus2);
  median(bus1), median(bus2);
  mean(bus1), mean(bus2);
  variance(bus1), variance(bus2);
  standarddeviation(bus1), standarddeviation(bus2);
```

We can perform similar calculations for the U.S. census data in Table 2.6.1. The results are given in Table 2.6.2.

```
> range(mpop), range(fpop), range(pop);
  median(mpop), median(fpop), median(pop);
  mean(mpop), mean(fpop), mean(pop);
  variance(mpop), variance(fpop), variance(pop);
  standarddeviation(mpop), standarddeviation(fpop),
  standarddeviation(pop);
```

Table 2.6.2   Summary for U.S. Age Distribution

	Range	Median	Mean	Standard Deviation
Male	0–84	29	31.7	21.16
Female	0–84	30	35.6	22.68
Total	0–84	29	33.8	22.10

The normal distribution is everywhere.

It is well known that histograms are often bell-shaped. This is especially true in the biological sciences. The mathematician Karl Gauss discovered the explanation for this, and it is now known as the *Central Limit Theorem* (see Hogg and Craig [6]).

Theorem 1

Central Limit Theorem. The accumulated result of many independent random outcomes, in the limit, tends to a Gaussian or normal distribution given by

$$G(x) = \frac{1}{\sqrt{2\pi}\sigma}e^{-1/2((x-\mu)/\sigma)^2}, \qquad -\infty < x < \infty,$$

where μ and σ are the mean and standard deviation of the distribution.

The normal distribution is a continuous distribution meaning that its resolution is infinitely fine; its histogram, given by $G(x)$, is smooth (see Figure 2.6.3). The two parameters mean μ and standard deviation σ completely determine the normal dis-

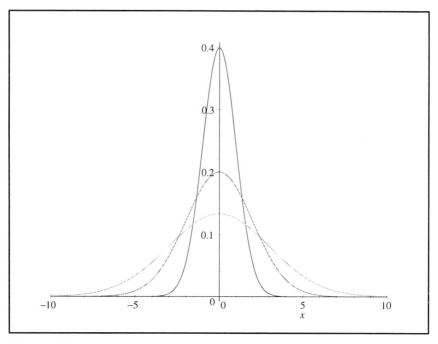

Figure 2.6.3

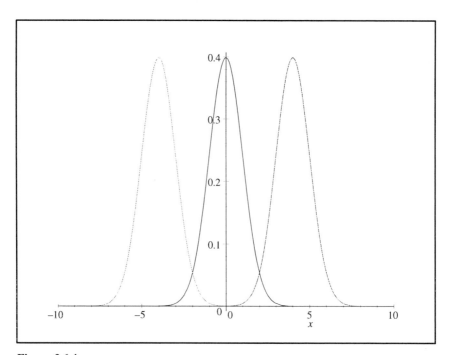

Figure 2.6.4

Table 2.6.3

Age range	Deaths per 100,000	Age range	Deaths per 100,000
0–1	1122.4	40–45	287.8
1–5	55.1	45–50	487.2
5–10	27.5	50–55	711.2
10–15	33.4	55–60	1116.9
15–20	118.4	60–65	1685.1
20–25	139.6	65–70	2435.5
25–30	158.0	70–75	3632.4
30–35	196.4	75–80	5300.0
35–40	231.0	80–85	8142.0
		85+	15279.0

tribution. Likewise, even though a given histogram is not Gaussian, nevertheless its description is often made in terms of just its mean and variance or standard deviation.

We show three curves in Figure 2.6.3 with the same mean and different standard deviations. In Figure 2.6.4, there are three curves with the same standard deviation and with different means.

```
> y:=(sigma,mu,x)->exp(-(x-mu)^2/(2*sigma^2))/ (sqrt(2*Pi)*sigma);
> plot({y(1,0,x),y(2,0,x),y(3,0,x)},x=-10..10);
> plot({y(1,-4,x),y(1,0,x),y(1,4,x)},x=-10..10);
```

Exercises

1. In the February 1994 Epidemiology Report published by the Alabama Department of Public Health, the following data was provided as Age-Specific Mortality. Make a histogram for this data. While the data is given over age ranges, determine a fit for the data so that one could predict the death rate for intermediate years. Find the median, mean, and standard deviation for the data.

```
> with(stats): with (plots): with(describe):
> Mort:=[1122.4, 55.1, 27.5, 33.4, 118.4, 139.6, 158.0, 196.4, 231.0,
        287.8, 487.2, 711.2, 1116.9, 1685.1, 2435.5, 3632.4, 5300.0, 8142.0, 15278.0]:
> MortRate:=[seq(Mort[i]/100000,i=1..19)];
> ranges=[seq(5*i..5*(i+1),i=1..17)];
> mortdata=[Weight(0..1,MortRate[1]), Weight(1..5,4*MortRate[2]),
        seq(Weight(ranges[i],5*MortRate[2+i]),i=1..17)]:
> statplots[histogram](mortdata);
```

a. a polynomial fit

```
> xcord:=[seq(3+5*(i-1),i=1..18)];
   mortrate:=[seq(MortRate[i+1], i=1..18)];
```

```
> plot([seq([xcord[i],mortrate[i]],i=1..18)],style=POINT, symbol=CROSS);
> fit[leastsquare[[x,y],y=a+b*x+c*x^2+d*x^3]]([xcord,mort rate]);
> approx:=unapply(rhs(" "),x);
  approx(30);
> plot(approx(x),x=0..90);
```

b. an exponential fit

```
> Lnmortrate:=map(ln,mortate);
> fit[leastsquare[[x,y],y=m*x+b]]([xcord,Lnmortrate]);
> k:=op(1,op(1,rhs(" "))); A:=op(2,rhs(" "));
> expfit:=t->exp(A)*exp(k*t);
  expfit(30)*10000;
> J:=plot(expfit(t),t=0..85):
  K:=plot([seq([xcord[i],MortRate[i+1]],i=1..18)],
    style=POINT,symbol=CROSS):
> display({J,K});
```

c. a linear spline for the data (see the discussion in this section)

```
> readlib(spline):
> linefit:=spline(xcord,mortrate,x,linear):
> y:='spline/makeproc'(linefit,x):
> J:=plot('y(t)','t'=0..85):
> display({J,K});
```

Give the range, median, mean, and standard deviation of the mortality rates. Note that the first entry is applicable to humans in an age group of width one year and the second is in a group of width four years. Each of the others applies to a span of five years. Thus we set up a weighted sum.

```
> summary:=[Weight(Mort[1],1),Weight(Mort[2],4),
    seq(Weight(Mort[i],5),i=3..19)];
> range(summary); median(summary); mean(summary);
    standarddeviation(summary);
```

2. What follows are data for the heights of a group of males. Determine a histogram for the data. Find the range, median, mean, and standard deviation for the data. Give a normal distribution with the same mean and standard deviation as the data. Plot the data and the distribution on the same graph.

Number of Males	2	1	2	7	10	14	7	5	2	1
Height (in.)	66	67	68	69	70	71	72	73	74	75

```
> with(stats): with(plots): with(describe):
> htinches:=[seq(60+i,i=1..15)];
  numMales:=[0,0,0,0,0,2,1,2,7,10,14,7,5,2,1];
  ranges:=[seq(htinches[i]..htinches[i]+1,i=1..15)];
  maledata:=[seq(Weight(ranges[i],numMales[i]),i=1..15)];
> statplots[histogram](maledata);
> range(maledata); median(maledata); mean(maledata);
  standarddeviation(maledata);
> 'The average height is',floor(" "/12),'feet and',floor(frac(" "/12)*12),'inches';
> 'The standard deviation is', floor(frac(" "/12)*12), 'inches';
```

In what follows, we give a normal distribution that has the same mean and standard deviation as the height data.

```
> mu:=mean(maledata); sigma:=standarddeviation(maledata);
> ND:=x->exp(-(x-mu)^2/(2*sigma^2))/(sigma*sqrt(2*Pi));
> J:=plot(mu*ND(x),x=60..76):
> lineplot:= seq([t,numMales[i],t=60+i..60+i+1],i=1..15);
> K:=plot(lineplot):
> plots[display]({J,K});
```

To the extent that the graph K is an approximation for the graph J, the heights are normally distributed about the mean.

3. Here are population data estimates for the United States (in thousands) as published by the U.S. Bureau of the Census, Population Division, release PPL-21 (1995).

Five Year Age Groups	1990	1995	Five Year Age Groups	1990	1995
0–5	18,849	19,662	50–55	11,368	13,525
5–10	18,062	19,081	55–60	10,473	11,020
10–15	17,189	18,863	60–65	10,619	10,065
15–20	17,749	17,883	65–70	10,077	9,929
20–25	19,133	18,043	70–75	8022	8816
25–30	21,232	18,990	75–80	6145	6637
30–35	21,907	22,012	80–85	3934	4424
35–40	19,975	22,166	85–90	2049	2300
40–45	17,790	20,072	90–95	764	982
45–50	13,820	17,190	95–100	207	257
			100+	37	52

Find the median and mean ages. Estimate the number of people at ages 21, 22, 23, 24, and 25 in 1990 and in 1995. Make a histogram for the percentages of the population in each age catagory for both population estimates.

4. In equation (2.6.3) we stated that the mean μ is defined as

$$\mu = \sum_{\text{all possible } x\text{'s}} x \cdot f(x),$$

where $f(x)$ is the fraction of all values equal to x. If these values are spread continuously over all numbers, μ can be conceived of as an integral. In this sense, this integral of the normal distribution given by equation (2.6.3) yields

$$\mu = \int_{-\infty}^{\infty} x \frac{1}{\sigma\sqrt{2\pi}} \exp\left(-\frac{1}{2}\left(\frac{x-\mu}{\sigma}\right)^2\right) dx.$$

In a similar manner,

$$\sigma^2 = \int_{-\infty}^{\infty} (x-\mu)^2 \frac{1}{\sigma\sqrt{2\pi}} \exp\left(-\frac{1}{2}\left(\frac{x-\mu}{\sigma}\right)^2\right) dx.$$

Here is a way to evaluate the integrals with *Maple*:

```
> f:=x->exp(-(x-mu)^2/(2*sigma^2))/(sigma*sqrt(2*Pi));
> assume(sigma > 0);
> int(x*f(x),x=-infinity..infinity);
> int((x-mu)^2*f(x),x=-infinity..infinity);
```

Section 2.7

Probability

The biosphere is a complicated place. One complication is its unpredictable events, such as when a tree will fall or exactly what the genome of an offspring will be. Probability theory deals with unpredictable events by making predictions in the form of relative frequency of outcomes. Histograms portray the distribution of these relative frequencies, and serve to characterize the underlying phenomenon.

Statistics deals with the construction and subsequent analysis of histograms retroactively, that is, from observed data. Probability deals with the prediction of histograms by calculation. In this regard, important properties to look for in calculating probabilities are independence, disjointness, and equal likelihood.

Probabilities and their distributions.

Probability theory applies mathematical principles to random phenomena in order to make precise statements and accurate predictions about seemingly unpredictable events. The probability of an event E, written $\Pr(E)$, is the fraction of times E occurs in an infinitely long sequence of trials. (Defining probability is difficult to do without being circular and without requiring experimentation. A definition requiring the performance of infinitely many trials is obviously undesirable. The situation is similar to that in geometry where the term "point" is necessarily left undefined; despite this, geometry has enjoyed great success.) For example, let an "experiment" consist of rolling a single die for which each of the six faces has an equal chance of landing up. Take event E to mean a 3 or a 5 lands up. Evidently the probability of E is then 1/3, $\Pr(E) = 1/3$, that is rolling a 3 or 5 will happen approximately one third of the time in a large number of rolls.

More generally, by an *event* E, in a probabilistic experiment, we mean some designated set of outcomes of the experiment. The number of outcomes, or *cardinality*, of E is denoted $|E|$. The set of all possible outcomes, or *universe*, for an experiment is denoted U. Here are some fundamental laws:

Principle of Universality   One of the possible outcomes of an experiment will occur with certainty

$$\Pr(U) = 1. \qquad (2.7.1)$$

Principle of Disjoint Events   If events E and F are *disjoint*, $E \cap F = \emptyset$, that is they have no outcomes in common, then the probability that E or F will occur (sometimes written $E \cup F$) is the sum

$$\Pr(E \text{ or } F) = \Pr(E) + \Pr(F). \qquad (2.7.2)$$

Principle of Equally Likelihood   Suppose each outcome in U has the same chance of occurring, i.e., is *equally likely*. Then the probability of an event E is the ratio of the number of outcomes making up E to the total number of outcomes,

$$\Pr(E) = \frac{|E|}{|U|}. \qquad (2.7.3)$$

To illustrate, consider the experiment of rolling a pair of dice, one red and one green. That any one of 6 numbers can come up on each die is equally likely, so the total number of possibilities is 36; the first possibility could be 1 on red and 1 on green; the second, 1 on red and 2 on green, and so on. In this scheme, the last would be 6 on red and 6 on green. So $|U| = 36$. There are two ways to roll an 11, a 5 on red and 6 on green or the other way around. So letting E be the event that an 11 is rolled, we have $\Pr(E) = 2/36 = 1/18$. Let S be the event that a 7

```
> with(stats): with(plots): with(describe):
> roll:=[seq(n,n=2..12)];
> prob:=[1/36,2/36,3/36,4/36,5/36,6/36,5/36,4/36,3/36,2/36,1/36];
> wtroll:=[seq(Weight(roll[i]-1/2..roll[i]+1/2,prob[i]),i=1..11)];
> statplots[histogram](wtroll);
```

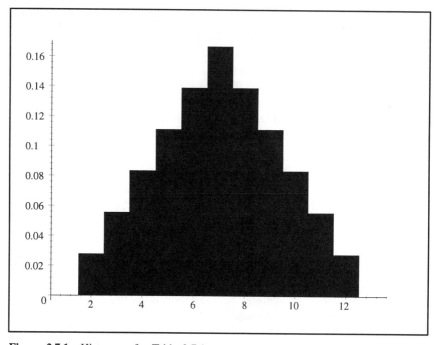

Figure 2.7.1   Histogram for Table 2.7.1

is rolled; this can happen in 6 different ways, so $\Pr(S) = 6/36 = 1/6$. Now the probability that a 7 or 11 is rolled is their sum

$$\Pr(S \cup E) = \Pr(S) + \Pr(E) = \frac{2+6}{36} = \frac{2}{9}.$$

Since probabilities are frequencies of occurrence, they share properties with statistical distributions. Probability distributions can be visualized by histograms and their mean and variance calculated. For example, let the variable X denote the outcome of the roll of a pair of dice, Table 2.7.1 gives the possible outcomes of X along with their probabilities. Figure 2.7.1 graphically portrays the table

Table 2.7.1   Probabilities for a dice roll

roll	2	3	4	5	6	7	8	9	10	11	12
probability	$\frac{1}{36}$	$\frac{2}{36}$	$\frac{3}{36}$	$\frac{4}{36}$	$\frac{5}{36}$	$\frac{6}{36}$	$\frac{5}{36}$	$\frac{4}{36}$	$\frac{3}{36}$	$\frac{2}{36}$	$\frac{1}{36}$

as a histogram. Just as in the previous section, the rectangle on x represents the fraction of times a dice roll will be x.

Equation (2.6.3) can be used to calculate the mean value $\bar{X}$ of the random variable X, also known as its *expected* value, $E(X)$,

$$\bar{X} = \sum_{\substack{\text{over all possible} \\ \text{values } x \text{ of } X}} x \cdot \Pr(X = x). \qquad (2.7.4)$$

From Table 2.7.1,

$$E(X) = 2\frac{1}{36} + 3\frac{2}{36} + 4\frac{3}{36} + 5\frac{4}{36} + 6\frac{5}{36} + 7\frac{6}{36}$$
$$+ 8\frac{5}{36} + 9\frac{4}{36} + 10\frac{3}{36} + 11\frac{2}{36} + 12\frac{1}{36} = 7.$$

> Sum('roll[i]*prob[i]','i'=1..11)=sum('roll[i]*prob[i]','i'=1..11);

Similarly, the variance is defined as

$$V(X) = E(X - \bar{X})^2 = \sum_{\substack{\text{over all possible} \\ \text{values } x \text{ of } X}} (x - \bar{X})^2 \cdot \Pr(X = x). \qquad (2.7.5)$$

For a dice roll

$$V(X) = (2-7)^2\frac{1}{36} + (3-7)^2\frac{2}{36} + (4-7)^2\frac{3}{36} + (5-7)^2\frac{4}{36}$$
$$+ (6-7)^2\frac{5}{36} + (7-7)^2\frac{6}{36} + (8-7)^2\frac{5}{36} + (9-7)^2\frac{4}{36}$$
$$+ (10-7)^2\frac{3}{36} + (11-7)^2\frac{2}{36} + (12-7)^2\frac{1}{36} = \frac{35}{6}.$$

> Sum('(roll[i]-7)^2*prob[i]','i'=1..11)=sum('(roll[i]-7)^2*prob[i]','i'=1..11);
> mean(wtroll);variance(wtroll);

Probability calculations can be simplified by decomposition and independence.

Consider the experiment of tossing a fair coin in the air four times and observing the result. Suppose we want to calculate the probability that heads will come up 3 of the 4 times. This grand experiment consists of four sub-experiments, namely, the four individual coin tosses. Decomposing a probability experiment into sub-experiments can often simplify making probability calculations. This is especially true if the sub-experiments, and therefore their events, are *independent*. Two events E and F are independent when the fact that one of them has or has not occurred has no bearing on the other.

Principle of Independence   If two events E and F are independent, then the probability that both will occur is the product of their individual probabilities

$$\Pr(E \text{ and } F) = \Pr(E) \cdot \Pr(F).$$

One way 3 heads in 4 tosses can occur by getting a head the first three tosses and a tail on the last one; we will denote this by $HHHT$. Since the four tosses are independent, to calculate the probability of this outcome, we just multiply the individual probabilities of an H the first, second and third times, and, on the fourth, a T; each of these has chance 1/2, hence

$$\Pr(HHHT) = \left(\frac{1}{2}\right)^4 = \frac{1}{16}.$$

There are three other ways that 3 of the 4 tosses will be H; $HHTH$, $HTHH$, and $THHH$. Each of these is also 1/16 probable, therefore, by the principle of disjoint events,

$$\Pr(3 \text{ heads out of 4 tosses}) = 4 \cdot \frac{1}{16} = \frac{1}{4}.$$

Permutations and combinations are at the core of probability calculations.

The previous example raises a question: By direct enumeration we found there are 4 ways to get 3 heads (or equivalently 1 tail) in 4 tosses of a coin, but how can we calculate, for example, the number of ways to get 8 heads in 14 coin tosses, or, in general, k heads in n coin tosses? This is the problem of counting *combinations*.

In answer, consider the following experiment: Place balls labeled 1, 2, and so on to n in a hat and select k of them at random to decide where to place the H's. For instance, if $n = 4$ and $k = 3$, the selected balls might be 3, then 4, then 1, signifying the sequence $HTHH$.

As a sub-question, in how many ways can the balls 1, 3, and 4 be selected? This is the problem of counting *permutations*, the various ways to order a set of objects. Actually, there are 6 permutations. They are $(1, 3, 4)$, $(1, 4, 3)$, $(3, 1, 4)$, $(3, 4, 1)$, $(4, 1, 3)$, and $(4, 3, 1)$. There are 3 choices for the first ball from the possibilities 1, 3, or 4. Having been made, there are two remaining choices for the second and, finally, only one possibility for the last. Hence the number of permutations of three objects $= 3 \cdot 2 \cdot 1 = 6$.

```
> with(combinat):
> permute([1,3,4]);
> numbperm(3);
```

More generally, the number of permutations of n objects is

$$\text{number of permutations of } n \text{ objects} = n \cdot (n-1) \cdot (n-2) \cdot \ldots \cdot 2 \cdot 1 = n!.$$

As indicated, this product is written $n!$ and called *n factorial*.

So, in a similar fashion, the number of ways to select k balls from a hat holding n balls is

$$n \cdot (n - 1) \cdot (n - 2) \cdot \ldots \cdot (n - k + 1).$$

As we said above, the labels on the selected balls signify when heads occur in the n tosses. But every such choice has $k!$ permutations, all of which also give k heads. Therefore, the number of ways of getting k heads in n tosses is

$$\frac{n(n - 1)(n - 2) \ldots (n - k + 1)}{k(k - 1) \ldots 2 \cdot 1}. \tag{2.7.6}$$

```
> with(combinat):
> numbcomb(6,3);
> binomial(6,3);
```

The value calculated by equation (2.7.6) is known as the number of combinations of n objects taken k at a time. This ratio occurs so frequently that there is a shorthand notation for it, $\binom{n}{k}$, called *n choose k*. An alternative form of $\binom{n}{k}$ is

$$\binom{n}{k} = \frac{n(n - 1) \ldots (n - k + 1)}{k(k - 1) \ldots 2 \cdot 1} = \frac{n!}{k!(n - k)!} \tag{2.7.7}$$

where the third member follows from the second by multiplying numerator and denominator by $(n - k)!$.

Some elementary facts about n choose k follow. For consistency in these formulas, zero factorial is defined to be 1:

$$0! = 1.$$

The first three combination numbers are

$$\binom{n}{0} = 1, \qquad \binom{n}{1} = n, \qquad \binom{n}{2} = \frac{n(n - 1)}{2}.$$

There is a symmetry,

$$\binom{n}{k} = \binom{n}{n - k} \qquad \text{for all } k = 0, 1, \ldots, n.$$

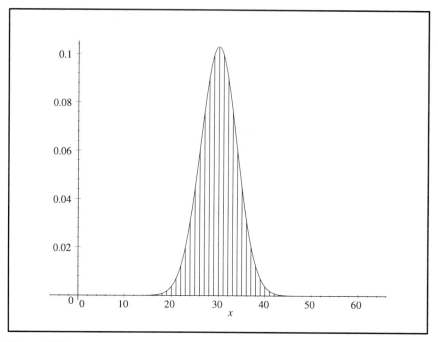

Figure 2.7.2

These numbers, n choose k, occur in the Binomial Theorem, which states, for any p and q,

$$\sum_{k=0}^{n} \binom{n}{k} p^k q^{n-k} = (p+q)^n. \qquad (2.7.8)$$

Finally, the probability of realizing k heads in n tosses of a fair coin, denoting it as $H_n(k)$, is

$$H_n(k) = \binom{n}{k} \left(\frac{1}{2}\right)^n, \quad k = 0, 1, \dots, n. \qquad (2.7.9)$$

The distribution $H_n(k)$ is shown in Figure 2.7.2 for $n = 60$. If the coin were not fair, with the probability of a heads being p and that of a tails being $q = 1 - p$, then $H_n(k)$ becomes

$$H_n(k) = \binom{n}{k} p^k q^{n-k}, \quad k = 0, 1, \dots, n. \qquad (2.7.10)$$

Continuous variations require continuous distributions.

In Figure 2.7.2 we show the heads distribution histogram $H_{60}(k)$ for 60 coin tosses. Notice that the distribution takes on the characteristic bell shape of the Gaussian distribution as predicted by the Central Limit Theorem discussed in the previous section:

$$G(x) = \frac{1}{\sqrt{2\pi\sigma}} e^{-1/2((x-\mu)/\sigma)^2}, \qquad -\infty < x < \infty,$$

where μ and σ are the mean and standard deviation. In Figure 2.7.2 we have superimposed the Gaussian over the histogram. In order to get the approximation right, we must match the means and variances of the two distributions. The mean of $H_n(k)$ for a biased coin, equation (2.7.9), is given by [7]

$$\mu = np. \tag{2.7.11}$$

And the variance of $H_n(k)$ is, (see Reference [6])

$$v = npq. \tag{2.7.12}$$

With $p = q = 1/2$ and $n = 60$, we get $\mu = 30$ and $\sigma^2 = 15$.

```
> n:= 60;
> flip:=[seq(binomial(n,i)*(1/2)^i*(1-1/2)^(n-i),i=0..n)]:
> wthd:=[seq(Weight(i-1,flip[i]),i=1..n+1)]:
> with(describe): mu:=mean(wthd); sigma:=standarddeviation(wthd);
```

Hence, Figure 2.7.2, shows the graph of

$$G(x) = \frac{1}{\sqrt{2 \cdot 15 \cdot \pi}} e^{-1/2(x-30)^2/15}.$$

```
> G:=x->exp(-(x-mu)^2/(2*sigma^2))/(sigma*sqrt(2*Pi));
> J:=plot(G(x),x=0..n):
> K:=statplots[histogram](wthd):
> plots[display]({J,K});
```

---

[7] Using the fact that $k\binom{n}{k} = n\binom{n-1}{k-1}$ and the Binomial Theorem,

$$\mu = \sum_{k=0}^{n} k\binom{n}{k} p^k q^{n-k} = \sum_{r=0}^{n-1} n\binom{n-1}{r} p^{r+1} q^{n-r} = np.$$

The normal distribution is an example of a continuous distribution. Any non-negative function, $f(x) \geq 0$, where

$$\int_{-\infty}^{\infty} f(x)dx = 1,$$

can define a probability distribution. The condition that the total integral be 1 is dictated by the universality principle (2.7.1). In this role, such a function f is called a *probability density function*. Probabilities are given as integrals of f. For example, let X denote the outcome of the probabilistic experiment governed by f. Then the probability that X lies between 3 and 5, say, is exactly

$$\Pr(3 \leq X \leq 5) = \int_{3}^{5} f(x)dx.$$

Similarly, the probability that an outcome will lie in a *very* small interval of width dx at the point x is[8]

$$\Pr(X \text{ falls in an interval of width } dx \text{ at } x) = f(x)\,dx. \qquad (2.7.13)$$

This shows that outcomes are more likely to occur where f is large and less likely to occur where f is small.

The simplest continuous distribution is the *uniform* distribution,

$$u(x) = \text{constant}.$$

Evidently, for an experiment governed by the uniform distribution, an outcome is just as likely to be at one place as another. For example, butterflies fly in a kind of random flight path that confounds their predators. As a first approximation, we might hypothesize that a butterfly makes its new direction somewhere within 45 degrees of its present heading uniformly. Let Θ denote the butterfly's directional change, Θ is governed by the uniform probability law

$$u(\Theta) = \begin{cases} \text{constant,} & \text{if } -45 \leq \Theta \leq 45 \\ 0, & \text{otherwise.} \end{cases} \qquad (2.7.14)$$

By the universality principle,

$$\int_{-45}^{45} u(\Theta)d\Theta = 1;$$

therefore the constant must be 1/90 in equation (2.7.14).

---

[8] This equation is interpreted in the same spirit as the concept "velocity at a point" in dynamics which is the ratio of infinitesimals $\frac{dx}{dt}$.

Exercises

1. An undergraduate student in mathematics wants to apply to three of six grad-
    uate programs in mathematical biology. She will make a list of three programs
    in the order of her preferences. Since the order is important, this is a problem
    of permutations. How many such choices can she make? There are three ways
    to answer this: Let *Maple* make a list of all permutations of six objects taken
    three at a time and let *Maple* count them, or let *Maple* compute the number
    of permutations of three objects where the objects in each permutation are
    chosen from a list of six, or use a formula.

    ```
    > with(combinat):
    > permute([a,b,c,d,e,f],3);nops(" );
    > numbperm(6,3);
    > 6!/3!;
    ```

    The student must send a list of three references to any school to which she
    applies. There are six professors who know her abilities well, of whom she
    must choose three. Since the order is not important, this is a problem of com-
    binations. How many such lists can she make? This question can be answered
    the same three ways as the preceding question.

    ```
    > with(combinat):
    > choose([a,b,c,d,e,f],3);nops(" );
    > numbcomb(6,3);
    > 6!/(3!*(6-3)!);
    ```

2. Five patients need heart transplants and three hearts for transplant surgery
    are available. How many ways are there to make a list of recipients? How
    many ways are there to make a list of the two of the five who must wait for
    further donors? (The answer to the previous two questions should be the same
    because for every three recipients, there are two nonrecipients.) How many
    lists can be made for the possible recipients in the order in which the surgery
    will be performed?

    ```
    > with(combinat):
    > numbcomb(5,3); numbcomb(5,2);
    > numbperm(5,3);
    ```

3. Choose an integer in the interval $[1, 6]$. If a single die is thrown 300 times,
    explain why one would expect to get the chosen number about 50 times. Do
    this experiment and record how often each face of the die appears. Compare
    how much this deviates from 50.

    ```
    > with(stats);
    > die:=rand(1..6);
    ```

```
> for i from 1 to 6 do
    count.i:=0
  od:
> for i from 1 to 300 do
    n:=die():
    count.n:= count.n + 1:
  od:
> for i from 1 to 6 do
    print(count.i);
  od;
```

4. Simulate throwing a pair of dice for 360 times using a random number gen-
    erator and complete the table using the sums of the top faces.

Sums	Predicted	Simulated
2	10	—————
3	20	—————
4	30	—————
5	40	—————
6	50	—————
7	60	—————
8	50	—————
9	40	—————
10	30	—————
11	20	—————
12	10	—————

Calculate the mean and standard deviation for your sample and compare them
with the outcome probabilities. Draw a histogram for the simulated throws
on the same graph as a continuous, normal distribution defined by equation
(2.13), where the mean and standard deviation in equation (2.13) are the same
as for the predicted sums.
        Maple syntax that will generate an answer follows.

```
> with(stats): with(describe):
> red:=rand(1..6):
  blue:=rand(1..6):
> for i from 2 to 12 do
    count.i:=0:
  od:
> for i from 1 to 360 do
    n:=red()+blue():
    count.n:=count.n + 1;
  od:
```

```
> for i from 2 to 12 do
    print(count.i);
    od;
> throws:=[seq(Weight(i,count.i),i=2..12)];
> mean(throws); standarddeviation(throws);
> theory:= [Weight(2,10), Weight(3,20),Weight(4,30),Weight(5,40),
    Weight(6,50),Weight(7,60),Weight(8,50),Weight(9,40),
    Weight(10,30),Weight(11,20),Weight(12,10)];
> mu:=mean(theory); sigma:=standarddeviation(theory);
> y:=x->360*exp(-(x- mu)^2/(2*sigma^2))/(sigma*sqrt(2*Pi));
> J:=statplots[histogram](throws):
  K:=plot([x,y(x),x=0..14]):
> with(plots): display({J,K});
```

5. This exercise is a study of independent events. Suppose a couple's genetic makeup makes the probability that they should have a brown-eyed child equal to 3/4. Assume that the eye color for two children is a pair of independent events.

   a. What is the probability that the couple will have two blue-eyed children? one blue-eyed and one brown-eyed? two brown-eyed children? What is the sum of these probabilities?

   ```
   > binomial(2,0)*1/4*1/4; binomial(2,1)*3/4*1/4; binomial(2,2)*3/4*3/4;
     sum(binomial(2,j)*(3/4)^j*(1/4)^(2-j),j=0..2);
   ```

   b. Suppose that the couple have five children. What is the probability that among the five, exactly two will have brown eyes?

   ```
   > binomial(5,2)*(3/4)^2*(1/4)^3;
   ```

   c. What is the probability that among the five children, there are at least two with brown eyes?

   ```
   > sum(binomial(5,j)*(3/4)^j*(1/4)^(5-j),j=2..5);
   ```

References and Suggested Further Reading

1. **AIDS cases in the U.S.:** HIV/AIDS Surveillance Report, U. S. Department of Health and Human Services, Centers for Disease Control, Division of HIV/AIDS, Atlanta, GA, July 1993.
2. **Cubic growth of AIDS:** S. A. Colgate, E. A. Stanley, J. M. Hyman, S. P. Layne and C. Qualls, "Risk-Behavior Model of the Cubic Growth of Acquired Immunodeficiency Syndrome in the United States," *Proc. Natl. Acad. Sci. USA*, vol. 86, pp. 4793–4797, 1989.

3. **Ideal height and weight:** Sue Rodwell Williams, *Nutrition and Diet Therapy*, 2nd ed., The C. V. Mosby Company, Saint Louis, p. 655, 1973.
4. **Georgia Tech Exercise Laboratory:** Philip B. Sparling, Melinda Millard-Stafford, Linda B. Rosskopf, Linda Dicarlo, and Bryan T. Hinson, "Body composition by bioelectric impedance and densitometry in black women," *American Journal of Human Biology* **5**, pp. 111–117, 1993.
5. **Classical differential equations:** E. Kamke, *Differentialgleichungen Lösungs-methoden und Lösungen*, Chelsea Publishing Company, New York, NY, 1948.
6. **The Central Limit Theorem:** Robert Hogg and Allen Craig, *Intro. to Math. Statistics*, Macmillan, New York, NY, 1965.
7. **Mortality tables for Alabama:** "Epidemiology Report," Alabama Department of Public Health, IX (number 2), February, 1994.

Chapter 3

Reproduction and the Drive for Survival

Introduction to this chapter

This chapter is an introduction to cell structure and biological reproduction and the effects that they have upon the survival of species according to the Darwinian model of evolution. The Darwinian model of evolution postulates that all living systems must compete for resources that are too limited to sustain all the organisms that are born. Those organisms possessing properties that are best suited to the environment can survive and may pass the favored properties to their offspring.

A system is said to be *alive* if it has certain properties. These life properties, e.g., metabolism, reproduction and response to stimuli, interact with each other, and, in addition, the interactions themselves must be part of the list of life-properties.

Cells contain organelles, which are subcellular inclusions dedicated to performing specific tasks such as photosynthesis and protein synthesis. Membranes are organelles that are components of other organelles and are functional in their own right—they regulate material transport into and out of cells. Prokaryotic organisms (bacteria and blue-green algae) lack most organelles; all other organisms, called eukaryotes, have a wide range of organelles.

A cell's genetic information is contained along the length of certain organelles called chromosomes. In asexual reproduction, or *mitosis*, genetic material of one cell is exactly replicated and the identical copies are partitioned among two daughter cells. Thus, the daughter cells end up with genetic information identical to that of the parent cell, a decided advantage if the environment is one in which the parent cell thrived. In multicellular organisms, certain genes may be "turned off" in mitosis; the result will be cells with different behaviors, which leads to the various tissues found in multicellular organisms. Genetic information is not lost in this way; it is merely inactivated, often reversibly. Mitosis also increases the surface-to-volume ratio of cells, which allows the cell to take up food and release waste more easily.

Sexual reproduction, the combining of genetic information from two parents into one or more offspring, leads to variations among the offspring. This is achieved by the production of novel combinations of genetic information and by complex interactions between genetic materials affecting the same property. The result is the possibility for immense variation, which is one of the empirical observations at the heart of the Darwinian model.

Left unchecked, populations would grow exponentially, but factors in the environment always control the sizes of populations.

Section 3.1

The Darwinian Model of Evolution

We introduce the Darwinian model of evolution, a model that ties all biology together. Finite resources of all kinds place limits on the reproduction and growth of organisms. All must compete for these resources and most will not get enough. Those that survive may pass their favorable properties to their offspring.

The diversity of organisms is represented by taxonomic categories.

A group of organisms is said to represent a species if there is a real or potential exchange of genetic material among its members and they are reproductively isolated from all other such groups. Thus, members of a single species are capable of interbreeding and producing fertile offspring. By inference, if individuals are very similar but reproductively isolated from one another, they are in different species. The definition above makes good sense in most cases: Horses and cows, different species, live in the same area but never mate; horses and donkeys, different species, may live in the same area and interbreed, but their offspring are sterile mules; lions and tigers, also different species, do not live in the same area, but have interbred in zoos to give sterile offspring. The definition also produces some odd results: St. Bernard dogs and chihuahuas would be in different species by the reproductive-isolation criterion, although both might be in the same species as, say, a fox terrier. English sparrows in the United States and in England would have to be put into different species, even though they are essentially identical. There are other, somewhat different definitions of species.

A group of species is a *genus*, and a group of genera is a *family*. Higher levels are *orders*, *classes*, *phyla* (called *divisions* in plants), and *kingdoms*. To identify an organism, its generic and specific names are usually given in the following format: *Homo sapiens* (humans), or *Acer rubrum* (red maple trees).

Living systems operate under a set of powerful constraints.

1. *Available space is finite.* Some organisms can survive a few kilometers into the air or under water and others live a few meters under the ground, but that does not change the basic premise: Our planet is a sphere of fixed surface area

and everything alive must share that area for all its needs, including nutrient procurement and waste disposal.

2. *The temperature range for life is very restricted.* Most living systems cannot function if their internal temperature is outside a range of about 0° to 50°C, the lower limitation being imposed by the destructive effect of ice crystals on cell membranes, and the upper limit being imposed by heat inactivation of large molecules. Some organisms can extend this range a bit with special mechanisms, e.g., antifreeze-like substances in their bodies, but this temperature limitation is generally not too flexible.

3. *Energetic resources are limited.* The only energy sources originating on earth are geothermal, radioactive and that which is available in some inorganic compounds. Some organisms, said to be chemoautotrophic, can use the latter compounds, but these organisms are exceptional. By far, the majority of the energy available for life comes from the sun. While the sun's energy is virtually inexhaustible, it tends not to accumulate in any long-term biological form on earth. This limitation lies in an empirical observation—the Second Principle of Thermodynamics—that energy becomes less useful as it undergoes a transformation from one form to another. The transformations that solar energy undergoes are described by a food chain: The sun's energy is captured and used by photosynthetic plants, which are eaten by herbivores, which are eaten by carnivores, which die and are broken down by decomposing organisms. At each step, much of the useful energy is lost irreversibly to the immediate creation of disorder and/or to heat, which radiates away and creates disorder elsewhere. Thus, the sun's radiant energy does not accumulate in living systems for longer than a single organism's lifetime, and must be constantly replenished. (See the reference by Yeargers [1] for further discussion.)

4. *Physical resources are finite.* Obviously, there is more mass to the non-organic world than to the organic one. The problem is that most of the earth's non-organic mass is not available to the organisms that inhabit the earth's surface. For example, only tiny fractions of our planet's inventory of such critical materials as carbon, oxygen and nitrogen are actually available to life. The rest is either underground or tied up in the form of compounds not chemically accessible to life.

The Darwinian model of evolution correlates biological diversity and the survival of species.

The four constraints listed above would not be so serious if living organisms were different from what they are. We might picture a world in which every organism was non-reproducing, had a constant size, and was immortal. Perhaps the organisms would be photosynthetic and would have unlimited supplies of oxygen, carbon dioxide, nitrogen and other important inorganic substances. They would have infinite sinks for waste materials or would produce little waste in the first place.

The biological world just described is, of course, just the opposite of the real one, where there is rapid reproduction and a resultant competition for space and resources. Charles Darwin formulated a model to describe the nature and effect of this competition on living systems. This model may be presented as two empirical observations and two conclusions.

Observation #1: More organisms are born than can survive to reproductive maturity.   The high death toll among the young, from primitive plants to humans, is plain to see. There simply are not enough resources or space to go around, and the young are among the first to be affected.

Observation #2: All organisms exhibit innate variability.   While we are easily able to spot differences between humans or even other mammals, it is not easy for us to identify differences between members of a group of daffodils or coral snakes. The differences are there nonetheless, and if we observe the plants and snakes carefully we will see that, because of the differences, some will thrive and others will not.

Conclusion #1: The only organisms that will survive and reproduce are those whose individual innate variations make them well suited to the environment. Note the importance of context here: An organism suited to one environment may be totally unsuited to another. Note also the importance of reproduction; it is not enough to live—one must pass one's genes on to subsequent generations. The ability to produce fertile offspring is called *fitness*. This combines the ability to attract a mate with the fertility of offspring. If Tarzan were sterile he would have zero fitness, in spite of his mate-attraction capabilities.

Conclusion #2:   Properties favored by selection can be passed on to offspring.   Selection winnows out the unfit, i.e., those individuals whose innate properties make them less competitive in a given environmental context. The survivors can pass on favored characteristics to their progeny.

Reproductive isolation can generate new species.

Suppose that a population, or large, freely interbreeding group, of a species becomes divided in half, such that members of one half can no longer breed with the other half. Genetic mutations and selection in one half may be independent of those in the other half, leading to a divergence of properties between the two halves. After enough time passes, the two groups may accumulate enough differences to become different species, as defined earlier in this section. This is the usual method for species creation. An example is found at the Grand Canyon; the squirrels at the north and south rims of the canyon have evolved into different species by virtue of their geographical separation.

The idea of reproductive isolation may suggest geographical separation, but many other forms of separation will work as well. For example, one part of the population may mate at night and the other during the day, even if they occupy the same geographical area. As a second example, we return to dogs: St. Bernards and chihuahuas are reproductively isolated from each other, even if they live in the same house.

Section 3.2

Cells

A cell is not just a bag of sap. It is a mass of convoluted membranes that separate the inside of a cell from the outside world. These membranes also form internal structures that perform specialized tasks in support of the entire cell. Certain primitive cells, e.g., bacteria and some algae, have not developed most of these internal structures.

Organelles are cellular inclusions that perform particular tasks.

A cell is not a bag of homogeneous material. High resolution electron microscopy shows that the interiors of cells contain numerous simple and complex structures, each functionally dedicated to one or more of the tasks that a cell needs carried out. The cell is thus analogous to a society, each different organelle contributing to the welfare of the whole. The sizes of organelles can range from about one-thousandth of a cell diameter to half a cell diameter, and the number of each kind can range from one to many thousands. The kinds of organelles that cells contain provide the basis for one of the most fundamental taxonomic dichotomies in biology: prokaryotes vs. eukaryotes.

Eukaryotes have many well-defined organelles and an extensive membrane system.

The group called the *eukaryotes*[1] include virtually all the kinds of organisms in our every-day world. Mammals, fish, worms, sponges, amoebas, trees, fungi, and most algae are in this group. As the name implies, they have obvious, membrane-limited, nuclei. Among their many other organelles, all formed from membranes, one finds an *endoplasmic reticulum* for partitioning off internal compartments of the cell, *chloroplasts* for photosynthesis, *mitochondria* to get energy from food, *ribosomes* for protein synthesis, and an external membrane to regulate the movement of materials into and out of the cell.

Prokaryotic cells have a very limited set of organelles.

The organisms called the *prokaryotes*[2] include only two groups, the bacteria and the blue-green algae. They lack a matrix of internal membranes and most other

---

[1] The word means "with true nuclei."
[2] Prokaryotes lack true nuclei.

organelles found in eukaryotes. They have genetic material in a more-or-less localized region, but it is not bounded by a membrane; thus, prokaryotes lack true nuclei. Prokaryotes have ribosomes for protein synthesis, but they are much simpler than those of eukaryotes. The function of prokaryotic mitochondria—getting energy from food—is performed in specialized regions of the plasma membrane, and the chlorophyll of photosynthetic prokaryotes is not confined to chloroplasts.

Section 3.3

Replication of Living Systems

Living systems can only be understood in terms of the integration of elemental processes into a unified whole. It is the organic whole that defines life, not the components.

Asexual reproduction can replace those members of a species that die. The new organisms will be genetically identical to the parent organism. To the extent that the environment does not change, the newly generated organisms should be well-suited to that environment.

Sexual reproduction results in offspring containing genetic material from two parents. It not only replaces organisms that die, but provides the new members with properties different from those of their parents. Thus, Darwinian selection will maximize the chance that some of the new organisms will have a better chance to fit into their environment than did their parents.

What do we mean by a "living system"?

To deal with this question, we need to back up conceptually and ask how we know whether something is alive in the first place. This question causes at least mild embarrassment to every thinking biologist. All scientists know that the solution of any problem must begin with clear definitions of fundamental terms, yet a definition of "life" is as elusive as quicksilver.

If we start with the notion that a definition of a "living system" must come before anything else in biology, then that definition should use only non-biological terms. However, one virtually always sees living systems defined by taking a group of things everyone has already agreed to be living things, and then listing properties they have in common. Examples of these life properties are organization, response to stimuli, metabolism, growth, evolution and, of course, reproduction. A system is said to be alive if it has these properties (and/or others) because other systems that have these properties are, by consensus, alive. Thus, living systems end up being defined in terms of living systems. This definition is a recursive one: The first case is simply given, and all subsequent cases are defined in terms of one or more preceding ones.

The list of life properties against which a putative living system would be compared is an interesting one, because no one property is sufficient. For example, a building is organized, dynamite responds to stimuli, many metabolic reactions can be carried out in a test tube, salt crystals grow, mountain ranges evolve, and many chemical reactions are autocatalytic, spawning like reactions. Of course, we could always insist that the putative system should exhibit two or more of the properties, but clever people will find a non-living exception.

In spite of these objections, definition-by-precedent, applied to living systems, has an appealing practicality and simplicity—most six-year-olds are quite expert at creating such definitions. At a more intellectual level, however, recursion always leaves us with the bothersome matter of the first case, which must be accepted as axiomatic—an idea foreign to biology—or accepted as a matter of faith, an idea that makes most scientists cringe.

One way out of this dilemma is to drop the pretense of objectivity. After all, almost everyone, scientist or lay person, will agree with each other that something is or isn't alive. One wag has said, "It's like my wife—I can't define her, but I always know her when I see her." There is, however, a more satisfying way to handle this problem, and that is to note the unity of the life-properties list: The listed properties are related to each other. For instance, only a highly organized system could contain enough information to metabolize and therefore to respond to stimuli. A group of organisms evolves and/or grows when some of its members respond to stimuli in certain ways, leading some to thrive and some not. Reproduction, which requires metabolism and growth, can produce variation upon which selection acts. Selection, in turn, requires reproduction to replace those organisms that were weeded out by selection.

We see then that living systems perform numerous elemental processes, none of which is unique to living systems. *What is unique to living systems is the integration of all these processes into a unified, smoothly functioning whole.* Any attempt to limit our focus to one process in isolation will miss the point; for example, we must view reproduction as one part of a highly interacting system of processes. This does not preclude discussion of the individual processes, but it is their *mutual interactions* that characterize life. In Chapter 9 we will further discuss the importance of organization to biological systems by considering biomolecular structure.

Why do living systems reproduce?

To try to answer this question we must first lay some groundwork by stating something that is obvious: Every organism is capable of dying. If an organism were incapable of any kind of reproduction, it would surely die at some point and would not be here for us to observe.[3] Reproduction is therefore required as part of any lifestyle that includes the possibility of death, i.e., it includes all living things.

The cause of an organism's death may be built-in, i.e., its life span may be genetically preprogrammed. Alternatively, the organism may wear out, a notion called the "wear-and-tear" theory, suggesting that we collect chemical and physical injuries until something critical in us stops working. Finally, some other

---

[3]This reasoning is analogous to the "anthropic principle" of cosmology, in response to the question "Why does our universe exist". The principle says that if any *other* kind of universe existed we would not be here to observe it. (But we do not wish to get too metaphysical here.)

organism, ranging from a virus to a grizzly bear, may kill the organism in the course of disease or predation.

A number of reproductive modes have evolved since life began, but they may be collected into two broad categories, asexual and sexual. Asexual reproduction itself is associated with three phenomena: First, there is the matter of a cell's surface-to-volume ratio, which affects the cell's ability to take up food and to produce and release waste; second, asexual reproduction allows the formation of daughter cells identical to the parent cell, thus providing for metabolic continuity under nonvarying environmental conditions; third, asexual reproduction allows individual multicellular organisms to develop physiologically different tissues by allowing genetic information to be switched on and off. This provides for organ formation.

Sexual reproduction, on the other hand, rearranges genetic information by combining genetic contributions from two parents in novel ways; this provides a range of variations in offspring upon which selection can act. In Chapter 11 we will describe the details of asexual and sexual reproduction in cells. Here we will restrict our discussion to general principles.

Simple cell division changes the surface-to-volume ratio (S/V) of a cell.

An interesting model connects asexual cell division to waste management. Consider a metabolizing, spherical cell of radius R: The amount of waste the cell produces ought to be roughly proportional to the mass, therefore to the volume, of the cell. The volume V of a sphere is proportional to R^3. On the other hand, the ability of the cell to get rid of waste ought to be proportional to the surface area of the cell, because waste remains in the cell until it crosses the outer cell membrane on the way out. The surface area S is proportional to R^2. As a result, the ratio (S/V), a measure of the cell's ability to get rid of its waste to the cell's production of waste, is proportional to R^{-1}. For each kind of cell there must be some minimum value permitted for the ratio $S/V = 1/R$, a value at which waste collects faster than the cell can get rid of it. This requires that the cell divide, thus decreasing R and increasing S/V. A similar model, describing the ability of a cell to take up and utilize food, should be obvious.

Asexual reproduction maintains the genetic material of a single parent
in its offspring.

In general, asexual reproduction leads to offspring that are genetically identical to the parent cell. This will be especially useful if the environment is relatively constant; the offspring will thrive in the same environment in which the parent thrived.

Most eukaryotic cells replicate asexually by a process called *mitosis*.[4] In mitosis a cell's genetic material is copied and each of two daughter cells gets one

---

[4]Bacteria reproduce asexually by a somewhat different process, called *binary fission.* We will not go into it.

of the identical copies. At the same time, the cytoplasm and its organelles are divided equally among the daughter cells. Single-celled organisms, like amoebas, divide asexually by mitosis, as do the individual cells of multicellular organisms like daisies and humans. The details of mitosis are spelled out in Chapter 11, where we will describe how the various cells of a multicellular organism get to be different, in spite of their generation by mitosis.

Entire multicellular organisms can reproduce asexually. A cut-up starfish can yield a complete starfish from each piece. Colonies of trees are generated by the spreading root system of a single tree. These and similar processes create offspring that are genetically identical to the parent.

The various tissues of multicellular organisms are created by turning genes on and off.

A human has dozens of physiologically and anatomically different kinds of cell types. Virtually all of them result from mitosis in a fertilized egg. Thus, we might expect them all to be identical because they have the same genes.[5]

The differences between the cells is attributable to different active gene sets. The active genes in a liver cell are not the same ones active in a skin cell. Nevertheless, the liver cell and the skin cell contain the same genes, but each cell type has turned off those not appropriate to that cell's function.

Sexual reproduction provides for variation in offspring.

Sexual reproduction is characterized by offspring whose genetic material is contributed by two different parents. The interesting thing about the two contributions is that they do not simply add to one another. Rather, they combine in unexpected ways to yield offspring that are often quite different from either parent. Further, each offspring will generally be different from the other offspring. We have only to compare ourselves to our parents and siblings to verify this.

The variations induced by sexual reproduction maximize the chance that at least a few progeny will find a given environment to be hospitable. Of course, this also means that many will die, but in nature that is no problem because those that die will serve as food for some other organism. Note the lack of mercy here— many variants are tried by sexual reproduction and most die. The few survivors perpetuate the species.

Sexual reproduction is found in a very wide variety of organisms, ranging from humans to single-celled organisms like amoebas and bacteria. In fact, organisms whose life cycles exclude sexual reproduction are so unusual that they are generally put into special taxonomic categories based solely on that fact. In simple organisms, sexual reproduction may not result in increased numbers, but the offspring will be different from the parent cells. Chapter 11 and references [2] and [3] contain detailed discussions of sexual reproduction and genetics.

---

[5]As always, there are notable exceptions. Mammalian red blood cells have nuclei when they are first formed, but lose them and spend most of their lives anucleate, therefore without genes.

Section 3.4

Population Growth and Its Limitations

We now combine the topics of the two previous sections of this chapter, namely the increase in an organism's numbers and the struggle among them for survival. The result is that, in a real situation, population growth is limited.

Unchecked growth of a population is exponential.

One of the observations of the Darwinian model of evolution is that more organisms are born than can possibly survive. (We use the word "born" here in a very broad sense to include all instances of sexual and asexual reproduction.) Let us suppose for a moment that some organism is capable of unchecked reproduction, doubling its numbers at each reproductive cycle. One would become 2, then 2 would become 4, 4 would become 8, etc. After N reproductive cycles, there would be 2^N organisms. If the organism's numbers increased M-fold at each reproductive cycle, there would be M^N organisms after N reproductive cycles. This kind of growth is *exponential,* and it can rapidly lead to huge numbers. Table 3.4.1 shows the numbers generated by an organism that doubles at each cycle.

Table 3.4.1

Number of Generations:	0	1	2	$\cdots$	10	$\cdots$	25	$\cdots$	40	$\cdots$	72
Number of Organisms:	1	2	4	$\cdots$	1024	$\cdots$	3.4×10^7	$\cdots$	1.1×10^{12}	$\cdots$	4.7×10^{21}

Many bacteria can double their numbers every 20 minutes. Therefore each cell could potentially generate 4.7×10^{21} cells per day. To put this number into perspective, a typical bacterium has a mass on the order of 10^{-12} grams, and a day of reproduction could then produce a mass of 4.7×10^9 grams of bacteria from each original cell. Assuming the cells have the density of water, 1 gm/cm^3, 4.7×10^9 grams is the mass of a solid block of bacteria about 1.6 meters on a side. Obviously, no such thing actually happens.

Real life: population growth meets environmental resistance

Every population has births (in the broad sense described above), and it has deaths. The net growth in numbers is (births − deaths). The *growth rate, r,* is defined by[6]

$$r = \frac{(\text{birth rate} - \text{death rate})}{\text{population size}}.$$

---

[6]The units of birth and death rates are *numbers of births per unit of time* and *number of deaths per unit of time.* The units of population are *numbers of individuals,* and r is in units of *time*$^{-1}$.

The maximum value that r can have for an organism is r_{max}, called the *biotic potential*. Estimates of r_{max} have been made by Brewer [4]. They range from about 0.03 per year for large mammals to about 10 per year for insects and about 10,000 per year for bacteria. These numbers are all positive, and we therefore expect organisms growing at their biotic potential to increase in numbers over time, not so dramatically as described by Table 3.4.1, but constantly nevertheless.

We must remember that r_{max} is the rate of natural increase under optimal conditions, which seldom exist. Under suboptimal conditions the birth rate will be low and the death rate high, and these conditions may even lead the value of r to be negative. In any case, the value of r will drop as inimical environmental factors make themselves felt. These factors are collectively called *environmental resistance*, and they are responsible for the fact that we are not waist-deep in bacteria or, for that matter, dogs or crabgrass.

From our discussion of evolution we now understand that some organisms have a higher tolerance for environmental resistance than do others. Those with the highest tolerance will prosper at the expense of those with low tolerance. Our experience, however, is that every species is ultimately controlled at *some* level by environmental resistance.

Section 3.5

The Exponential Model for Growth and Decay

Assuming a constant per capita growth rate leads to exponential growth. Despite its simplicity, most populations do in fact increase exponentially at some time over their existence. There are two parameters governing the process, initial population size and the per capita growth rate. Both or either may be easily determined from experimental data by least squares.

If the growth rate parameter is negative, then the process is exponential decay. Although populations sometimes collapse catastrophically, they can also decline exponentially. Moreover exponential decay pertains to other phenomena as well, such as radioactive decay. In conjunction with decay processes, it is customary to recast the growth rate as a half-life.

The exponential model approximates population growth in its early stage.

We will discuss logistic growth in the next chapter, after the biological discussion has been extended. Here we will discuss the mathematics of exponential growth only.

The size of a population, and its trend, has vital significance for that population, for interacting populations, and for the environment. It is believed that the Polynesian population of Easter Island grew too large to be supported by the island's resources with disastrous consequences for most of the flora and fauna of the island. A large sea gull population living near a puffin rookery spells high chick losses for the puffins. At the height of a disease, the pathogen load on the victim can reach 10^9 organisms per milliliter.

By a population we mean an interbreeding subpopulation of a species. Often this implies geographical localization; for example, the Easter Island community, or a bacterial colony within a petri dish. The first published model for predicting population size was by Malthus in 1798, in which it was assumed that the growth rate of a population is proportional to their numbers y, that is,

$$\frac{dy}{dt} = ry, \tag{3.5.1}$$

where r is the constant of proportionality. By dividing equation (3.5.1) by y, we see that r is the per capita grow rate,

$$\frac{1}{y}\frac{dy}{dt} = r,$$

with units of time, e.g., per second. Hence Malthus' Law assumes the per capita growth rate to be constant. For a no-growth, replacement-only colony, r will be zero.

Malthus' model is a vast oversimplification of survival and reproduction. Although population size can only be integer-valued in reality, by incorporating the derivative $\frac{dy}{dt}$, y is necessarily a continuous variable in this model, it can take on any non-negative value. Further, the parameter r must be taken as an average value over all population members. Therefore equation (3.5.1) is a continuum model and does not apply to extremely small populations. Nevertheless, it captures a germ of truth about population dynamics and is mathematically tractable. It is a significant first step from which better models emerge and to which other models are compared.

This simple differential equation can be solved by separating the y and t variables and integrating,

$$\frac{dy}{y} = r\,dt \quad \text{or} \quad \int \frac{dy}{y} = \int r\,dt.$$

These integrals evaluate to

$$\ln y = rt + c$$

where c is the constant of integration. Exponentiate both sides to obtain

$$y = e^{rt+c} \quad \text{or} \quad y = y_0 e^{rt} \tag{3.5.2}$$

where $y_0 = e^c$. In this, the parameter y_0 is the value of y when $t = 0$, cf. Section 2.4.

Under Malthus' Law equation (3.5.2), a population increases *exponentially*, the solid curve in Figure 3.5.5. While exponential growth cannot continue indefinitely, it is observed for populations when resources are abundant and population density is low, compare the data designated by the circles in Figure 3.5.5. Under these conditions, populations approximate their biotic potential, r_{max}, cf. Section 3.4.

Per capita growth rate parameters r are often given in terms of *doubling time*. Denote by T_2 the time when the population size reaches twice its initial value, then using equation (3.5.2)

$$2y_0 = y_0 e^{rT_2}.$$

Divide out y_0 and solve for T_2 by first taking the logarithm of both sides,

$$\ln 2 = rT_2,$$

and then dividing by r; the doubling time works out to be

$$T_2 = \frac{\ln 2}{r}. \tag{3.5.3}$$

Thus the per capita growth rate and doubling time are inversely related, a higher per capita growth rate makes for a smaller doubling time, and conversely. Rearranging equation (3.5.3) gives the per capita growth rate in terms of doubling time

$$r = \frac{\ln 2}{T_2}. \tag{3.5.4}$$

Growth rate parameters are not always positive. In the presence of serious adversity, a population can die off exponentially. To make it explicit that the parameter is negative, the sign is usually written out,

$$y = y_0 e^{-\mu t} \tag{3.5.5}$$

where $\mu > 0$. Such an exponential decay process is characterized by its *half-life* $T_{1/2}$ given by

$$\frac{1}{2}y_0 = y_0 e^{-\mu T_{1/2}}$$

or

$$T_{1/2} = \frac{\ln 2}{\mu}. \tag{3.5.6}$$

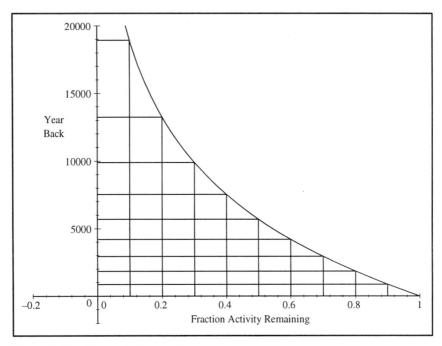

Figure 3.5.1 Decay of Carbon-14.

Exponential growth and decay apply to other natural phenomena as well as biological processes. One of these is radioactive decay, where emissions occur in proportion to the amount of radioactive material remaining. The activity of a material in this regard is measured in terms of half-life, for example, the half-life of Carbon-14 is about 5700 years. Radioactive decay is the scientific basis behind various artifact dating methods using different isotopes. Figure 3.5.1 is a chart for ^{14}C. Locate the fraction of ^{14}C remaining in an artifact relative to the environment on the horizontal axis and read its age on the vertical axis.

As observed above, Malthus' assumption of immediate reproduction embodied in equation (3.5.1) hardly seems accurate. Mammals, for instance, undergo a lengthy maturation period. Further, since no real population grows unboundedly, the assumption of constant per capita growth breaks down eventually for all organisms. Nevertheless, there is often a phase in the growth of populations, even populations of organisms with structured life cycles, when exponential growth is observed. This is referred to as the *exponential growth phase* of the population.

It is possible to mathematically account for a maturation period and hence more accurately model population growth. This is done by the incorporation of a *delay*, τ, between the time offspring are born and the time they reproduce. In differential equation form, we have

$$\left.\frac{dy}{dt}\right|_{t} = r \cdot y(t - \tau);$$ (3.5.7)

in words, the growth rate at the present time is proportional to the population size τ time units ago (births within the last τ periods of time do not contribute offspring). Equation (3.5.7) is an example of a *delay differential equation*. An initial condition for the equation must prescribe $y(t)$ for $-\tau \le t \le 0$. As an illustration, let $r = 1$, $\tau = 1$ and $y(t) = e^{t/10}$ for $-1 \le t \le 0$ as an initial condition. Begin by setting $f_0(t)$ to this initial function and solving

$$\frac{dy}{dt} = f_0(t - 1)$$

for y on the interval $[0, 1]$, that is for $0 \le t \le 1$. In this, solving means integrating since the right hand side is a given function of t. Define this solution to be $f_1(t)$, and repeat the procedure to get a solution on $[1, 2]$. Continue in this way to move ahead by steps of length 1.

Here is a computational procedure that produces a graph of a solution for equation (3.5.7):

```
> f0:=t->exp(t/10);
> dsolve({diff(y(t),t)=f0(t-1),y(0)=f0(0)},y(t));
> f1:=unapply(rhs("),t);
  dsolve({diff(y(t),t)=f1(t-1),y(1)=f1(1)},y(t));
> f2:=unapply(rhs("),t);
  dsolve({diff(y(t),t)=f2(t-1),y(2)=f2(2)},y(t));
> f3:=unapply(rhs("),t);
```

We see that this delay population model still follows an exponential-like growth law. The extent to which this is "exponential" is examined in the exercises.

Growth parameters can be determined from experimental data.

Exponential growth entails two parameters, initial population size y_0 and growth rate r. Given n experimental data values, (t_1, y_1), (t_2, y_2), ..., (t_n, y_n), we would like to find the specific parameter values for the experiment. As discussed in Section 2.2, this is done by the method of least squares. The equations are, cf equations (2.2.3),

$$\ln y_0 = \frac{\sum_{i=1}^n t_i^2 \sum_{i=1}^n \ln y_i - \sum_{i=1}^n t_i \sum_{i=1}^n t_i \ln y_i}{n \sum_{i=1}^n t_i^2 - (\sum_{i=1}^n t_i)^2}$$

and

$$r = \frac{n \sum_{i=1}^n t_i \ln y_i - \sum_{i=1}^n t_i \sum_{i=1}^n \ln y_i}{n \sum_{i=1}^n t_i^2 - (\sum_{i=1}^n t_i)^2}.$$

A slightly different problem presents itself when we are sure of the initial population size, y_0, and only want to determine r by fit. If there were no experimental error, only one data value (t_1, y_1), besides the starting one, would be needed for this, thus

$$y_1 = y_0 e^{rt_1}$$

so

$$r = \frac{\ln y_1 - \ln y_0}{t_1}.$$

Unfortunately, however, experimental error invariably affects data, and performing this calculation using two data values will likely result in two different (but close) values of r. Given n data values (beyond the starting one), (t_1, y_1), (t_2, y_2), ..., (t_n, y_n), there will be n corresponding calculations of r. Which is the right one?

To solve this, we use a specialization of the least squares method. As before (Section 2.2), use the logarithm form of equation (3.5.2); the squared error is then given by

$$E = \sum_{i=1}^{n} [\ln y_i - (\ln y_0 + rt_i)]^2. \qquad (3.5.8)$$

As before, differentiate E with respect to r and set the derivative to zero:

$$2 \sum_{i=1}^{n} [\ln y_i - (\ln y_0 + rt_i)] (-t_i) = 0.$$

Now solve this for r and get

$$r = \frac{\sum_{i=1}^{n} t_i(\ln y_i - \ln y_0)}{\sum_{i=1}^{n} t_i^2}. \qquad (3.5.9)$$

Alternatively, we can let the computer algebra system derive equation (3.5.9).

The reader should understand the importance of (3.5.9); this is not a result that should be memorized. Indeed, a computer algebra system will be able to compute this result.

Suppose the starting value is known, $y(0) = A$, and we have data given symbolically as

$$\{[a[1], b[1]], [a[2], b[2]], [a[3], b[3]]\}.$$

We find the value of r given in (3.5.9) for this general problem in the following manner:

```
> with(linalg):
> xval:=[seq(a[i],i=1..3)];
> yval:=[seq(b[i],i=1..3)];
> lny:=map(ln,yval);
> with(stats);
> fit[leastsquare[[x,y],y=r*x+ln(A),{r}]]([xval,lny]);
> coeff(rhs(" "),x);
> combine(simplify(" "));
```

This computer calculation yields the same results as equation (3.5.9).

Example 3.5.1

The U.S. Census Data

To illustrate these ideas, we determine the per capita growth rate for the U.S. over the years 1790 to 1990. In Table 3.5.1 we give the U.S. census for every 10 years, the period required by the Constitution.

Table 3.5.1   U.S. Population Census

1790	3,929,214	1860	31,433,321	1930	122,775,046
1800	5,308,483	1870	39,818,449	1940	131,669,275
1810	7,239,881	1880	50,155,783	1950	151,325,798
1820	9,638,453	1890	62,947,714	1960	179,323,175
1830	12,866,020	1900	75,994,575	1970	203,302,031
1840	17,069,453	1910	91,972,266	1980	226,545,805
1850	23,191,876	1920	105,710,620	1990	248,709,873

SOURCE: Statistical Abstracts of the United States: 1993, 113th Edition, U.S. Department of Commerce, Bureau of the Census, Washington, DC.

First, we plot the data to note that it does seem to grow exponentially. We read in population data as millions of people and plot the data in order to see that the population appears to be growing exponentially.

```
> tt:=[seq(1790+i*10,i=0..20)];
> pop:=[ 3.929214, 5.308483, 7.239881, 9.638453, 12.866020,
   17.069453, 23.191876, 31.433321, 39.818449, 50.155783,
   62.947714, 75.994575, 91.972266, 105.710620, 122.775046,
   131.669275, 151.325798,179.323175, 203.302031, 226.545805,
   248.709873];
> data:= [seq([tt[i],pop[i]],i=1..21)];
> plot(data,style=POINT,symbol=CROSS,tickmarks=[4,5]);
```

In order to make the data manageable, we rescale the time data by taking 1790 as year zero.

```
> scaledata:= [seq([(i-1)*10,pop[i]],i=1..21)];
> plot(scaledata,style=POINT,symbol=CROSS,tickmarks=[4,5]);
```

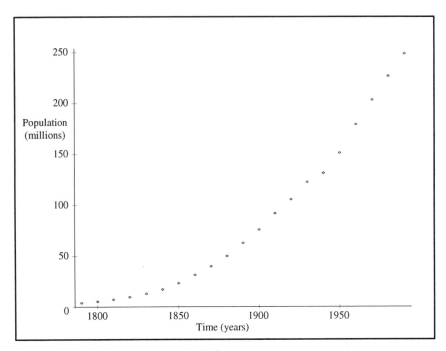

Figure 3.5.3   Population data for the U.S.

It appears that the growth of the U.S. population is approximately exponential with a deviation in the 1940's. We will try to get an exponential fit between 1790 and 1930.

We take the logarithm of the data. The plot of the logarithm of the data should be approximately a straight line.

Since this linear fit is the logarithm of the population, its exponential will approximate the data. Recall that these techniques were used in Chapter 2.

Figure 3.5.5 shows the fit which is reasonably good up to 1910. We return to this data in the next exercise and in the exercises for Chapter 4.

Exercises

1. Repeat Example 3.5.1 with all the U.S. population data, instead of just 15 points. Which fit is better for the data up to 1930, the partial fit or the total fit? Using an *error* similar to the one in equation (2.2.1) of Chapter 2, give a quantitative response. (If this fit for the U.S. population data interests you, note that we will return to it again in the exercises for Section 4.3.)

```
> semilnpop:= [seq([(i-1)*10,ln(pop[i])],i=1..15)];
> plot(semilnpop,style=POINT,symbol=CIRCLE);
```

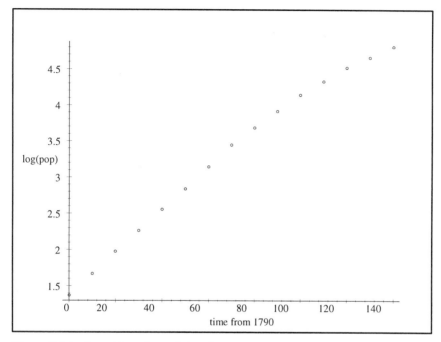

Figure 3.5.4 Logarithm of population data

2. We present below the expected number of deaths per 1000 people as a function of increasing age. Surprisingly, an exponential fit approximates this data well. Find an exponential fit for the data. The set DR is the death rate at the ages in the set yrs.

```
> yrs:=([9,19,29,39,49,59,69,79,89]);
  DR:=([.3,1.5, 1.9, 2.9, 6.5, 16.5, 37, 83.5, 181.9]);
> pts:=[seq([yrs[i],DR[i]], i=1..9)];
> plot(pts,style=POINT, symbol=CROSS);
> lnpts:=[seq([yrs[i],ln(DR[i]) ], i=1..9)];
> plot(lnpts,style=POINT,symbol=CIRCLE);
> with(stats):
> lnDR:=map(ln,DR);
> fit[leastsquare[[t,y],y=a*t+b]]([yrs,lnDR]);
> a:=op(1,op(1,rhs(" "))); b:=op(2,rhs(" "));
> death:=t->exp(a*t+b);
> J:=plot(pts,style=POINT, symbol=CROSS):
  K:=plot(death(t),t=0..90):
  plots[display]({J,K});
```

```
> with(stats):
> tzeroed:=[seq((i-1)*10,i=1..15)];
  lnpop:=[seq(ln(pop[i]),i=1..15)];
> fit[leastsquare[[t,y],y=m*t+b]]([tzeroed,lnpop]);
> y:=unapply(rhs("),t);
> J:=plot(exp(y(t-1790)),t=1790..1930,tickmarks=[4,5]):
> K:=plot(data,style=POINT,symbol=CROSS,tickmarks=[4,5]):
> plots[display]({J,K});
```

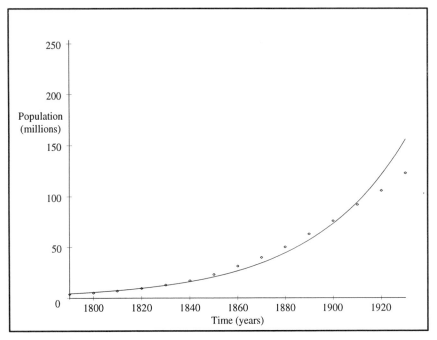

Figure 3.5.5   Exponential growth data fit between 1790 and 1930

3. Using the least square methods of this section, and by sampling nine data points on the interval $[0, 3]$, determine if the growth of the solution for the delay equation (3.5.7) is exponential.

4. In Section 2.2 we gave a cubic polynomial fit for the cumulative number of AIDS cases in the U.S. Find an exponential fit for that data. Determine which fit has the smaller error—the cubic polynomial fit or the exponential fit.

```
> AIDS:=([97, 206, 406, 700, 1289, 1654, 2576, 3392,
    4922, 6343, 8359, 9968,12990,14397, 16604, 17124,
    19585, 19707, 21392, 20846, 23690, 24610,26228,
    22768, 4903]);
> CAC:=[seq(sum(AIDS[j]/1000.0,j=1..i),i=1..24)];
> Time:=[seq(1981+(i-1)/2,i=1..24)]:
```

```
> pts:=[seq([Time[i],CAC[i]],i=1..24)]:
> LnCAC:=map(ln,CAC);
  Times:=[seq((i+1)/2/10,i=1..24)];
> with(stats):
  fit[leastsquare[[x,y],y=m*x+b]]([Times,LnCAC]);
> k:=op(1,op(1,rhs(" "))); A:=op(2,rhs(" "));
> y:=t->exp(A)*exp(k*t);
> J:=plot(y((t-1980)/10),t=1980..1992):
  K:=plot(pts,style=POINT,symbol=CIRCLE):
  plots[display]({J,K});
```

Section 3.6

Questions for Thought and Discussion

1. What is the surface-to-volume (S/V) ratio of a spherical cell with a radius of
    2? What is the radius of a spherical cell with $S/V = 4$? A spherical cell with
    $S/V = 3$ divides exactly in two. What is the S/V ratio of each of the daughter
    cells?
2. Name some factors that might prevent a population from reaching its biotic
    potential.
3. Variations induced by sexual reproduction generally lead to the early deaths
    of many, if not most, of the organisms. What could be advantageous about
    such a ruthless system?

References and Suggested Further Reading

1. **Biophysics of living systems:** Edward Yeargers, *Basic Biophysics for Biology*,
    CRC Press, Boca Raton, FL, 1992.
2. **Cell division and reproduction:** William Keeton and James Gould, *Biological
    Science*, 5th ed., W. W. Norton and Company, New York, 1993.
3. **Cell division and reproduction:** William S. Beck, Karel F. Liem and George
    Gaylord Simpson, *Life – An Introduction to Biology*, 3rd ed., Harper-Collins Pub-
    lishers, New York, 1991.
4. **Population biology:** Richard Brewer, *The Science of Ecology*, 2nd ed., Saunders
    College Publishing Co., Fort Worth, TX, 1988.

Chapter 4

Interactions Between Organisms and Their Environment

Introduction to this chapter

This chapter is a discussion of the factors that control the growth of populations of organisms.

Evolutionary fitness is measured by the ability to have fertile offspring. Selection pressure is due to both biotic and abiotic factors and is usually very subtle, expressing itself over long periods of time. In the absence of constraints, the growth of populations would be exponential, rapidly leading to very large population numbers. The collection of environmental factors that keep populations in check is called environmental resistance, which consists of density-independent and density-dependent factors. Some organisms, called r-strategists, have short reproductive cycles marked by small prenatal and postnatal investments in their young and by the ability to capitalize on transient environmental opportunities. Usually, their numbers increase very rapidly at first but then decrease very rapidly when the environmental opportunity disappears. Their deaths are due to climatic factors that act independently of population numbers.

A different life style is exhibited by K-strategists, who spend a lot of energy caring for their relatively infrequent young, under relatively stable environmental conditions. As the population grows, density-dependent factors such as disease, predation and competition act to maintain the population at a stable level. A moderate degree of crowding is often beneficial, however, allowing mates and prey to be located. From a practical standpoint, most organisms exhibit a combination of r- and K-strategic properties.

The composition of plant and animal communities often changes over periods of many years, as the members make the area unsuitable for themselves. This process of succession continues until a stable community, called a climax community, appears.

Section 4.1

How Population Growth Is Controlled

In Chapter 3 we saw that uncontrolled growth of a biological population is exponential. In natural populations, however, external factors control growth. We can distinguish two extremes of population growth kinetics, depending on the nature of these external factors, although most organisms are a blend of the two. First, r-strategists exploit unstable environments and make a small investment in the raising of their young. They produce many offspring, which often are killed off in large numbers by climatic factors. Second, K-strategists have few offspring, and invest heavily in raising them. Their numbers are held at some equilibrium value by factors that are dependent on the density of the population.

An organism's environment includes biotic and abiotic factors.

An *ecosystem* is a group of interacting living and nonliving elements. Every real organism sits in such a mixture of living and non-living elements, interacting with them all at once. Biologist Barry Commoner has summed this up with the observation "Everything is connected to everything else." Living components of an organism's environment include those organisms that it eats, those that eat it, those that exchange diseases and parasites with it, and those that try to occupy its space. The non-living elements include the many compounds and structures that provide the organism with shelter, that fall on it, that it breathes and that poison it. (See References [1–4] for discussions of environmental resistance, ecology and population biology.)

Density-independent factors regulate the populations of r-strategists.

Figure 4.1.1 shows two kinds of population growth curves, in which an initial increase in numbers is followed by either a precipitous drop (curve a) or a period of zero growth (curve b). The two kinds of growth curves are generated by different kinds of enviromental resistance.[1]

Organisms whose growth kinetics resemble curve (a) of Figure 4.1.1 are called *r-strategists*, and the environmental resistance that controls their numbers is said to be *density-independent*.[2] This means that the organism's numbers are limited by factors that do not depend upon the organism's population density. Climatic factors, such as storms or bitter winters, earthquakes and volcanoes, are density-independent factors, in that they exert their effects on dense and sparse populations alike.

---

[1]Note that the vertical axis in Figure 4.1.1 is the total number of individuals in a population; thus, it allows for births, deaths, and migration.

[2]The symbol "r" indicates the importance of the rate of growth, which is also symbolized by "r."

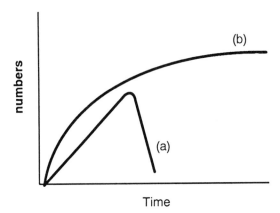

Figure 4.1.1   A graph of the number of individuals in a population vs. time, for an idealized r-strategist (a) and an idealized K-strategist (b). r-strategists suffer rapid losses when density-independent factors like the weather change. K-strategists' numbers tend to reach a stable value over time because density-dependent environmental resistance balances birth rate.

Two characteristics are helpful in identifying r-strategists:

1. Small parental investment in their young.   The concept of "parental investment" combines the energy and time dedicated by the parent to the young in both the prenatal and the postnatal period. Abbreviation of the prenatal period leads to the birth of physiologically vulnerable young, while abbreviation of postnatal care leaves the young unprotected. As a result, an r-strategist must generate large numbers of offspring, most of whom will not survive long enough to reproduce themselves. Enough, however, will survive to continue the population. Figure 4.1.2 is a *survivorship curve* for an r-strategist; it shows the number of survivors from a group as a function of time.[3] Note the high death rate during early life.

Because of high mortality among its young, an r-strategist must produce many offspring, which makes death by disease and predation numerically unimportant, inasmuch as the dead ones are quickly replaced. On the other hand, the organism's short life span ensures that the availability of food and water do not become limiting factors either. Thus, density-dependent factors like predation and resource availability do not affect the population growth rates of r-strategists.

2. The ability to exploit unpredictable environmental opportunities rapidly.   It is common to find r-strategists capitalizing on transient environmental opportunities. The mosquitoes that emerge from one discarded, rain-filled beer can are capable of making human lives in a neighborhood miserable for months. Dan-

---

[3]Note that the vertical axes in Figures 4.1.2 and 4.1.4 are the numbers of individuals surviving from an initial, fixed group; thus, they allow only for deaths.

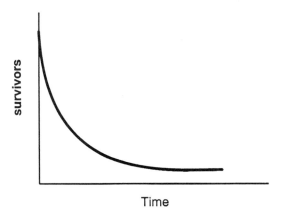

Figure 4.1.2 An idealized survivorship curve for a group of r-strategists. The graph shows the number of individuals surviving as a function of time, beginning with a fixed number at time $t = 0$. Lack of parental investment and an opportunistic lifestyle lead to a high mortality rate among the young.

delions can quickly fill up a small patch of disturbed soil. These mosquitoes and dandelions have exploited situations that may not last long; therefore a short, vigorous reproductive effort is required. Both organisms, in common with all r-strategists, excel in that regard.

We can now interpret curve (a) of Figure 4.1.1 by noting the effect of environmental resistance, i.e., density-independent factors. Initial growth is rapid and results in a large population increase in a short time, but a population "crash" follows. This crash is usually the result of the loss of the transient environmental opportunity because of changes in the weather: Drought, cold weather or storms can bring the growth of the mosquito or dandelion population to a sudden halt. By this time, however, enough offspring have reached maturity to propagate the population.

Density-dependent factors regulate the populations of K-strategists.

Organisms whose growth curve resembles that of curve (b) of Figure 4.1.1 are called *K-strategists*, and their population growth rate is regulated by population *density-dependent* factors. As with r-strategists, the initial growth rate is rapid but, as the density of the population increases, certain resources such as food and space become scarce, predation and disease increase, and waste begins to accumulate. These negative conditions generate a feedback effect: Increasing population density produces conditions that slow down population growth. An equilibrium situation results, in which the population growth curve levels out; this long-term, steady-state population is the *carrying capacity* of the environment.

The carrying capacity of a particular environment is symbolized by "K," hence the name "K-strategist" refers to an organism that lives in the equilibrium

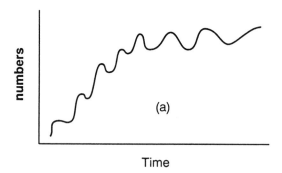

Figure 4.1.3 A more realistic growth curve of a population of K-strategists. The numbers fluctuate around an idealized curve, as shown. Compare this with Figure 4.1.1(b).

situation. The growth curve of a K-strategist, shown as (b) in Figure 4.1.1, is called a *logistic curve*. Figure 4.1.3 is a logistic curve for a more realistic situation.

Two characteristics are helpful in identifying K-strategists:

1. Large parental investment in their young. K-strategists reproduce slowly, with long gestation periods, to increase physiological and anatomical development of the young, who therefore must be born in small broods. After birth, the young are tended until they can reasonably be expected to fend for themselves. One could say that K-strategists put all their eggs in one basket and then watch that basket very carefully!

Figure 4.1.4 is an idealized survivorship curve for a K-strategist. Note that infant mortality is low (compared to the infant mortality rate of r-strategists; see Figure 4.1.2)

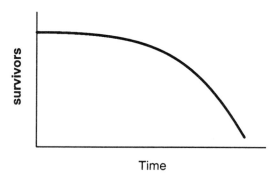

Figure 4.1.4 An idealized survivorship curve for a group of K-strategists. The graph shows the number of individuals surviving as a function of time, beginning with a fixed number at time $t = 0$. High parental investment leads to a low infant mortality rate.

2. The ability to exploit stable environmental situations. Once the population of a *K*-strategist has reached the carrying capacity of its environment, the population size stays relatively constant. This is nicely demonstrated by the work of H. N. Southern, who studied mating pairs of tawny owls in England [5]. The owl pairs had adjacent territories, with each pair occupying a territory that was the right size to provide it with nesting space and food (mainly rodents). Every year some adults in the area died, leaving one or more territories that could be occupied by new mating pairs. Southern found that while the remaining adults could have more than replaced those who died, only enough owlets survived in each season to keep the overall numbers of adults constant. The population control measures at work were failure to breed, reduced clutch size, death of eggs and chicks, and emigration. These measures ensured that the total number of adult owls was about the same at the start of each new breeding season.

As long as environmental resistance remains the same, population numbers will also remain constant. But, if the environmental resistance changes, the carrying capacity of the environment will, also. For example, if the amount of food is the limiting factor, a new value of *K* is attained when the amount of food increases. This is shown in Figure 4.1.5.

The density-dependent factors that maintain a stabilized population, in conjunction with the organism's reproductive drive, are discussed in the next section. In a later section we will discuss some ways that a population changes its own environment, and thereby changes that environment's carrying capacity.

Some density-dependent factors exert a negative effect on populations and can thus help control K-strategists.

There are many environmental factors that change with the density of populations. This section is a discussion of several of them.

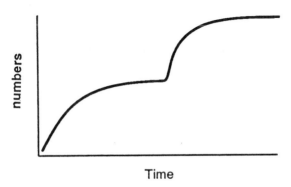

Figure 4.1.5 The growth of a population of animals, with an increase in food availability midway along the horizontal axis. The extra food generates a new carrying capacity for the environment.

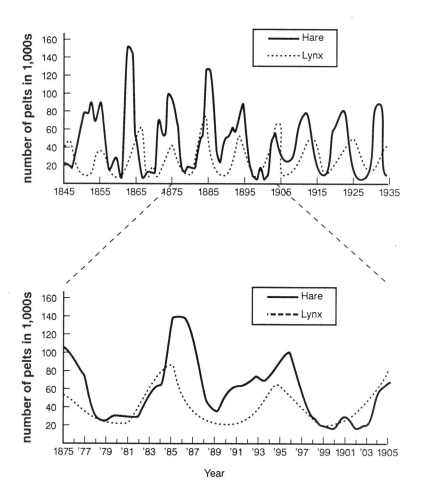

Figure 4.1.6 The Hudson's Bay Company data. The curve shows the number of predator (lynx) and prey (hare) pelts brought to the company by trappers over a 90-year period. Note that, from 1875 to 1905, changes in the number of lynx pelts sometimes precede changes in the number of hare pelts. (Redrawn from "Sunspots and Abundance of Animals," by D. A. McLulich; *The Journal of the Royal Astronomical Society of Canada*, Volume 30, 1936, page 233. Used with permission.)

Predation: The density of predators, free-living organisms that feed on the bodies of other organisms, would be expected to increase or decrease with the density of prey populations. Figure 4.1.6 shows some famous data, the number of hare and lynx pelts brought to the Hudson Bay Company in Canada over a period of approximately 90 years. Over most of this period, changes in the number of hare pelts led changes in the number of lynx pelts, as anticipated. After all, if the density of hares increased we would expect the lynx density to follow suit. A detailed study of the data, however, reveals that things were not quite that simple, because

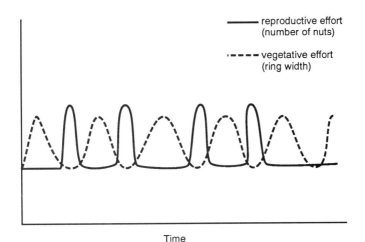

legend:
——— reproductive effort (number of nuts)

·—— — vegetative effort (ring width)

Time

Figure 4.1.7 This idealized graph shows the amount of sexual (reproductive) effort and asexual (vegetative) effort expended by many trees as a function of time. Sexual effort is measured by nut (seed) production and asexual effort is measured by tree ring growth. Note that the tree periodically switches its emphasis from sexual to asexual and back again. Some related original data can be found in the reference by Harper [2].

in the cycles beginning in 1880 and 1900 the lynxes led the hares. Analysis of this observation can provide us with some enlightening information.

Most importantly, prey population density may depend more strongly on its own food supplies than on predator numbers. Plant matter, the food of many prey species, varies in availability over periods of a year or more. For example, Figure 4.1.7 shows how a tree might partition its reproductive effort (represented by nut production) and its vegetative effort (represented by the size of its annual tree rings). Note the cycles of abundant nut production (called *mast years*) alternating with periods of vigorous vegetative growth; these alternations are common among plants. We should expect that the densities of populations of prey, which frequently are herbivores, would increase during mast years and decrease in other years, independently of predator density (see Reference [2]).

There are some other reasons why we should be cautious about the Hudson Bay data: First, in the absence of hares, lynxes might be easier to catch because, being hungry, they would be more willing to approach baited traps. Second, the naive interpretation of Figure 4.1.6 assumes equal trapping efficiencies of prey and predator. Third, to be interpreted accurately, the hares whose pelts are enumerated in Figure 4.1.6 should consist solely of a subset of all the hares that could be killed by lynxes, and the lynxes whose pelts are enumerated in the figure should consist solely of a subset of all the lynxes that could kill hares. The problem here is that very young and very old lynxes, many of whom would have contributed pelts to the study, may not kill hares at all (e.g., because of infirmities they may subsist on carrion).

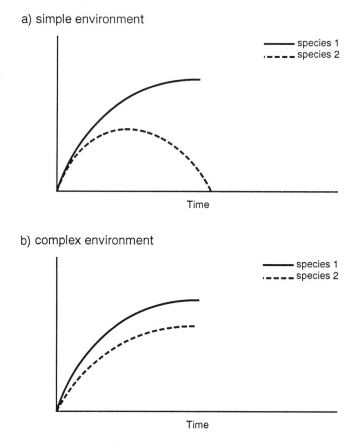

Figure 4.1.8   Graphs showing the effect of environmental complexity on interspecies relationships. The data for (a) are obtained by counting the individuals of two species in a pure growth medium. The data for (b) are obtained by counting the individuals of the two species in a mechanically complex medium where, for example, pieces of broken glass tubing provide habitats for species 2. The more complex environment supports both species, while the simpler environment supports only one species.

Parasitism.   *Parasitism* is a form of interaction in which one of two organisms benefits while the other is harmed but not generally killed. A high population density would be unfavorable for a parasite's host. For example, many parasites, e.g., hookworms and roundworms, are passed directly from one human host to another. Waste accumulation is implicated in both cases, because these parasites are transmitted in fecal contamination. Other mammalian and avian parasites must go through intermediate hosts between their primary hosts, but crowding is still required for effective transmission.

Disease.   The ease with which diseases are spread goes up with increasing population density. The spread of colds through school populations is a good example.

An important aggravating factor in the spread of disease would be the accumulation of waste. For example, typhoid fever and cholera are easily carried between victims by fecal contamination of drinking water.

Interspecific competition.   Every kind of organism occupies an *ecological niche*, which is the functional role that organism plays in its community. An organism's niche would include a consideration of all of its behaviors, their effects on the other members of the community, and the effects of the behaviors of other members of the community on the organism in question.

An empirical rule in biology, *Gause's Law*, states that no two species can long occupy the same ecological niche. What will happen is that differences in fitness, even very subtle ones, will eventually cause one of the two species to fill the niche, eliminating the other species. This concept is demonstrated by Figure 4.1.8. When two organisms compete in a uniform habitat, one of the two species always becomes extinct. The "winner" is usually the species having a numerical advantage at the outset of the experiment. (Note the role of luck here – a common and decisive variable in Darwinian evolution.) On the other hand, when the environment is more complex, both organisms can thrive because each can fit into its own special niche.

Intraspecific competition.   As individuals die, they are replaced by new individuals who are presumably better suited to the environment than their predecessors. The general fitness of the population thus improves because it becomes composed of fitter individuals.

The use of antibiotics to control bacterial diseases has contributed immeasurably to the welfare of the human species. Once in a while, however, a mutation occurs in a bacterium that confers on it resistance to that antibiotic. The surviving bacterium can then exploit its greater fitness to the antibiotic environment by reproducing rapidly, making use of the space and nutritional resources provided by the deaths of the antibiotic-sensitive majority. Strains of the bacteria that cause tuberculosis and several sexually transmitted diseases have been created that are resistant to most of the available arsenal of antibiotics. Not unexpectedly, a good place to find such strains is in the sewage from hospitals, from which they can be dispersed to surface and ground water in sewage treatment plant effluent.

This discussion of intraspecific competition would not be complete without including an interesting extension of the notion of *biocides*, as suggested by Garrett Hardin. Suppose the whole human race practices contraception to the point that there is zero population growth. Now suppose that some subset decides to abandon all practices that contribute to zero population growth. Soon that subset will be reproducing more rapidly than everyone else, and will eventually replace the others. This situation is analogous to that of the creation of an antibiotic-resistant bacterium in an otherwise sensitive culture. The important difference is that antibiotic resistance is genetically transmitted and a desire for population growth is not. But, as long as each generation continues to teach the next to ignore population control, the result will be the same.

Some density-dependent factors exert positive effects on populations.

The effect of increasing population density is not always negative. Within limits, increasing density may be beneficial, a phenomenon referred to as the "Allee Effect."[4] For example, if a population is distributed too sparsely it may be difficult for mates to meet; a moderate density, or at least regions in which the individuals are clumped into small groups, can promote mating interactions (think "singles bars").

An intimate, long-term relationship between two organisms is said to be *symbiotic*. Symbiotic relationships require at least moderate population densities to be effective. Parasitism, discussed earlier, is a form of symbiosis in which one participant benefits and the other is hurt, although it would be contrary to the parasite's interests to kill the host. The closeness of the association between parasite and host is reflected in the high degree of parasite-host specificity. For instance, the feline tapeworm does not often infect dogs, nor does the canine tapeworm often infect cats.

Another form of symbiosis is *commensalism*, in which one participant benefits and the other is unaffected. An example is the nesting of birds in trees: The birds profit from the association, but the trees are not affected.

The third form of symbiosis recognized by biologists is *mutualism*, in which both participants benefit. An example is that of termites and certain microorganisms that inhabit their digestive systems. Very few organisms can digest the cellulose that makes up wood; the symbionts in termite digestive systems are rare exceptions. The termites provide access to wood and the microorganisms provide digestion. Both can use the digestive products for food, so both organisms profit from the symbiotic association.

It would be unexpected to find a pure K-strategist or a pure r-strategist.

The discussions above, in conjunction with Figure 4.1.1, apply to idealized K- or r- strategists. Virtually all organisms are somewhere in between the two, being controlled by a mixture of density-independent and dependent factors. For example, a prolonged drought is non-discriminatory, reducing the numbers of both mosquitoes and rabbits. The density of mosquitoes might be reduced more than that of rabbits, but both will be reduced to some degree. On the other hand, both mosquitoes and rabbits serve as prey for other animals. There are more mosquitoes in a mosquito population than rabbits in a rabbit population, and the mosquitoes reproduce faster, so predation will affect the rabbits more. Still, both animals suffer from predation to some extent.

Density-independent factors may control a population in one context and density-dependent factors may control it in another context. A bitter winter could reduce rodent numbers for a while and then, as the weather warms up, predators,

---

[4]named for a prominent population biologist

arriving by migration or arousing from hibernation, might assume control of the numbers of rodents. Even the growth of human populations can have variable outcomes, depending on the assumption of the model (see Reference [6]).

The highest sustainable yield of an organism is obtained during the period of most rapid growth.

Industries like lumbering or fishing have, or should have, a vested interest in sustainable maintenance of their product sources. The key word here is "sustainable." It is possible to obtain a very high initial yield of lumber by clear-cutting a mature forest or by seining out all the fish in a lake. Of course, this is a one-time event and is therefore self-defeating. A far better strategy is to keep the forest or fish population at its point of maximal growth, i.e., the steepest part of the growth curve (b) in Figure 4.1.1. The population, growing rapidly, is then able to replace the harvested individuals. Any particular harvest may be small, but the forest or lake will continue to yield products for a long time, giving a high long-term yield. The imposition of bag limits on duck hunters, for instance, has resulted in the stable availability of wild ducks, season after season. Well-managed hunting can be viewed as a density-dependent population-limiting factor that replaces predation, disease and competition, all of which would kill many ducks anyway.

Section 4.2

Community Ecology

There is a natural progression of plant and animal communities over time in a particular region. This progression occurs because each community makes the area less hospitable to itself and more hospitable to the succeeding community. This succession of communities will eventually stabilize into a climax community that is predictable for the geography and climate of that area.

Continued occupation of an area by a population may make that region less hospitable to them and more hospitable to others.

Suppose that there is a *community* (several interacting populations) of plants in and around a small lake in north Georgia. Starting from the center of the lake and moving outward, we might find algae and other aquatic plants in the water, marsh plants and low shrubs along the bank, pine trees farther inland and, finally, hardwoods well removed from the lake. If one could observe this community for a hundred or so years, the pattern of populations would be seen to change in a predictable way.

As the algae and other aquatic plants died, their mass would fill up the lake, making it hostile to those very plants whose litter filled it. Marsh plants would start growing in the center of the lake, which would now be boggy. The area that once rimmed the lake would start to dry out as the lake disappeared and small shrubs and pine trees would take up residence on its margins. Hardwoods would move into the area formerly occupied by the pine trees. This progressive change, called *succession*, would continue until the entire area was covered by hardwoods, after which no further change would be seen. The final, stable, population of hardwoods is called the *climax community* for that area. Climax communities differ from one part of the world to another, e.g., they may be rain forests in parts of Brazil and tundra in Alaska, but they are predictable.

If the hardwood forest described above is destroyed by lumbering or fire, a process called *secondary succession* ensues: Grasses take over, followed by shrubs, then pines and then hardwoods again. Thus, both primary and secondary succession lead to the same climax community.

Succession applies to both plant and animal populations and, as the above example demonstrates, it is due to changes made in the environment by its inhabitants. The drying of the lake is only one possible cause of succession; for instance, the leaf litter deposited by trees could change the pH of the soil beneath the trees, thus reducing mineral uptake by the very trees that deposited the litter. A new population of trees might then find the soil more hospitable, and move in. Alternatively, insects might drive away certain of their prey, making the area less desirable for the insects and more desirable for other animals.

Section 4.3

Environmentally Limited Population Growth

Real populations do not realize constant per capita growth rates. By engineering the growth rate as a function of the population size, finely structured population models can be constructed. Thus, if the growth rate is taken to decrease to zero with increasing population size, then a finite limit, the carrying capacity, is imposed on the population. On the other hand, if the growth rate is assigned to be negative at small population sizes, then small populations are driven to extinction.

Along with the power to tailor the population model in this way comes the problem of its solution and the problem of estimating parameters. However, for one-variable models simple sign considerations predict the asymptotic behavior and numerical methods can easily display solutions.

Logistic growth stabilizes a population at the environmental carrying capacity.

As discussed in Sections 3.1 and 4.1, when a biological population becomes too large, the per capita growth rate diminishes. This is because the individuals interfere with each other and are forced to compete for limited resources. Consider the model, due to Verhulst in 1845, wherein the per capita growth rate decreases linearly with population size y (see Figure 4.3.1),

$$\frac{1}{y}\frac{dy}{dt} = r\left(1 - \frac{y}{K}\right). \qquad (4.3.1)$$

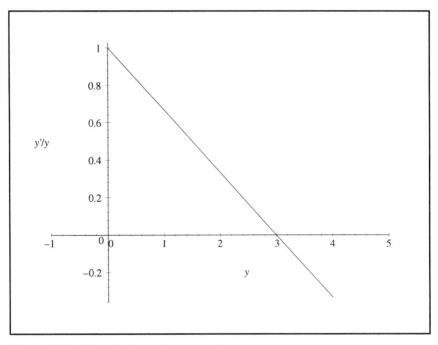

Figure 4.3.1   Linearly decreasing per capita growth rate

This differential equation is known as the *logistic (differential) equation*; two of its solutions are graphed in Figure 4.3.2. Multiplying equation (4.3.1) by y yields the alternate form

$$\frac{dy}{dt} = ry\left(1 - \frac{y}{K}\right). \tag{4.3.2}$$

In this form we see that the derivative $\frac{dy}{dt}$ is zero when $y = 0$ or $y = K$. These are the *stationary points* of the equation (see Section 2.4). The stationary point $y = K$, at which the per capita growth rate becomes zero, is called the *carrying capacity* (of the environment).

When the population size y is small, the term $\frac{y}{K}$ is nearly zero and the per capita growth rate is approximately r as before. Thus for small population size (but not so small that the continuum model breaks down), the population increases exponentially. Hence solutions are repelled from the stationary point $y = 0$. But as the population size approaches the carrying capacity K, the per capita growth rate decreases to zero and the population ceases to change in size. Further, if the population size ever exceeds the carrying capacity for some reason, then the per capita growth rate will be negative and the population size will decrease to K. Hence solutions are globally attracted to the stationary point $y = K$.

From the form (4.3.2) of the logistic equation we see that it is non-linear, with a quadratic non-linearity in y. Nevertheless, it can be solved by the separation of variables (see Section 2.4). Rewrite equation (4.3.1) as

$$\frac{dy}{y(1 - \frac{y}{K})} = r\,dt.$$

The fraction on the left-hand side can be expanded by *partial fraction decomposition* and written as the sum of two simpler fractions

$$\left(\frac{1}{y} + \frac{\frac{1}{K}}{(1 - \frac{y}{K})} \right) dy = r\,dt.$$

The solution is now found by integration. Since the left-hand side integrates to

$$\int \left(\frac{1}{y} + \frac{\frac{1}{K}}{(1 - \frac{y}{K})} \right) dy = \ln y - \ln(1 - \frac{y}{K}),$$

we get

$$\ln y - \ln(1 - \frac{y}{K}) = rt + c \qquad (4.3.3)$$

where c is the constant of integration. Combining the logarithms and exponentiating both sides we get

$$\frac{y}{1 - y/K} = Ae^{rt} \qquad (4.3.4)$$

where $A = e^c$, A is not the $t = 0$ value of y. Finally, we solve equation (4.3.4) for y. First divide the numerator and denominator of the left-hand side by y and reciprocate both sides; this gives

$$\frac{1}{y} - \frac{1}{K} = \frac{1}{Ae^{rt}}$$

or, isolating y,

$$\frac{1}{y} = \frac{1}{Ae^{rt}} + \frac{1}{K}. \qquad (4.3.5)$$

Now reciprocate both sides of this and get

$$y = \frac{1}{\frac{1}{Ae^{rt}} + \frac{1}{K}},$$

or, equivalently,

$$y = \frac{Ae^{rt}}{1 + \frac{A}{K}e^{rt}}. \tag{4.3.6}$$

Equation (4.3.6) is the solution of the logistic equation (4.3.1). To emphasize that it is the concept of "logistic growth" that is important here, and not these preceeding procedures, we illustrate how a solution for equation (4.3.1) with initial value $y(0) = y_0$ might be found.

```
> dsolve({diff(y(t),t) = r*y(t)*(1-y(t)/k),y(0)=y0},y(t));
```

The output of this computation is

$$y(t) = \frac{k}{1 + \frac{e^{-rt}(k-y_0)}{y_0}}.$$

It is an exercise to reduce this solution found by the computer to

$$y(t) = \frac{ke^{rt}y_0}{y_0(e^{rt} - 1) + k}.$$

Three members of the family of solutions (4.3.6) are shown in Figure 4.3.2 for different starting values y_0. We take $r = 1$, $K = 3$ and find solutions for equation (4.3.2) with $y_0 = 1$, or 2, or 4.

Logistic parameters can sometimes be estimated by least squares.

Unfortunately, the logistic solution, equation (4.3.6), is not linear in its parameters A, r, and K. Therefore there is no straightforward way to implement least squares. However, if the data values are separated by fixed time periods, τ, then it is possible to remap the equations so least squares will work.

Suppose the data points are (t_1, y_1), (t_2, y_2), ..., (t_n, y_n) with $t_i = t_{i-1} + \tau$, $i = 2, \ldots, n$. Then $t_i = t_1 + (i-1)\tau$ and the predicted value of $1/y_i$, from equation (4.3.5), is given by

$$\frac{1}{y_i} = \frac{1}{Ae^{rt_1}e^{(i-1)r\tau}} + \frac{1}{K}$$

$$= \frac{1}{e^{r\tau}} \left[\frac{1}{Ae^{rt_1}e^{(i-2)r\tau}} + \frac{e^{r\tau}}{K} \right] \tag{4.3.7}$$

By rewriting the term involving K as

```
> r:=1; k:=3;
> dsolve({diff(y(t),t) = r*y(t)*(1-y(t)/k),y(0)=1},y(t));
  y1:=unapply(rhs("),t);
> dsolve({diff(y(t),t) = r*y(t)*(1-y(t)/k),y(0)=2},y(t));
  y2:=unapply(rhs("),t);
> dsolve({diff(y(t),t) = r*y(t)*(1-y(t)/k),y(0)=4},y(t));
  y4:=unapply(rhs("),t);
> plot({y1(t),y2(t),y4(t)},t=0..5,y=0..5);
```

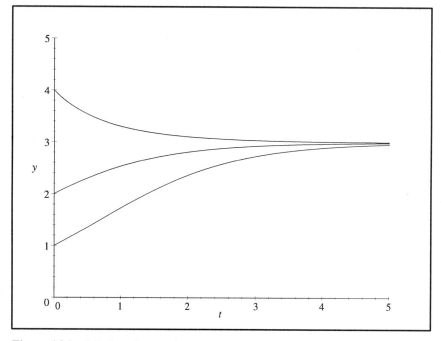

Figure 4.3.2 Solutions for equation (4.3.1)

$$\frac{1}{K} + \frac{e^{rt} - 1}{K},$$

and using equation (4.3.5) again, equation (4.3.7) becomes

$$\frac{1}{y_i} = \frac{1}{e^{rt}}\left[\frac{1}{y_{i-1}} + \frac{e^{rt} - 1}{K}\right].$$

Now put $z = 1/y$ and we have

$$z_i = e^{-rt}z_{i-1} + \frac{1 - e^{-rt}}{K} \quad \text{where } y = 1/z. \tag{4.3.8}$$

A least squares calculation is performed on the points (z_1, z_2), (z_2, z_3), ..., (z_{n-1}, z_n) to determine r and K. With r and K known, least squares can be performed on, say, equation (4.3.5) to determine A.

In the exercises we will illustrate this method and suggest another, for U.S. population data.

Non-linear per capita growth rates allow more complicated population behavior.

Real populations are in danger of extinction if their size falls to a low level. Predation might eliminate the last few members completely, finding mates becomes more difficult, and a lack of genetic diversity renders the population susceptible to epidemics. By constructing a per capita growth rate that is actually negative below some critical value, θ, there results a population model that tends to extinction if the population size falls too low. Such a per capita growth rate is given as the right-hand side of the following modification of the logistic equation

$$\frac{1}{y}\frac{dy}{dt} = r\left(\frac{y}{\theta} - 1\right)\left(1 - \frac{y}{K}\right), \qquad (4.3.9)$$

where $0 < \theta < K$. This form of the per capita growth rate is pictured in Figure 4.3.3 using the specific parameters: $r = 1$, $\theta = 1/5$, and $K = 1$. It is sometimes referred to as the *predator pit*.

We draw the graph in Figure 4.3.3 with these parameters:

```
> r:=1; theta:=1/5; K:=1;
> plot([y,r*(y/theta-1)*(1-y/K),y=0..1],-.2..1,-1..1);
```

The stationary points of equation (4.3.9) are $y = 0$, $y = \theta$, and $y = K$. But now $y = 0$ is asymptotically stable; that is, if the starting value y_0 of a solution is near enough to 0, then the solution will tend to 0 as t increases. This follows because the sign of the right-hand side of equation (4.3.9) is negative for $0 < y < \theta$ causing $\frac{dy}{dt} < 0$. Hence y will decrease. On the other hand, a solution starting with $y_0 > \theta$ tends to K as t increases. This follows because when $\theta < y < K$, the right-hand side of equation (4.3.9) is positive, so $\frac{dy}{dt} > 0$ also, hence y will increase even more. As before, solutions starting above K decrease asymptotically to K.

Some solutions to equation (4.3.9) are shown in Figure 4.3.4 with the following syntax:

```
> r:=1; theta:=1/5; K:=1;
> inits:={[0,.05],[0,.1],[0,0.3],[0,.5],[0,1],[0,0.7],[0,1.5]};
> with(DEtools):
> DEplot1(diff(y(t),t)=r*y*(y/theta-1)*(1-y/K),y(t),t=0..3,inits,
      arrows=NONE,stepsize=0.1);
```

As our last illustration, we construct a population model that engenders little population growth for small populations, rapid growth for intermediate ones, and

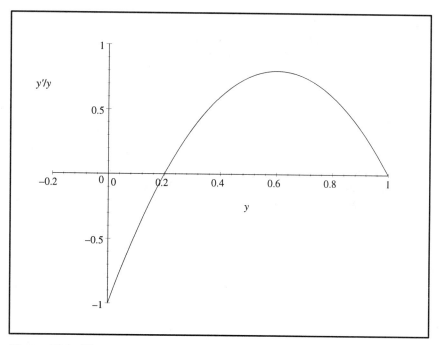

Figure 4.3.3   The predator pit per capita growth rate function

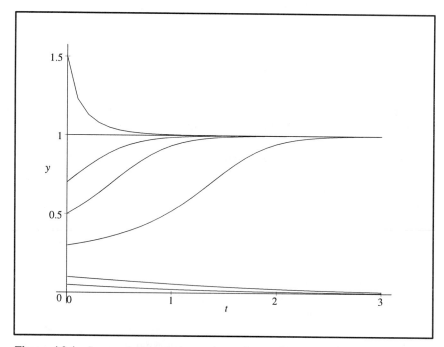

Figure 4.3.4   Some solutions to the predator pit equation

low growth again for large populations. This is achieved by the quadratic per capita growth rate and given as the right-hand side of the following differential equation:

$$\frac{1}{y}\frac{dy}{dt} = ry\left(1 - \frac{y}{K}\right). \tag{4.3.10}$$

Exercises

1. At the meeting of the Southeastern Section of the Mathematics Association of America, Terry Anderson presented a *Maple* program that determined a logistic fit for the U.S. population data. His fit is given by

$$\text{U.S. Population} \approx \frac{\alpha}{1 + \beta e^{-\delta t}},$$

where $\alpha = 387.9802$, $\beta = 54.0812$, and $\delta = 0.0270347$. Here, population is measured in millions and $t = $ time since 1790. (Recall the population data of Example 3.5.1.)

   a. Show that the function given by the Anderson Fit satisfies a logistic equation of the form

$$\frac{dy}{dt} = \delta y(t)\left(1 - \frac{y(t)}{\alpha}\right),$$

   with

$$y(0) = \frac{\alpha}{1 + \beta}.$$

   b. Plot the graphs of the U.S. population data and this graph superimposed. Compare the exponential fits from Chapter 3.
   c. If population trends continue, what is the long-range fit for the U.S. population level?

```
> Anderfit:=t->alpha/(1+beta*exp(-delta*t));
> dsolve({ diff(y(t),t)-delta*y(t)*(1-y(t)/alpha)=0,
      y(0)=alpha/(1+beta)},y(t));
> alpha:=387.980205; beta:=54.0812024; delta:=0.02270347337;
> J:=plot(Anderfit(t),t=0..200):
> tt:=[seq(i*10,i=0..20)];
> pop:=[ 3.929214, 5.308483, 7.239881, 9.638453, 12.866020,
      17.069453, 23.191876, 31.433321, 39.818449, 50.155783,
      62.947714, 75.994575, 91.972266, 105.710620, 122.775046,
      131.669275, 151.325798,179.323175, 203.302031, 226.545805,
      248.709873];
```

```
> data:= [seq([tt[i],pop[i]],i=1..21)];
> K:=plot(data,style=POINT,symbol=CROSS):
> plots[display]({J,K});
> expfit:=t->exp(0.02075384393*t+1.766257672);
> L:=plot(expfit(t),t=0..200):
> plots[display]({J,K,L});
> plot(Anderfit(t),t=0..350);
```

2.   Using the method of equation (4.3.8), get a logistic fit for the U.S. population.
     Use the data in Example 3.5.1.
3.   Suppose that the spruce budworm, in the absence of predation by birds, will
     grow according to a simple logistic equation of the form

$$\frac{dB}{dt} = rB\left(1 - \frac{B}{K}\right).$$

Budworms feed on the foliage of trees. The size of the carrying capacity, K,
would depend on the amount of foliage on the trees. We take it to be constant
for this model.

a.   Draw graphs for how the population might grow if r were 0.48 and k
     were 15. Use several initial values.
b.   Introduce predation by birds into this model in the following manner:
     Suppose that for small levels of worm population there is almost no
     predation, but for larger levels birds are attracted to this food source.
     Allow for a limit to the number of worms that each bird can eat. A model
     for predation by birds might have the form

$$P(B) = a\frac{B^2}{b^2 + B^2},$$

where a and b are positive (see Reference [7]). Sketch the graph for
the level of predation of the budworms as a function of the size of the
population. Take a and b to be 2.
c.   A model for the budworm population size in the presence of predation
     could be modeled as

$$\frac{dB}{dt} = rB\left(1 - \frac{B}{K}\right) - a\frac{B^2}{b^2 + B^2}.$$

To understand the delicacy of this model and the implications for the
care that needs to be taken in modeling, investigate graphs of solutions
for this model with parameters $r = 0.48 \,(= 48/100)$, $a = b = 2$, and
$K = 15$ or $K = 17$.
d.   Verify that in one case, there are two positive, attracting, steady state
     solutions and in the other, there is only one.

The significance of the graph with $K = 17$ is that the worm population can rise to a high level. With $K = 15$ only a low level for the size
of the budworms is possible. The birds will eat enough of the budworms
to save the trees!

Here is the syntax for making the study with $K = 15$:

```
> k:= 15;
> plot({.48*(1-mu/k),2*mu/(4+mu^2)},mu=0..20);
> with(share): readshare(ODE,plots):
> h:=(t,P)->.48*P*(1-P/k)-2*P^2/(4+P^2);
> inits:={[0,1], [0,2], [0,4], [0,5], [0,6], [0,8], [0,10], [0,12],
    [0,14], [0,16]};
> directionfield(h,0..30,0..18,inits,grid=[0,0]);
```

Section 4.4

A Brief Look at Multiple Species Systems

Without exception biological populations interact with populations of other
species. Indeed, the web of interactions is so pervasive that the entire field of
Ecology is devoted to it. Mathematically the subject began about 70 years ago
with a simple two-species, predator–prey differential equation model. The central
premise of this Lotka–Volterra model is a mass action–interaction term. While
community differential equation models are difficult to solve exactly, they can
nonetheless be analyzed by qualitative methods. One tool for this is to linearize
the system of equations about their stationary solution points and to determine the
eigenvalues of the resulting interaction, or community, matrix. The eigenvalues in
turn predict the stability of the web. The Lotka–Volterra system has neutral stability at its non-trivial stationary point which, like Malthus' unbounded population
growth, is a shortcoming that indicates the need for a better model.

Interacting population models utilize a mass action interaction term.

Lotka (1925) and, independently, Volterra (1926) proposed a simple model for the
population dynamics of two interacting species (see Reference [8]). The central
assumption of the model is that the degree of interaction is proportional to the
numbers, x and y, of each species and hence to their product, that is,

$$\text{degree of interaction} = (\text{constant})xy.$$

The Lotka–Volterra system is less than satisfactory as a serious model because it
entails neutral stability (see below). However, it does illustrate the basic principles
of multi-species models and the techniques for their analysis. Further, like the
Malthusian model, it serves as a point of departure for better models. The central
assumption stated above is also used as the interaction term between reactants in

the description of chemical reactions. In that context it is called the *Mass Action Principle*. The principle implies that encounters occur more frequently in direct proportion to their concentrations.

The original Lotka–Volterra equations are

$$\frac{dx}{dt} = rx - axy$$
$$\frac{dy}{dt} = -my + bxy$$

(4.4.1)

where the positive constants r, m, a, and b are parameters. The model was meant to treat predator–prey interactions. In this, x denotes the population size of the prey, and y the same for the predators. In the absence of predators, the equation for the prey reduces to $dx/dt = rx$. Hence the prey population increases exponentially with rate r in this case, see Section 3.5. Similarly, in the absence of prey, the predator equation becomes $dy/dt = -my$, dictating an exponential decline with rate m.

The sign of the interaction term for the prey, $-a$, is negative, indicating that interaction is detrimental to them. The parameter a measures the average degree of the effect of one predator in depressing the per capita growth rate of the prey. Thus a is likely to be large in a model for butterfies and birds but much smaller in a model for Caribou and wolves. In contrast, the sign of the interaction term for the predators, $+b$, is positive, indicating that they benefit by the interaction. As above, the magnitude of b is indicative of the average effect of one prey on the per capita predator growth rate.

Besides describing predator–prey dynamics, the Lotka–Volterra system describes a host–parasite interaction as well. Furthermore, by changing the signs of the interaction terms, or allowing them to be zero, the same basic system applies to other kinds of biological interactions such as mutualism, competition, commensalism, and amensalism.

Mathematically, the Lotka–Volterra system is not easily solved. Nevertheless solutions may be numerically approximated and qualitatively described. Since the system has two dependent variables, a solution consists of a pair of functions $x(t)$ and $y(t)$ whose derivatives satisfy equations (4.4.1). Figure 4.4.1 is the plot of the solution to equations (4.4.1) with $r = a = m = b = 1$ and initial values $x(0) = 1.5$ and $y(0) = 0.5$.

Figure 4.4.1 is drawn with the following syntax:

```
> with(plots): with(DEtools):
> predprey:=[diff(x(t),t)=r*x-a*x*y,diff(y(t),t)=-m*y+b*x*y];
> r:=1; a:=1; m:=1; b:=1;
> J:=DEplot2(predprey,[x,y],0..10,{[0,3/2,1/2]},stepsize=.1,
    scene=[t,x]):
  K:=DEplot2(predprey,[x,y],0..10,{[0,3/2,1/2]},stepsize=.1,
    scene=[t,y]):
  display({J,K});
```

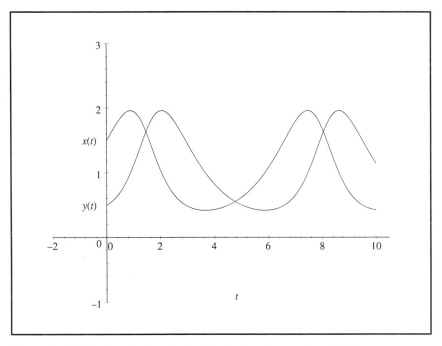

Figure 4.4.1   Graphs of $x(t)$ and of $y(t)$, solutions for equation (4.3.1)

Notice that the prey curve leads the predator curve.[5] We discuss this next.

Although there are three variables in a Lotka–Volterra system, t is easily eliminated by dividing dy/dt by dx/dt, thus

$$\frac{dy}{dx} = \frac{-my + bxy}{rx - axy}.$$

This equation does not contain t and can be solved exactly as an implicit relation between x and y:[6]

```
> dsolve(diff(y(x),x)=(-y+x*y)/(x-x*y),y(x));
```

$$-\ln(y(x)) + y(x) - \ln(x) + x = C.$$

---

[5]In Section 4.1 we discussed a number of biological reasons why, in a real situation, this model is inadequate.

[6]*Implicit* means that neither variable x nor y is solved for in terms of the other.

```
> inits:={[0,3/2,1/2],[0,4/5,3/2]};
> phaseportrait(predprey,[x,y],0..10,inits,stepsize=.1);
```

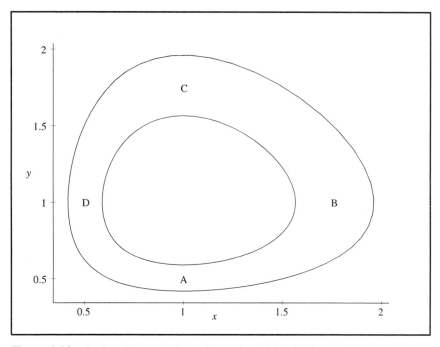

Figure 4.4.2   A plot of two solutions of equation (4.4.1) in the x, y plane

This solution gives rise to a system of closed curves in the x,y plane called the *phase plane* of the system. These same curves, or *phase portraits*, can be generated from a solution pair $x(t)$ and $y(t)$ as above by treating t as a parameter. In Figure 4.4.2, we show the phase portrait of the solution pictured in Figure 4.4.1. (Recall the discussion of phase portraits in Section 2.4.[7])

Let us now trace this phase portrait. Start at the bottom of the curve, region A, with only a small number of prey and predators. With few predators, the population size of the prey grows almost exponentially. But as the prey size becomes large, the interaction term for the predators, bxy, becomes large, and their numbers y begin to grow. Eventually the product ay first equals and then exceeds r, in

---

[7]Syntax similar to that for drawing Figure 4.4.1 and producing a picture similar to Figure 4.4.2 (without the direction field) is the following.

```
> with(DEtools):
> predprey:=[diff(x(t),t)=r*x-a*x*y,diff(y(t),t)=-m*y+b*x*y];
> r:=1; a:=1; m:=1; b:=1;
> DEplot2(predprey,[x,y],0..10,{[0,3/2,1/2],[0,4/5,3/2]},stepsize=.1);
```

the first of equations (4.4.1), at which time the population size of the prey must decrease. This takes us to point B on the figure.

However, the number of prey are still large, so predator size y continues to grow, forcing prey size x to continue declining. This is the upward and leftward section of the portrait. Eventually the product bx first equals and then falls below m in the second of equations (4.4.1) whereupon the predator size now begins to decrease. This is point C in the figure.

At first, the predator size is still at a high level, so the prey size will continue to decrease until reaching its smallest value. But with few prey around, predator numbers y rapidly decrease until finally the product ay falls below r. Then the prey size starts to increase again. This is point D in the figure. But the prey size is still at a low level, so the predator numbers continue to decrease bringing us back to point A and completing one cycle.

Thus the phase portrait is traversed counter-clockwise and, as we have seen in the above narration, the predator population cycle qualitatively follows that of the prey population cycle but lags behind it.

Of course the populations won't change at all if the derivatives $\frac{dx}{dt}$ and $\frac{dy}{dt}$ are both zero in the Lotka–Volterra equations (4.4.1). Setting them to zero and solving the resulting algebraic system locates the stationary points,

$$0 = x(r - ay)$$
$$0 = y(-m + bx).$$

Thus if $x = m/b$ and $y = r/a$, the populations remain fixed. Of course, $x = y = 0$ is also a stationary point.

Stability determinations are made from an eigen-analysis of the community matrix.

Consider the stationary point $(0, 0)$. What if the system starts close to this point, that is, y_0 and x_0 are both very nearly 0? We assume these values are so small that the quadratic terms in equations (4.4.1) are negligible, and we discard them. This is called *linearizing the system* about the stationary point. Then the equations become

$$\frac{dx}{dt} = rx$$

$$\frac{dy}{dt} = -my.$$
(4.4.4)

Hence x will increase and y will further decrease (but not to zero) and a phase portrait will be initiated as discussed above. The system will *not*, however, return to $(0, 0)$. Therefore this stationary point is unstable.

We can come to the same conclusion by rewriting the system (4.4.4) in matrix form and examining the eigenvalues of the matrix on the right-hand side. This matrix is

$$
\begin{bmatrix} r & 0 \\ 0 & -m \end{bmatrix},
$$
(4.4.5)

and its eigenvalues are $\lambda_1 = r$ and $\lambda_2 = -m$. Since one of these is real and positive, the conclusion is the stationary point $(0,0)$ is unstable.

Now consider the stationary point $x = m/b$ and $y = r/a$ and linearize about it as follows. Let $\xi = x - m/b$ and $\eta = y - r/a$. In these new variables the first equation of the system (4.4.1) becomes

$$
\frac{d\xi}{dt} = r\left(\xi + \frac{m}{b}\right) - a\left(\xi + \frac{m}{b}\right)\left(\eta + \frac{r}{a}\right)
$$
$$
= -\frac{am}{b}\eta - a\xi\eta.
$$

Again discarding the quadratic term, this yields

$$
\frac{d\xi}{dt} = -\frac{am}{b}\eta.
$$

The second equation of the system becomes

$$
\frac{d\eta}{dt} = -m\left(\eta + \frac{r}{a}\right) + b\left(\xi + \frac{m}{b}\right)\left(\eta + \frac{r}{a}\right)
$$
$$
\frac{d\eta}{dt} = \frac{br}{a}\xi + b\xi\eta.
$$

Discarding the quadratic term gives

$$
\frac{d\eta}{dt} = \frac{br}{a}\xi.
$$

Thus equations (4.4.1) become

$$
\frac{d\xi}{dt} = -\frac{am}{b}\eta
$$
$$
\frac{d\eta}{dt} = \frac{br}{a}\xi.
$$
(4.4.6)

The right-hand side of equation (4.4.5) can be written in matrix form

$$
\begin{bmatrix} 0 & -\frac{am}{b} \\ \frac{br}{a} & 0 \end{bmatrix}\begin{bmatrix} \xi \\ \eta \end{bmatrix}.
$$
(4.4.7)

This time the eigenvalues of the matrix are imaginary, $\lambda = \pm i\sqrt{mr}$. This implies that the stationary point is *neutrally stable*.

Determining the stability at stationary points is an important problem. Linearizing about these points is a common tool for studying this stability, and has been formalized into a computational procedure. In the exercises, we give more applications that utilize the above analysis and use a computer algebra system. Also, we give an example where the procedure incorrectly predicts the behavior at a stationary point. The text by Steven H. Strogatz [9] explains conditions to guarantee when the procedure works.

To illustrate a computational procedure for this predator–prey model, first make the vector function V:

```
> restart:
> with(linalg):
> V:=vector([r*x-a*x*y,-m*y+b*x*y]);
```

Find the critical points of (4.2.1) by asking where this vector-valued function is zero:

```
> solve({V[1]=0,V[2]=0}, {x,y});
```

This investigation provides the solutions $\{0,0\}$ and $\{m/b, r/a\}$, as we stated above. We now make the linearization of V about $\{0,0\}$ and about $\{m/b, r/a\}$:

```
> jacobian(V,[x,y]);
> subs({x=0,y=0},");
> subs({x=m/b,y=r/a},"");
```

The output is the matrix in equation (4.4.5) and the matrix as in equation (4.4.7). Note that in matrix form

$$\begin{pmatrix} \frac{dx}{dt} \\ \frac{dy}{dt} \end{pmatrix} = \begin{pmatrix} r & 0 \\ 0 & -m \end{pmatrix} \begin{pmatrix} x-0 \\ y-0 \end{pmatrix} + \begin{pmatrix} -a \\ b \end{pmatrix}(x-0)(y-0) \tag{4.4.8}$$

and

$$\begin{pmatrix} \frac{dx}{dt} \\ \frac{dy}{dt} \end{pmatrix} = \begin{pmatrix} 0 & -\frac{am}{b} \\ \frac{br}{a} & 0 \end{pmatrix} \begin{pmatrix} x-\frac{m}{b} \\ y-\frac{r}{a} \end{pmatrix} + \begin{pmatrix} -a \\ b \end{pmatrix}\left(x-\frac{m}{b}\right)\left(y-\frac{r}{a}\right).$$

Finally, we compute the eigenvalues for the linearization about each of the critical points.

```
> eigenvals(" "); eigenvals(" ");
```

The result is the same as the result after equation (4.2.5) and after equation (4.2.7).

Exercises

1. The following *competition model* is provided in Reference [9]. Imagine rabbits and sheep competing for the same limited amount of grass. Assume a logistic growth for the two populations, that rabbits reproduce rapidly, and that the sheep will crowd out the rabbits. Assume these conflicts occur at a rate proportional to the size of each population. Further, assume that the conflicts reduce the growth rate for each species, but make the effect more severe for the rabbits by increasing the coefficient for that term. A model that incorporates these assumptions is

$$\frac{dx}{dt} = x(3 - x - 2y)$$

$$\frac{dy}{dt} = y(2 - x - y)$$

where $x(t)$ is the rabbit population and y is the sheep population. (Of course, the coefficients are not realistic, but are chosen to illustrate the possibilities.) Find four stationary points and investigate the stability of each. Show that one of the two populations is driven to extinction.

2. Imagine a *three-species predator–prey* problem, which we identify with grass, sheep, and wolves. The grass grows according to a logistic equation in the absence of sheep. The sheep eat the grass and the wolves eat the sheep. (See McLaren [10] for a three-species population in observation.) We model this with the equations that follow. Here x represents the wolf population, y represents the sheep population, and z represents the area in grass:

$$\frac{dx}{dt} = -x + xy,$$

$$\frac{dy}{dz} = -y + 2yz - xy,$$

$$\frac{dz}{dt} = 2z - z^2 - yz.$$

What would be the steady state of grass, with no sheep or wolves present? What would be the steady state of sheep and grass, with no wolves present? What is the revised steady state with wolves present? Does the introduction of wolves benefit the grass? This study can be done as follows:

```
> with(share): readshare(ODE,plots):
> rsx:=(t,x,y,z)–>-x+x*y;
  rsy:=(t,x,y,z)–>-y+2*y*z-x*y;
  rsz:=(t,x,y,z)–>2*z-z^2-y*z;
```

For just grass:

```
> init:=[0,0,0,1.5];
> output:=rungekuttahf([rsx,rsy,rsz],init,0.1,200):
> plot({makelist(output,1,4),makelist(output,1,3),
    makelist(output,1,2)});
```

For grass and sheep:

```
> init:=[0,0,.5,1.5];
> output:=rungekuttahf([rsx,rsy,rsz],init,0.1,200):
> plot({makelist(output,1,4),makelist(output,1,3),
    makelist(output,1,2)});
```

For grass, sheep, and wolves:

```
> init:=[0,.2,.5,1.5];
> output:=rungekuttahf([rsx,rsy,rsz],init,0.1,200):
> plot({makelist(output,1,4),makelist(output,1,3),
    makelist(output,1,2)});
```

3. J.M.A. Danby [11] has a collection of interesting population models in his delightful text. The following *predator–prey model with child care* is included. Suppose that the prey $x(t)$ is divided into two classes: $x_1(t)$ and $x_2(t)$ of young and adults. Suppose that the young are protected from predators $y(t)$. Assume the young increase proportional to the number of adults and decrease due to death or to moving into the adult class. Then

$$\frac{dx_1}{dt} = ax_2 - bx_1 - cx_1.$$

The number of adults is increased by the young growing up and decreased by natural death and predation, so that we model

$$\frac{dx_2}{dt} = bx_1 - dx_2 - ex_2y.$$

Finally, for the predators, we take

$$\frac{dy}{dt} = -fy + gx_2y.$$

Investigate the structure for the solutions of this model. Parameters that might be used are

$$a = 2, \quad b = c = d = 1/2, \quad \text{and} \quad e = f = g = 1.$$

4. Show that linearization of the system

$$\frac{dx}{dt} = -y + ax(x^2 + y^2)$$

$$\frac{dy}{dt} = x + ay(x^2 + y^2)$$

predicts that the origin is a center for all values of a, whereas in fact the origin is a stable spiral if $a < 0$ and an unstable spiral if $a > 0$. Draw phase portraits for $a = 1$ and $a = -1$.

5. Suppose there is a small group of individuals who are infected with a contagious disease and who have come into a larger population. If the population is divided into three groups, the susceptible, the infected, and the recovered, we have what is known as a classical S–I–R problem. The susceptible class consists of those who are not infected, but who are capable of catching the disease and becoming infected. The infected class consists of the individuals who are capable of transmitting the disease to others. The recovered class consists of those who have had the disease, but are no longer infectious.

A system of equations that is used to model such a situation is often described as follows:

$$\frac{dS}{dt} = -rS(t)I(t),$$

$$\frac{dI}{dt} = rS(t)I(t) - aI(t),$$

$$\frac{dR}{dt} = aI(t),$$

for positive constants r and a. The proportionality constant r is called the infection rate, and the proportionality constant a is called the removal rate.

a. Rewrite this model as a matrix model and recognize that the problem forms a closed compartment model. Conclude that the total population remains constant.

b. Draw graphs for solutions. Observe that the susceptible class decreases in size and that the infected class increases in size and later decreases.

```
> readlib(spline);
> r:=1; a:=1;
> sol:=dsolve({diff(SU(t),t)=-r*SU(t)*IN(t),
    diff(IN(t),t)=r*SU(t)*IN(t)-a*IN(t),
    diff(R(t),t)=a*IN(t),
      SU(0)=2.8,
      IN(0)=0.2,
      R(0)=0},
    {SU(t),IN(t),R(t)}, numeric, output=listprocedure):
> f:=subs(sol,SU(t)): g:=subs(sol,IN(t)): h:=subs(sol,R(t));
> plot({f,g,h}, 0..20);
```

c. Suppose, now, that the recovered do not receive permanent immunity. Rather, we suppose that after a delay of one unit of time, those who have recovered lose immunity and move into the susceptible class. The system of equations changes to the following:

$$\frac{dS}{dt} = -rS(t)I(t) + R(t-1),$$

$$\frac{dI}{dt} = rS(t)I(t) - aI(t),$$

$$\frac{dR}{dt} = aI(t) - R(t-1).$$

Draw graphs for solutions to this system. Observe the possibility of oscillating solutions. How do you explain these oscillations from the perspective of an epidemiologist? (Note: The program has a long run time.)

```
> readlib(spline);
> r:=1; a:=1;
> f.0:=t->2.8; g.0:=t->0.2; h.0:=t->0;
    J.0:=plot({f.0,g.0,h.0}, -1..0):
> x0seq:=[seq(j/5,j=0..5)]: N:=5;
> for n from 1 to N do
    sol:=dsolve({diff(SU(t),t)=-r*SU(t)*IN(t)+'h.(n-1)(t-1)',
      diff(IN(t),t)=r*SU(t)*IN(t)-a*IN(t),
      diff(R(t),t)=a*IN(t)-'h.(n-1)(t-1)',
        SU(n-1)=f.(n-1)(n-1),
        IN(n-1)=g.(n-1)(n-1),
        R(n-1)=h.(n-1)(n-1)},
      {SU(t),IN(t),R(t)}, numeric, output=listprocedure):
    f:=subs(sol,SU(t)): g:=subs(sol,IN(t)): h:=subs(sol,R(t)):
    xseq:=map(t->t+n-1,x0seq):
    fxseq:=map(f,xseq): gxseq:=map(g,xseq): hxseq:=map(h,xseq):
      F:=spline(xseq,fxseq,t,cubic):
      G:=spline(xseq,gxseq,t,cubic):
```

```
    H:=spline(xseq,hxseq,t,cubic):
    f.n:='spline/makeproc'(F,t):
    g.n:='spline/makeproc'(G,t):
    h.n:='spline/makeproc'(H,t):
    J.n:= plot({f.n,g.n,h.n}, (n-1)..n):
    od:
  > plots[display]({J.1,J.2,J.3,J.4,J.5});
```

Section 4.5

Questions for Thought and Discussion

1. Name and discuss four factors that affect the carrying capacity of an environment for a given species.
2. Draw and explain the shape of survivorship and population growth curves for an r-strategist.
3. Draw and explain the shape of survivorship and population growth curves for a K-strategist.
4. Define carrying capacity and environmental resistance.
5. Discuss the concept of parental investment and its role in r- and K-strategies.

References and Suggested Further Reading

1. **Environmental resistance:** William T. Keeton and James L. Gould, *Biological Science*, 5th. ed., W. W. Norton and Company, New York, 1993.
2. **Partitioning of resources:** John L. Harper, *Population Biology of Plants*, Academic Press, New York, 1977.
3. **Population ecology:** Richard Brewer, *The Science of Ecology*, Saunders College Publishing, Ft. Worth, 2nd ed., 1988.
4. **Ecology and public issues:** Barry Commoner, *The Closing Circle—Nature, Man and Technology*, Alfred A. Knopf, New York, 1971.
5. **Natural population control:** H. N. Southern, *J. Zool., Lond.* vol. 162, pp. 197–285, 1970.
6. **A doomsday model:** David A. Smith; "Human Population Growth: Stability or Explosion," *Mathematics Magazine*, vol. 50, no. 4, Sept. 1977, pp. 186–197.
7. **Budworm, balsalm fir, and birds:** D. Ludwig, D. D. Jones and C. S. Holling; "Qualitative Analysis of Insect Outbreak Systems: The Spruce Budworm and Forests," *J. Animal Ecology*, vol. 47, 1978, pp. 315–332.
8. **Predator or prey:** J. D. Murray, "Predator–Prey Models: Lotka–Volterra Systems," Section 3.1 in *Mathematical Biology,* Springer-Verlag, 1990.
9. **Linearization:** Steven H. Strogatz, *Nonlinear Dynamics and Chaos, with Applications to Physics, Biology, Chemistry, and Engineering*, Addison-Wesley Publishing Company, New York, 1994.

10. **A matter of wolves:** B. E. McLaren, R. O. Peterson, "Wolves, Moose and Tree Rings on Isle Royale," *Science*, vol. 266, 1994, pp. 1555–1558.
11. **Predator–prey with child care, cannibalism, and other models:** J. M. A. Danby, *Computing Applications to Differential Equations*, Reston Publishing Company, Inc., Reston, VA, 1985.

Chapter 5

Age-Dependent
Population Structures

Introduction to this chapter

This chapter presents an analysis of the distribution of ages in a population. We begin with a discussion of the aging process itself and then present some data on the age-structures of actual populations. We finish with a mathematical description of age-structures. Our primary interest is in humans, but the principles we present will apply to practically any mammal and perhaps to other animals as well.

Section 5.1

Aging and Death

The notion of aging is not simple. One must consider that oak trees, and perhaps some animals like tortoises, seem to have unlimited growth potential, that a Pacific salmon mates only once and then ages rapidly, and that humans can reproduce for many years. In each case a different concept of aging may apply.

The reason that aging occurs, at least in mammals, is uncertain. The idea that the old must die to make room for the new gene combinations of the young is in considerable doubt. An alternative hypothesis is that organisms must partition their resources between the maintenance of their own bodies and reproduction, and that the optimal partitioning for evolutionary fitness leaves much damage unrepaired. Eventually the unrepaired damage kills the organism. We present several hypotheses about how and why damage can occur.

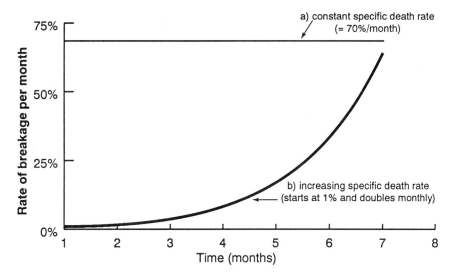

Figure 5.1.1   Death rate, modeled on the breakage of test tubes. Curve (a) is obtained by assuming a specific death (breakage) rate of 70% of survivors per month of test tubes surviving to that point. This is equivalent to assuming that there is no aging, because the probability of death (breakage) is independent of time. Curve (b) is obtained by assuming that the specific death rate is 1% of the survivors in the first month and then doubles each month thereafter. This is equivalent to assuming than the test tubes age, because the probability of death (breakage) increases with time.

What is meant by "aging" in an organism?

We will use a simple definition of aging, or *senescence*:[1] It is a series of changes that accelerate with age and eventually result in the death of an organism. This definition is a loose one because it does not specify the source of the changes—the only requirement is that they accelerate. We will adopt a common approach and not regard predation, injury and disease caused by parasites, e.g., microorganisms, as causes of aging, even though their incidence may increase with age.

The effect of aging on survival is demonstrated in Figure 5.1.1 for a simple model system of test tubes. Suppose that a laboratory technician buys 1000 test tubes and that 70% of all surviving test tubes are broken each month. Curve (a) of Figure 5.1.1 shows the specific rate of breakage of the tubes—a constant 70% per month.[2] Note that a test tube surviving for three months would have the same chance of breakage in the fourth month as would one at the outset of the experiment (because aging has not occurred). Alternatively, suppose that the test tubes broke more easily as time passed. A tube surviving for three months would

---

[1] There is much argument about definitions in the study of aging and we wish to avoid being part of the dispute. Our simplification may have the opposite effect!

[2] The specific death (= breakage) rate is the number dying per unit time *among those of a specific age*. This is to be distinguished from the simple death rate, which is the death rate irrespective of age. In this experiment, of course, all the test tubes are of the same age.

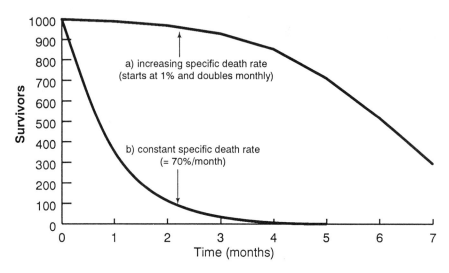

Figure 5.1.2   (a) A survivorship curve for a non-aging system, using the data of Figure 5.1.1(b). (b) A survivorship curve for a system that exhibits aging, using the data of Figure 5.1.1(a). Both curves assume an initial cohort of 1000 test tubes at time $t = 0$. Note the similarity of curves (a) and (b) to Figures 4.1.2 and 4.1.4, which are survivorship curves for r-strategists and K-strategists, respectively.

have a much greater chance of breakage during the fourth month than would one at the outset of the experiment (because the older one has aged). Curve (b) shows the rate of breakage for these tubes (doubling each month in this example).

Figure 5.1.2 shows survivorship curves for the two cases whose specific death rates are described by Figure 5.1.1. You should compare them to Figures 4.1.2 and 4.1.4, which are survivorship curves for r-strategists and K-strategists, respectively. It should be clear that r-strategists do not show aging (because they are held in check by climatic factors, which should kill a constant fraction of them, regardless of their ages).[3] The situation with regard to K-strategists is a bit more complex: Mammals, for instance, are held in check by density-dependent factors. If they live long enough, aging will also reduce their numbers. Both density-dependent factors and aging become more important as time passes. Thus, the survivorship curve for a mammalian K-strategist should look somewhat like that shown in Figures 4.1.4 and 5.1.2.

Why do organisms age and die?

When asking "why" of any biological process as profound as senescence we should immediately look to the Darwinian model of evolution for enlightenment and seek a positive selective value of aging to aspecies. A characteristic confer-

---

[3]admittedly an approximation

ring a positive advantage is called an *adaptation* and, as we shall see, the adaptation we seek may not exist.

A simple adaptive explanation for senescence is that the Darwinian struggle for survival creates new organisms to fit into a changing environment. Thus, the previous generation must die to make space and nutrients available for the new generation. Thomas Kirkwood has made two objections to this hypothesis [1]. The first objection is posed in the question "How can aging have a positive selective value for a species when it can kill all the members of the species?" Besides, many organisms show the most evident aging only after their reproductive lives have ended. If the organism should show genetically programmed deterioration in its old age, that would have minimal (or no) selective value because the organism's reproductive life would have already ended anyway.

Kirkwood's second objection is that most organisms live in the wild and almost always die from disease and predation. Thus, there is no need for selection based on aging in most organisms—they die too soon from other causes.

There is another way to answer the question "Why do organisms age?," a non-adaptive way in that aging does not have a positive selective value. First, recall that in Section 4.1 we discussed how trees can partition each year's energetic resources and physical resources between asexual and sexual reproduction. For a year or two a tree would add thick trunk rings (asexual growth) at the expense of reduced nut production (sexual reproduction). Then, for a year or two the tree would reverse the situation and produce lots of nuts at the expense of vegetative growth. There is a hypothesis about aging that generalizes this situation; it is called the *disposable soma model*.[4]

Kirkwood assumes that the organisms whose aging is of interest to us must partition their finite resources between reproduction and the maintenance of the soma, i.e., the body. In particular, somatic maintenance means the repair of the many insults and injuries that are inflicted on the body by factors like ordinary wear and tear, toxin production, radiation damage and errors in gene replication and expression. The two needs, reproduction and somatic maintenance, thus compete with one another. If excessive resources are put into somatic maintenance there will be no reproduction, and the species will die out. If excessive resources are devoted to reproduction there will be insufficient somatic maintenance, and the species will die out. We thus assume that there is an optimal partitioning of resources between somatic maintenance and reproduction. The disposable soma model postulates that this optimal partitioning is such that some somatic damage must go unrepaired and that the organism eventually dies because of it. Thus, the organism has a finite lifetime, one marked by an increasing rate of deterioration, i.e., aging.

The disposable soma model is non-adaptive in that aging is a harmful process. It is, however, an essential process because it is a measure of the resources diverted into reproduction. In a way, aging is a side effect, but, of course, it has powerful consequences to the individual organism.

---

[4]"Soma" means "body."

Aging of cells can provide insight into organismal aging.

The death of the only cell comprising an amoeba has consequences that are quite different from those associated with the death of a single skin cell of a person; thus, we will have to distinguish between aging in single-celled and multicellular organisms.

It is fine to study the processes that lead to the death of a cell, but what if that cell is only one of many in an organ of a multicellular organism? To answer this question we must first understand that cell death is a natural part of the life and development of organisms. Our hands are initially formed with interdigital webbing, perhaps suggesting our aquatic ancestry. This webbing is removed *in utero* by the death of the cells that comprise it. There are many other examples of cell death as a natural consequence of living: Our red blood cells live only about three months, and our skin cells peel off constantly, and both are quickly replaced.

We can now return to the question of what happens if one, or even a small fraction, of the cells in an organ dies. Usually nothing, of course—we see that it happens all the time. But, if that cell dies for a reason connected to the possible deaths of other cells, then the study of the one cell becomes very important. Thus, the study of aging in cells can contribute greatly to our knowledge of aging in multicellular organisms.

How do organisms become damaged?

Whether we accept Kirkwood's disposable soma model or not, it is clear that our cells age, and we must suggest ways that the relevant damage occurs. Numerous mechanisms have been proposed but no single one has been adequate, and in the end it may be that several will have to be accepted in concert. Some examples of damage mechanisms that have been proposed are:

a. *Wear and tear*: A cell accumulates "insults," until it dies. Typical insults result from the accumulation of wastes and toxins, as well as from physical injuries like radiation damage and mechanical injury. These are all well-known causes of cell death. Cells have several mechanisms by which insults can be repaired, but it may be that these repair systems themselves are subject to damage by insults.

b. *Rate of living*: This is the "live fast, die young" hypothesis. In general, the higher a mammal's basal metabolic rate, the shorter its lifespan is. Perhaps some internal cellular resource is used up, or wastes accumulate, resulting in cell death.

c. *Preprogrammed aging*: Our maximum lifespan is fixed by our genes. While the average lifespan of humans has increased over the past few decades, the maximum lifespan seems fixed at 100–110 years. Noncancerous mammalian cell lines in test tube culture seem capable of only a fixed number of divisions. If, halfway through that fixed number of divisions, the cells are frozen in liquid nitrogen for ten years and then thawed, they will complete only the remaining half of their allotted divisions.

Cell reproduction seems to have a rejuvenating effect on cells.

It is a common observation that cells which reproduce often tend to age more slowly than cells that divide infrequently. This effect is seen in both asexual and sexual reproduction. Cancer cells divide rapidly and will do so in culture forever. Cells of our pancreas divide at a moderate rate and our pancreas seems to maintain its function well into old age. Brain cells never divide and brain function deteriorates noticeably in old age. Even single-celled organisms can exhibit this effect: They may show obvious signs of senescence until they reproduce, at which point those signs disappear.

Section 5.2

The Age-Structure of Populations

Age-structure diagrams show the frequency distribution of ages in a population. The data for males and females are shown separately. The shape of these diagrams can tell us about the future course of population changes: The existence of a large proportion of young people at any given time implies that there will be large proportions of individuals of child-bearing age 20 years later and of retirees 60 years later. The shapes of age-structure diagrams are also dependent on migration into and out of a population. Comparison of data for males and females can tell us about the inherent differences between the genders and about society's attitude toward the two genders.

Age-structure diagrams are determined by age-specific rates of birth, death and migration

Figure 5.2.1 is a set of age-structure diagrams for the United States for 1955, 1985, 2015 (projected) and 2035 (projected) (see also Reference [2]). They show how the population is, or will be, distributed into age groups. Data are included for males and females. These diagrams can convey a great deal of information. For example, look at the data for 1955 and note the 20–30 year-old *cohort*.[5] There are relatively fewer people in this group because the birth rate went down during the Great Depression. On the other hand, the birth rate went up dramatically after World War II and the 20–40 year-old cohort in 1985 (the "baby-boomers") shows clearly. Both of these cohorts can be followed in the projected data. Note also how the population of elderly people, especially women, is growing.

---

[5]A *cohort* is a group of people with a common characteristic. Here the characteristic they share is that they were born in the same decade.

■ Male

□ Female

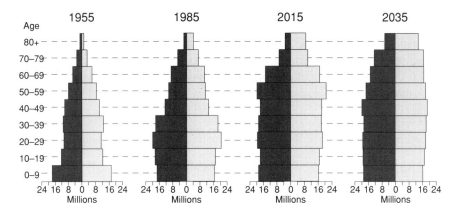

Figure 5.2.1   Past and future (projected) age-structure diagrams for the United States. Note the growing proportion of elderly, compared to young, people. The cohort of "baby-boomers" is evident at the base of the 1955 data. That group moves up the 1985 and 2015 diagrams. (Redrawn from "Age and Sex Composition of the U.S. Population," in "U.S. Population: Charting the Change—Student Chart Book," Population Reference Bureau, Inc. 1988. Used with permission.)

Figure 5.2.2 shows recent data for four countries—Kenya, China, the United States and Russia. Future population growth can be estimated by looking at the cohort of young people, i.e., the numbers of people represented by the bottom part of each diagram. In a few decades these people will be represented by the middle part of age-structure diagrams *and* will be having babies. Thus, we can conclude that the population of Russia will remain steady or even decrease, those of the United States and China will grow slowly to moderately and that of Kenya will grow rapidly.

Another factor besides births and deaths can change an age-structure diagram: Migration into and out of a population may change the relative numbers of people in one age group. Figure 5.2.3 shows data for Sheridan County, North Dakota, and for Durham County, North Carolina, for 1990. Rural areas of the Great Plains have suffered a loss of young people due to emigration, and the data for Sheridan County demonstrates it clearly. On the other hand, Durham County is in the North Carolina Research Triangle, the site of several major universities and many research industries. It is therefore a magnet for younger people, and its age-structure diagram shows that fact.

Some populations have more men than women.

We are accustomed to the idea that there are more women than men in our country. That (true) fact can be misleading, however. While the sex ratio at conception is

Male

Female

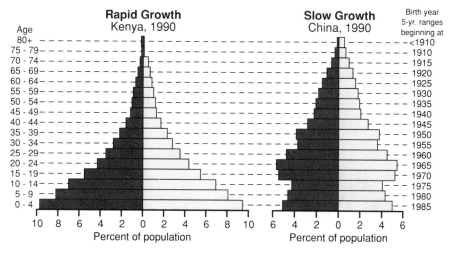

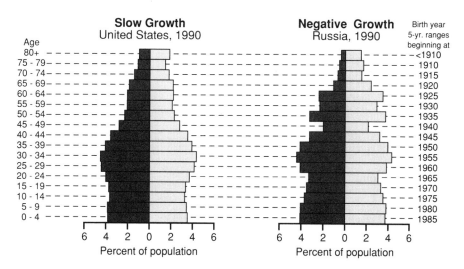

Figure 5.2.2 Age-structure diagram for four countries for 1990. Each is labelled according to its expected future growth rate. For instance, Kenya has a high proportion of young people, so we expect its future growth rate to be high. (Redrawn from "Patterns of Population Change" in "World Population – Toward the Next Century," page 5; Population Reference Bureau, Inc. 1994. Used with permission.)

not known, there is evidence that a disproportionate number of female fetuses are spontaneously aborted in the first trimester of pregnancy. On the other hand, in the second and third trimesters more male than female fetuses are lost. The ratio of sexes at birth in the United States is about 106 males to every 100 females. The specific death rate for males is higher than for women and by early adolescence the sex ratio is 100:100. You can refer to Figure 5.2.1 to see the effect of males' higher death rate on the relative numbers of males and females in later life.

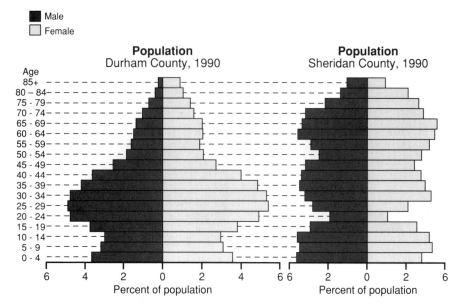

Figure 5.2.3   An age-structure diagram showing the effects of migration. Many young people in the 20–45 year-old age group have moved into Durham County, North Carolina, and many young people in the 20–30 year-old age group have moved out of Sheridan County, North Dakota. (Redrawn from "Age and Sex Profiles of Sheridan and Durham Counties, 1990," in "Americans on the Move," Population Bulletin, vol. 48, no. 3, page 25; Population Reference Bureau, Inc. 1993. Used with permission.)

The fact that there are more females than males in the United States might lead us to be surprised by the data of Figure 5.2.4, an age-structure diagram for the United Arab Emirates. The unbalanced sex ratio, heavily tilted toward males, arises from immigration: U.A.E. has brought in many men from other countries to work in its oil fields, and the men seldom bring their families.

Another feature of gender ratios can be noted in age-structure diagrams of certain countries. In the late 1980s, the ratio of men to women in advanced countries was about 94:100; in developing countries it was about 104:100.

Section 5.3

Predicting the Age-Structure of a Population

A graph of population size P as a function of age y visually documents the age-structure or profile of a population. Over time a population profile can change due to periodic environmental conditions which may be favorable or unfavorable to the population, and to occasional events such as natural diasters and epidemics. For human populations, medical improvements have gradually increased the representation in the higher age brackets.

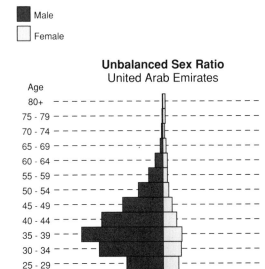

Figure 5.2.4 Age-structure diagram showing an unbalanced sex-ratio, from the United Arab Emirates. The gender imbalance, males outnumbering females, is due to the importation of males to work in the oil fields: These males are not accompanied by their families. (Redrawn from "Unbalanced Sex Ratio: United Arab Emirates, 1985" in "Population—A Lively Introduction," Population Bulletin, vol. 46, no. 2, page 25; Population Reference Bureau, Inc. 1991. Used with permission.)

But much greater use can be made of the population density function P. With a knowledge of survival rates by age, $\ell(y)$, the trend in P can be predicted. It can be shown that if survival rates are relatively constant over time, then the age-structure of a population tends to a fixed profile within which the overall size of the population may nonetheless increase or decrease.

Age-structure is the distribution of a population by age.

The age-structure of a population can be described by means of a function $P(y)$ giving the size of the population in the yth age group for a set of groups covering all possible ages. Table 5.3.1 shows the age distribution of the U.S population in 1990 refined to 20-year age brackets. Mathematically it is more common to use 1-year age brackets so that $P(0)$ is the number of newborns less than one year of

Table 5.3.1   U.S. Population, 1990

Age Bracket	Number (in millions)
0–20	71.8
20–40	103.4
40–60	60.3
60–80	20.9
80–100	.209
100+	.001

age, $P(1)$ counts the one-year-olds, and so on. We shall refer to $P(y)$ as the *age density* function. The total size of a population is calculated from its density by summing,

$$P = \sum_{y=0}^{\infty} P(y). \qquad (5.3.1)$$

The use of infinity as the upper limit of this sum is a simplifying measure; for some age (maybe $y_{max} = 115$), $P(y) = 0$ for $y > y_{max}$, so the indicated infinite sum is in reality only from 0 to 115.

The age-structure of the United States has gradually evolved over the last half of the 20th century, as seen in Figure 5.2.1. On the other hand, any of several catastrophes can bring about rapid change to an age-structure. We account for these possibilities by regarding the age density function as dependent on calendar time t as well as age y, and in deference to these dual dependencies we write $P(y, t)$. In addition, by including the reference to time, a mechanism is provided for describing births year by year, namely $P(0, t)$. This is the birth rate of a population. If the birth rate is down in some year, say $t = t_0$, this affects the population in subsequent years as well, as we have seen above. To begin with, the population of 1-year-olds cannot exceed the population of newborns in the previous year:

$$P(1, t_0 + 1) \leq P(0, t_0),$$

assuming no immigration into the population, of course. This is generally true for any age bracket, thus, under the condition of no immigration,

$$P(y + 1, t + 1) \leq P(y, t) \qquad \text{for } y \geq 0 \text{ and for all } t. \qquad (5.3.2)$$

While the population in an age bracket cannot increase in the following year, it can decrease due to deaths that occur during the year. Let $\mu(y)$ denote the death rate, or *mortality*, experienced by the population of age y. The death rate is dimensionless, being the fraction of deaths per individual, or, since it is usually a number in the thousandths, it is frequently given as deaths per 1000 individuals.

Table 5.3.2   U.S. Mortality Table for 1991

Age	Deaths (%)
0–10	1.2
10–20	.57
20–30	1.2
30–40	1.8
40–50	3.1
50–60	7.2
60–70	16.4
70+	100

SOURCE: U.S. Dept. of Health and Human Services, Hyattsville, MD.

The actual number of deaths that occur among the segment of the population of age y in year t is the product of the death rate and the number of individuals at risk,

$$\mu(y)P(y,t),$$

(μ must be deaths per individual here, or P must be population in thousands).

Virtually all natural populations experience very high pre-adult mortality rates. Insect populations and other unnurtured species (r-strategists; cf Chapter 4) experience death rates similar to that shown in curve (a) of Figure 5.1.1. Notice that the newly hatched young suffer the highest mortality rates with improvement as the animal ages. By contrast, nurtured species (K-strategists), such as mammals, experience much lower pre-adult mortality rates, as seen in curve (b) of Figure 5.1.1.

A mortality table for the United States is given in Table 5.3.2. In most species, mortality rates are lowest during the middle adult years.

Returning to equation (5.3.2), taking deaths into account yields the equality

$$P(y+1,t+1) = P(y,t) - \mu(y)P(y,t) \tag{5.3.3}$$

provided there is no immigration or emigration. But this equation ignores the effect of external events, which may play havoc with death rates. For example, due to a catastrophic epidemic, death rates in the youth age groups may be high during the calendar year in which the epidemic strikes. On the other hand, the U.S. population has experienced a gradually decreasing death rate over this century as a result of improved medical care (see Table 5.3.3). To account for these and other factors unrelated to age, we must regard μ as a function of time as well as age. Thus equation (5.3.3) becomes

$$\begin{aligned} P(y+1,t+1) &= P(y,t) - \mu(y,t)P(y,t) \\ &= \ell(y,t)P(y,t) \end{aligned} \tag{5.3.4}$$

where $\ell(y,t) = 1 - \mu(y,t)$ is the fraction of the population of age y which will live through year t. These factors $\ell(\cdot,\cdot)$ are called *survival rates*.

In the absence of external events, populations evolve to a stable age distribution.

While survival rates depend on calendar time in general, here we are interested in predicting the population structure in the absence of external events. Consequently we will regard μ (and ℓ) as a function of age only.

Knowing yearly birth rates $P(0,t)$ and age-specific survival rates, $\ell(y)$, equation (5.3.4) allows the calculation of the course of the population through time including its age distribution and size. We also need to know the present age distribution, $P(y,0)$, where we may regard the present time as $t = 0$. Usually the calculation is done for the female population of the species, since birth rates depend largely on the number of females while being somewhat independent of the number of males. The birth rates given will therefore pertain to the birth of females.

We illustrate this calculation for a K-strategist, specifically, for the grey seal, whose (female) fecundity and survival rates are given in Table 5.3.4.

Table 5.3.4 Grey Seal Fecundity and Survival Rates

Age	0	1	2	3	4	5	5+
Fecundity	0	0	0	0	0.08	0.28	0.42
Survival	0.657	0.930	0.930	0.930	0.935	0.935	0

SOURCE: D. Brown and P. Rothery, *Models in Biology: Mathematics, Statistics and Computing*, John Wiley & Sons Ltd., Chirchester, 1993.

To get it started, we make the assumption that the present population has uniform age density. Actually this assumption about the starting population is not important in the long term, as we will see in the exercises. The key values are the birth and survival rates in the table. Since the survival rate for age 0 is 0.657, from equation (5.3.4) we have

$$P(1,t+1) = 0.657P(0,t), \qquad \text{for all } t \geq 0.$$

Similarly, for $y = 1,2,3$,

$$P(y+1,t+1) = 0.930P(y,t), \qquad \text{for all } t \geq 0.$$

And for $y = 4,5$,

$$P(y+1,t+1) = 0.935P(y,t), \qquad \text{for all } t \geq 0.$$

In this we take $5 + 1$ to be 5+. Since there is no catagory beyond "5+", the survival rate $\ell(5+)$ is 0. The birth rate calculation uses the fecundity entries and is only slightly more complicated:

$$P(0,t+1) = 0.08P(4,t) + 0.28P(5,t) + 0.42P(5+,t).$$

It is convenient to write the calculation in matrix form. Let $\mathbf{p}(t)$ be the vector whose components are $P(y,t)$,

$$\mathbf{p}(t) = \begin{bmatrix} P(0,t) \\ P(1,t) \\ P(2,t) \\ P(3,t) \\ P(4,t) \\ P(5,t) \\ P(5+,t) \end{bmatrix}.$$

Then, $\mathbf{p}(1)$ is given as the matrix product

$$\mathbf{p}(1) = \begin{pmatrix} 0 & 0 & 0 & 0 & 0.08 & 0.28 & 0.42 \\ .657 & 0 & 0 & 0 & 0 & 0 & 0 \\ 0 & .930 & 0 & 0 & 0 & 0 & 0 \\ 0 & 0 & .930 & 0 & 0 & 0 & 0 \\ 0 & 0 & 0 & .930 & 0 & 0 & 0 \\ 0 & 0 & 0 & 0 & .935 & 0 & 0 \\ 0 & 0 & 0 & 0 & 0 & .935 & 0 \end{pmatrix} \begin{pmatrix} P(0,0) \\ P(1,0) \\ P(2,0) \\ P(3,0) \\ P(4,0) \\ P(5,0) \\ P(5+,0) \end{pmatrix}$$

$$= L\mathbf{p}(0). \tag{5.3.5}$$

Denote by L the 7×7 matrix indicated. The first row reflects the births coming from various age groups and has non-zero terms indicated by these births. Except for the first row, the only non-zero terms are the principal subdiagonal entries, the survival rates $\ell(y)$. Such a matrix is called a *Leslie matrix,* and it always has the same form:

$$L = \begin{pmatrix} a_1 & a_2 & a_3 & \cdots & a_n \\ b_1 & 0 & 0 & \cdots & 0 \\ 0 & b_2 & 0 & \cdots & 0 \\ \cdots & \cdots & \cdots & \cdots & 0 \\ 0 & 0 & 0 & \cdots & 0 \end{pmatrix}.$$

To be specific, assume a starting density $\mathbf{p}(0)$. The new density $\mathbf{p}(1)$ in equation (5.3.5) can be computed by inspection, or by using the computer:

```
> with(linalg):
> el:=matrix(7,7);
> for i from 1 to 7 do
     for j from 1 to 7 do
         el[i,j]:= 0:
  od
  od:
> el[1,5]:=2/25: el[1,6]:=7/25: el[1,7]:=21/50:
  el[2,1]:=657/1000: el[3,2]:=93/100: el[4,3]:=93/100:
  el[5,4]:=93/100: el[6,5]:=935/1000: el[7,6]:=935/1000:
> evalm(el &* [P0,P1,P2,P3,P4,P5,P6]);
```

Either way we get

$$\mathbf{p}(1) = \begin{pmatrix} 0.08P(4,0) + 0.28P(5,0) + 0.42P(5+,0) \\ .657P(0,0) \\ .930P(1,0) \\ .930P(2,0) \\ .930P(3,0) \\ .935P(4,0) \\ .935P(5,0) \end{pmatrix}.$$

Furthermore, the population size after one time period is simply the sum of the components of $\mathbf{p}(1)$.

The beauty of this formulation is that advancing to the next year is just another multiplication by L. Thus

$$\mathbf{p}(2) = L\mathbf{p}(1) = L^2\mathbf{p}(0), \qquad \mathbf{p}(3) = L\mathbf{p}(2) = L^3\mathbf{p}(0), \qquad \text{etc.}$$

The powers of a Leslie matrix have a special property which we illustrate. For example, we compute L^{10}.

```
> el10:=evalf(evalm(el^10)):
> Digits:=2;
  evalf(evalm(el10));
  Digits:=10;
```

The result, accurate to three places, is

$$L^{10} = \begin{pmatrix} .0018 & .018 & .058 & .094 & .71 & 0 & 0 \\ 0 & .0018 & .013 & .041 & .067 & .050 & 0 \\ 0 & 0 & .0018 & .013 & .041 & .066 & .050 \\ .11 & 0 & 0 & .0018 & .013 & .031 & .033 \\ .073 & .16 & 0 & 0 & .0018 & .0063 & .0094 \\ .021 & .10 & .16 & 0 & 0 & 0 & 0 \\ 0 & .030 & .10 & .16 & 0 & 0 & 0 \end{pmatrix}. \qquad (5.3.6)$$

Remarkably, the power L^n can be easily approximated as predicted by the Perron-Frobenius Theorem [3]. Letting λ be the largest eigenvalue of L and letting V be the corresponding normalized eigenvector, $LV = \lambda V$, then

$$L^n p(0) \approx c\lambda^n V$$

where c is a constant determined by the choice of normalization, see (5.3.7). This approximation improves with increasing n. The importance of this result is that the long range forecast for the population is predictable in form. That is, the ratios between the age classes are independent of the initial distribution and scale as powers of λ.

The number λ is a real, positive eigenvalue of L (recall Section 2.5). This eigenvalue for L can be found rather easily by the computer algebra system. The eigenvector can also be found by the computer. It is shown in Reference [4] that the eigenvector has the simple form

$$V = \begin{pmatrix} 1 \\ b_1/\lambda \\ b_1 b_2/\lambda^2 \\ \vdots \\ b_1 b_2 b_3 \cdots b_n/\lambda^n \end{pmatrix}. \tag{5.3.7}$$

To illustrate this property of Leslie matrices, we find λ, V, and L^{10} for this example. We explore other models in the exercises.

```
> fel:=evalf(evalm(el)):
> vel:=eigenvects(fel);
```

This syntax finds all 7 eigenvalues and their corresponding eigenvectors. By inspection we see which is the largest, for instance, the seventh here.

```
> B:=convert(vel[7][3],list);
> vec:=B[7]; type(vec,vector);
> V:=[seq(vec[i]/vec[1],i=1..7)];
```

We find the eigenvalue and eigenvector as

$$\lambda \approx .8586, \quad \text{and} \quad V \approx \begin{pmatrix} 1.0 \\ .765 \\ .829 \\ .898 \\ .972 \\ 1.06 \\ 1.15 \end{pmatrix} \tag{5.3.8}$$

normalized to have first component equal to 1. The alternate formula (5.3.7) for computing **V** can be used to check this result.

```
> chk:=[1,el[2,1]/lambda, el[2,1]*el[3,2]/lambda^2,
  el[2,1]*el[3,2]*el[4,3]/lambda^3,
  el[2,1]*el[3,2]*el[4,3]*el[5,4]/lambda^4,
  el[2,1]*el[3,2]*el[4,3]*el[5,4]*el[6,5]/lambda^5,
  el[2,1]*el[3,2]*el[4,3]*el[5,4]*el[6,5]*el[7,6]/lambda^6];
```

This gives the same vector **V** as (5.3.8). We illustrate the approximation of the iterates for this example taking the intial value to be uniform, say, 1. Then, making calculations,

```
> evalf(evalm(el10 &* [1,1,1,1,1,1,1]));
> evalm(lambda^10 * V);
```

$$
p(0) = \begin{pmatrix} 1 \\ 1 \\ 1 \\ 1 \\ 1 \\ 1 \\ 1 \end{pmatrix} \qquad L^{10}p(0) = c \begin{pmatrix} .24 \\ .17 \\ .17 \\ .19 \\ .25 \\ .28 \\ .29 \end{pmatrix} \approx c\lambda^{10}\mathbf{V} = c \begin{pmatrix} .22 \\ .17 \\ .18 \\ .19 \\ .21 \\ .23 \\ .25 \end{pmatrix}.
$$

One implication of this structure is that the total population is stable if $\lambda = 1$, and increases or decreases depending on the comparative size of λ to 1.

Continuous population densities provide exact population calculations.

Any table of population densities, such as $P(y,t)$ for $y = 0, 1, \ldots$, as above, will have limited resolution, in this case one-year brackets. Alternatively, an age distribution can be described with unlimited resolution by a continuous age density function, which we also denote by $P(y,t)$, such as we have shown in Figure 4.1.2 and Figure 4.1.4.

Given a continuous age density $P(y,t)$, to find the population size in any age group, just integrate. For instance, the number in the group 17.6 to 21.25 is

$$
\text{number between age 17.6 and 21.25} = \int_{17.6}^{21.25} P(y,t)dy.
$$

This is the area under the density curve between $y = 17.6$ and $y = 21.25$. The total population at time t is

$$
P = \int_0^{\infty} P(y,t)\,dy,
$$

which is the analog of equation (5.3.1). For a narrow range of ages at age y, for example, y to $y + \Delta y$ with Δy small, there is a simpler formula. Population size is approximately given by the product

$$P(y) \cdot \Delta y$$

because density is approximately constant over a narrow age bracket.

The variable y in an age density function is a continuous variable. The period of time an individual is exactly 20, for instance, is infinitesimal; so what does $P(20, t)$ mean? In general, $P(y, t)$ is the limit as $\Delta y \to 0$ of the number of individuals in an age bracket of size Δy that includes y, divided by Δy,

$$P(y, t) = \lim_{\Delta y \to 0} \frac{\text{population size between } y \text{ and } y + \Delta y}{\Delta y}.$$

As above, the density is generally a function of time as well as age, and is written $P(y, t)$ to reflect this dependence.

Table 2.6.3 gives the mortality rate for Alabama in 1990. From the table, the death rate for 70-year-olds, i.e., someone between 70.0 and 70.999 . . ., is approximately 40 per 1000 individuals over the course of the year. Over one-half the year it is approximately 20 per 1000, and over Δt fraction of the year the death rate is approximately $\mu(70, 1990) \cdot \Delta t$ in deaths per 1000, where $\mu(70, 1990)$ is 40. To calculate the actual number of deaths we must multiply by the population size of the 70-year-olds in thousands. On January 1st 1990, there were $\int_{70}^{71} P(y, 1990) \, dy / 1000$ numbers of such individuals. Thus the number of deaths among 70-year-olds over a small fraction Δt of time at the beginning of the year 1990 is given by

$$\mu(70, 1990) \Delta t \int_{70}^{71} P(y, 1990) \, dy / 1000. \qquad (5.3.9)$$

A calculation such as equation (5.3.9) works, provided the death rate is constant over the year and the time interval Δt is less than one year. But, in general, death rates vary continuously with age. In Figure 5.3.1 we show an exponential fit to the data of Table 2.6.3. The approximate continuously varying death rate is

$$\mu(y, t) = A e^{by},$$

which is drawn using the methods of Exercise 1, Section 2.6. This equation assumes the death rate is independent of time; but, as we have seen, it can depend on time as well as age.

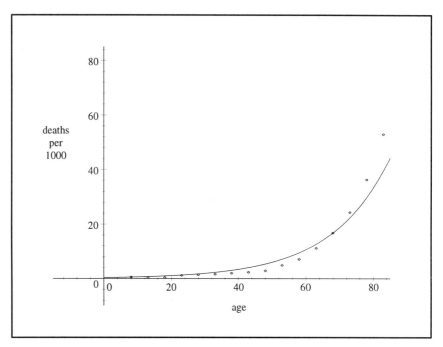

Figure 5.3.1   Least square fit to the Death Rate Table 2.6.3

To calculate a number of deaths accurately, we must account for the changing death rate as well as the changing density. The equation for calculating the number of deaths to individuals of exact age y at time t over the interval of time Δt is

$$P(y,t)\mu(y,t)\Delta t. \tag{5.3.10}$$

The number of deaths among those individuals who are between y and $y + \Delta y$ years old over this same period of time is

$$[P(y,t)\Delta y]\,\mu(y,t)\Delta t.$$

Suppose we want to do the calculation for those between the ages of a_1 to a_2 over the calendar time t_t to t_2. The approximate answer is given by the double sum of such terms,

$$\sum\sum \mu(y,t)P(y,t)\Delta y\Delta t,$$

over a grid of small rectangles $\Delta y\Delta t$ covering the range of ages and times desired. In the limit as the grid becomes finer this double sum converges to the double integral

$$\int_{t_1}^{t_2} \int_{a_1}^{a_2} \mu(y,t)P(y,t)\, dy dt. \tag{5.3.11}$$

Return to equation (5.3.10) which calculates the loss of population, ΔP, in the exact age group y over the time interval Δt:

$$\Delta P = -\mu(y,t)P(y,t)\Delta t.$$

But by definition the change in population is

$$\Delta P = P(y+\Delta y, t+\Delta t) - P(y,t).$$

Equate these two expressions for ΔP, incorporating the fact that as time passes, the population ages at the same rate, that is $\Delta y = \Delta t$. Therefore we have the continuous analog of equation (5.3.3)

$$P(y+\Delta t, t+\Delta t) - P(y,t) = -\mu(y,t)P(y,t)\Delta t.$$

Add and subtract the term $P(y, t+\Delta t)$ to the lefthand side and divide by Δt,

$$\frac{P(y+\Delta t, t+\Delta t) - P(y, t+\Delta t)}{\Delta t} + \frac{P(y, t+\Delta t) - P(y,t)}{\Delta t} = -\mu(y,t)P.$$

Finally, take the limit as $\Delta t \to 0$ and get

$$\frac{\partial P}{\partial y} + \frac{\partial P}{\partial t} = -\mu(y,t)P. \tag{5.3.12}$$

This is referred to as the *Von Foerster equation*. Its solution for $y > t$ is

$$P(y,t) = P(y-t,0)e^{-\int_0^t \mu(y-t+u,u)du}$$

as can be verified by direct substitution.

```
> P:=(y,t)->h(y-t)*exp(-int(mu(y-t+u,u),u=0..t));
> diff(P(y,t),t)+diff(P(y,t),y)+mu(y,t)*P(y,t);
> simplify(");
```

This solution does not incorporate new births, however. As in the computation of equation (5.3.5), we must use experimental data to determine $P(0,t)$ as a function of $P(y,t)$, $y > 0$.

Exercises

1. Consider this discrete population model, using equation (5.3.1): Suppose the
    initial population is given by

    $$P(n, 0) = (100 - n) \cdot (25 + n), \qquad n = 0, \ldots, 100.$$

    Take the birth rate to be 1.9 children per couple in the ten-year age bracket
    from 21 through 30 years of age, so that

    $$P(0, n) = \frac{1.9}{2} \sum_{i=21}^{30} \frac{P(n - 1, i)}{10}, \qquad n = 1, \ldots, 20.$$

    Take the death rate to be given by the exponential approximation

    $$\mu(n) = \exp(0.0756 \cdot t - 1.6134), \qquad n = 0, \ldots, 100.$$

    Advance the population for three years, keeping track of the total population:

    $$\text{total}(n) = \sum_{i=0}^{100} P(n, i).$$

    Does the total population increase?

```
> P0:=t->(100-t)*(25+t):
> sum('P0(p)','p'=0..100);
> birth0:= 1.9*sum('P0(p)','p'=21..30)/10;
> plot(P0,1..100);
> Death:=t->exp(.0756*t-1.6134);
> P1:=proc(p) if p <= 1 then birth0 else
     if 1 < p and p < 100 then P0(p-1)*(1-Death(p-1)/1000) else
     if p = 100 then 0
     fi fi fi end;
  sum('P1(p)','p'=0..100);
  birth1:= 1.9*sum('P1(p)','p'=21..30)/20;
> plot(P1,1..100);

> P2:=proc(p) if p ¡= 1 then birth1 else
     if 1 < p and p < 100 then P1(p-1)*(1-Death(p-1)/1000) else
     if p = 100 then 0
     fi fi fi end;
  sum('P2(p)','p'=0..100);
  birth2:= 1.9*sum('P2(p)','p'=21..30)/10;
> plot(P2,1..100);
```

2. For the following two Leslie matrices find λ and V as given in equation (5.3.4). What is the ratio of the ages of associated populations?

$$L_1 = \begin{pmatrix} 1 & 2/3 \\ 1/2 & 0 \end{pmatrix} \qquad L_2 = \begin{pmatrix} 0 & 4 & 3 \\ 1/2 & 0 & 0 \\ 0 & 1/4 & 0 \end{pmatrix}.$$

Section 5.4

Questions for Thought and Discussion

1. Draw age-structure diagrams for the three cases of populations whose maximum numbers are young, middle-aged and elderly people. In each case draw the age-structure diagram to be expected thirty years later if birth and death rates are equal and constant, and if there is no migration.
2. Repeat Question #1 for the situation where the birth rate is larger than the death rate, and there is no migration.
3. Repeat Question #1 for the situation where the birth and death rates are constant, but there is a short-but-extensive incoming migration of middle-aged women at the outset.

References and Suggested Further Reading

1. **Aging:** T. B. L. Kirkwood, "The Nature and Causes of Ageing," in *Research and the Ageing Population,* **134**, 193, 1988.
2. **Aging in humans:** Ricki L. Rusting, "Why do we age?" *Scientific American*, December, pp. 130–141, 1992.
3. **Perron-Frobernius Theorem:** E. Seneta, *Non-negative Martices and Markov Chains*, Springer-Verlag, New York, pp. 22, 1973.
4. **Leslie matrices:** H. Anton and C. Rorres, *Elementary Linear Algebra*, John Wiley and Sons, New York, pp. 653, 1973.

Chapter 6

Random Movements
in Space and Time

Introduction to this chapter

Many biological phenomena, at all levels of organization, can be modeled by treating them as random processes, behaving much like the diffusion of ink in a container of water. In this chapter we discuss some biological aspects of random processes, namely the movement of oxygen across a human placenta and the spread of infectious diseases. While these processes might seem to be quite different at first glance, they actually act according to very similar models.

We begin with a description of biological membranes, structures that regulate the movement of material into, out of, and within the functional compartments of a cell. At the core of a membrane is a layer of water-repelling molecular moieties. This layer has the effect of restricting the free transmembrane movement of any substance that is water-soluble, although water itself can get past the layer. The transmembrane movement of the normal water-soluble compounds of cellular metabolism is regulated by large biochemical molecules that span the membrane. They are called *permeases,* or transport proteins. Permeases have the ability to select the materials that cross a membrane. Other membranes anchor critical cellular components that promote chemical reactions through catalysis.

A human fetus requires oxygen for its metabolic needs. This oxygen is obtained from its mother, who breathes it and transfers it via her blood to the placenta, an organ that serves as the maternal-fetal interface. Because the blood of mother and child do not mix, material exchange between them must take place across a group of membranes. The chemical that transports the oxygen is hemoglobin, of which there are at least two kinds, each exhibiting a different strength of attachment to oxygen molecules. Further, chemical conditions around the hemoglobin also affect its attachment to oxygen. The conditions at the placenta are such that there is a net transmembrane movement of oxygen from maternal hemoglobin to fetal hemoglobin.

Diseases seem to diffuse, as anyone who has watched the progress of a flu epidemic can testify. The diffusion of an infectious disease, however, is subject to numerous perturbations and variables; examples are variations in infective ability of the infective agent and in the resistance of the hosts and the frequent need for vectors to carry the disease.

This chapter also serves as an introduction to the discussions of the blood vascular system in Chapter 7, of biomolecular structure in Chapter 9 and of HIV in Chapter 10.

Section 6.1

Biological Membranes

Biological membranes do much more than physically separate the interior of cells from the outside world. They provide organisms with control over the substances that enter, leave and move around their cells. This is accomplished by selective molecules that can recognize certain smaller molecules whose transmembrane movement is required by the cell. A waterproof layer in the membrane otherwise restricts the movement of these smaller compounds. In addition, membranes maintain compartments inside a cell, allowing the formation of specific chemical environments in which specialized reactions can take place.

The molecular structure of a substance determines its solubility in water.

Distinctions between oil and water are everywhere. We have all seen how salad oil forms spherical globules when it is mixed with watery vinegar. Likewise, we say that two hostile people "get along like oil and water." These experiences seem to suggest that all materials are either water-soluble or not. This is an oversimplification: Ethyl alcohol is infinitely soluble in water (gin is about half alcohol, half water), but isopropanol (rubbing alcohol), table salt and table sugar all have moderate water-solubility. Salt and sugar have very low solubility in gasoline and benzene (erstwhile dry-cleaning fluid). On the other hand, benzene will easily dissolve in gasoline and in fatty substances.

The electronic basis for water-solubility will be described in Chapter 9, but for now it is sufficient that we recognize that the ability of a substance to dissolve in water is determined by its electronic structure. Further, an appropriate structure is found in ions (like sodium and chlorine from salt) and in molecules with oxygen and nitrogen atoms (like sugars and ammonia). Such substances are said to be *hydrophilic*, or *polar*. Hydrophilic structures are not found in electrically neutral atoms, nor in most molecules lacking oxygen and nitrogen. This is especially true when the latter molecules have a very high proportion of carbon and hydrogen (e.g., benzene, gasoline and fatty substances). These latter materials are said to be *hydrophobic*, or *nonpolar*.

Both faces of a membrane are attracted to water, but the interior of the membrane repels water.

The biological world is water-based.[1] Therefore, cells face a bit of a problem in that water is a major component of the external world, which could lead to too much interaction between a cell's contents and its environment. To deal with this problem, cells are surrounded by a water-proofing, or hydrophobic, membrane layer. We should be glad for this structural feature of our bodies—it keeps us from dissolving in the shower!

Figure 6.1.1 shows a model of a cell membrane. The side facing the cellular interior is hydrophilic because it must interact with the cell's internal environment; the outside is also hydrophilic because it interacts with the external world.[2] The interior of the membrane, however, is strongly hydrophobic, being a kind of *hydrocarbon* (constructed from hydrogen and carbon only). This arrangement is thermodynamically favorable because there are no direct interactions between hydrophilic and hydrophobic groups.[3] Attached to, and sometimes piercing, the membrane are complicated biological molecules called proteins, which will be described in more detail in Chapter 9.

No material can enter or leave the cell unless it negotiates its way past this membrane, because the membrane completely envelops the cell. Clearly, the efficiency of transmembrane movement of a substance will be determined by the ability of the substance to interact with the membrane's components, especially the interior hydrophobic layer of the membrane.

Only a few kinds of substances can diffuse freely across the hydrophobic layer of a membrane.

The substances that can move across the hydrophobic layer in response to a concentration gradient fall into two groups. The first group, surprisingly, contains water and carbon dioxide, a fact that seems contrary to our earlier discussion. What seems to happen is that water and CO_2 molecules are small enough that they can slip past the large hydrocarbon fragments in the membrane. A nice demonstration of this is seen by placing human red blood cells into distilled water. The interior of the cell contains materials that cannot go through the membrane, so the water there is at a *lower* concentration than in the surroundings. Thus, water moves into the cell and eventually bursts it like a balloon. This movement of water (or any other solvent) is called *osmosis*.

The second kind of material that easily passes through the membrane hydrocarbon layer is a hydrocarbon. Of course, our cells are almost never exposed to hydrocarbons, so this material is of little interest. We will, however, point out in

---

[1]Our bodies must resort to special tricks to solubilize fats that we eat. Our livers produce a detergent-like substance, called bile, that allows water to get close to the fats. The hydrocarbon-metabolizing microorganisms that are useful in dealing with oil spills often use similar methods.

[2]It might be easiest here to picture a single-celled organism in a pond.

[3]The arrangement of molecules in the membrane of Figure 6.1.1 is called a *bilayer* because it consists of two leaflets of molecules, arranged back-to-back.

Chapter 7 that one route of lead into our bodies is through our skin: If we spill leaded gasoline on ourselves, the hydrocarbons of the gasoline can carry the lead right across our hydrophobic membrane barriers and into our blood stream.

Selective channels control the passive and active movements of ions and large molecules across membranes.

Many relatively large hydrophilic particles, like ions and sugars, can pass through membranes—after all, these particles are essential components of cellular systems. They do not move directly through the bulk of the membrane. Rather, their movement is regulated by large proteins that completely penetrate the membrane and act like specialized channels, choosing which substances get past the membrane (see Figure 6.1.1). These proteins are called *permeases*, or *transport proteins*, and they are very selective: The substitution of a single atom in a transportable molecule with molecular weight of several hundred can cause the

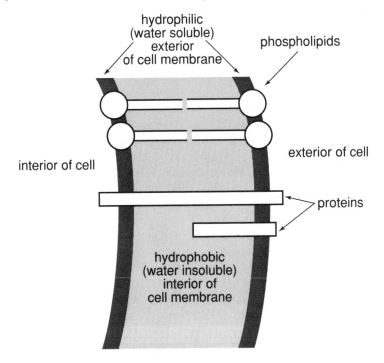

Figure 6.1.1 A model of a cell membrane, showing the hydrocarbon (hydrophobic, water insoluble) interior and hydrophilic (water soluble) exterior of the membrane. This dual nature of the membrane is the result of the orientation of many phospholipid molecules, only four of which are actually shown in the figure. Figure 9.2.5 in Chapter 9 will show how the chemical nature of a phospholipid leads to hydrophobic and hydrophilic parts of the membrane. Two proteins are also shown to demonstrate that some span the membrane completely and others only pierce the outside halfway through (on either side of the membrane).

molecule to be excluded by its normal permease. Permeases thus act like selective gates to control material transport into and out of a cell.

Materials can move across membranes via permeases by two different mechanisms, both often called *facilitated transport*. First, the *passive* movement of a material in response to a concentration gradient (diffusion) is usually facilitated by permeases. The point is that only those substances recognized by a permease will behave this way. Any other substances will diffuse up to the membrane and then be stopped by the hydrophobic layer of the membrane.

Second, many materials are pumped *against* a concentration gradient past a membrane. This process, called *active transport*, requires energy because it is in the opposite direction to the usual, spontaneous movement of particles. Active transport also requires a facilitating permease.

Facilitated transport is discussed further in Chapter 8, in the text by Beck et al. [1], and in the reference by Yeargers [2].

Some cellular membranes face the outside world and regulate intercellular material movement.

The day-to-day processes that a cell must perform require that nutrients and oxygen move into the cell and that wastes and carbon dioxide move out. In other words, the cell must maintain constant, intimate communication with its external environment. The cell membrane provides the interface between the cell and the outside world, and membrane permeases, because of their selectivity, control the transmembrane movement of most of the substances whose intercellular transport is required.

What about water? It moves across membranes irrespective of permeases and would therefore seem to be uncontrollable. In fact, cells can regulate water movement, albeit by indirect means. They accomplish this by regulating other substances and that, in turn, affects water. For example, a cell might pump sodium ions across a membrane to a high concentration. Water molecules will then follow the sodium ions across the membrane by simple diffusion, to dilute the sodium.

Some cellular membranes are inside the cell and regulate intracellular material movements.

Students are sometimes surprised to learn that the interior of a cell, exclusive of the nucleus, is a labyrinth of membranes. A mechanical analog can be obtained by combining some confetti and a sheet of paper, crumpling up the whole mess into a wad and then stuffing it into a paper bag. The analogy cannot be pushed too far; membranes inside the cell often have very regular structures, lying in parallel sheets, or forming globular structures (like the nucleus). In short, the interior of a cell is a very complicated place.

Many thousands of different biochemical reactions occur in a mammalian cell. If these reactions were not coordinated in space and time the result would be chaos. Membranes provide coordinating mechanisms in several ways: First, large

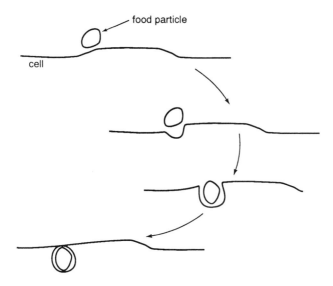

food particle

cell

Figure 6.1.2 A schematic diagram showing the process of pinocytosis. A small indentation forms at the particle's point of contact, and the particle is then drawn into the cell's interior.

biochemical molecules are always assembled step-wise, beginning with small structures and ending up with large ones. All of the fragments to be added must be close to the nascent biomolecule so that they can be added at the right time. Intracellular membranes provide compartmentalization to keep the reactants and products of related reactions in close proximity to one another. Second, the efficiencies of different cellular biochemical reactions are dependent on environmental conditions, e.g., pH and salt concentration. The specialized environmental needs of each reaction, or set of reactions, are maintained by membrane compartmentalization. Thus, a cell is partitioned off into many small chambers, each with a special set of chemical conditions. A third point, related to the first two, is that virtually all chemical reactions in a cell are catalyzed by special proteins, and these catalysts often work only when they are attached to a membrane. Refer to Figure 6.1.1, and note that many of the proteins do not pierce the membrane, but rather are attached to one side or the other. These proteins represent several of the membrane-bound protein catalysts of a cell. You will read more about these catalysts in Chapter 9.

Large objects move into and out of a cell by special means.

Neither simple diffusion nor facilitated transport can move particulate objects like cellular debris, food, bacteria and viruses into or out of a cell; those require completely different routes. If a cell is capable of amoeboid movement it can surround the particle with *pseudopods* and draw it in by *phagocytosis*. If the cell is not amoeboid, it can form small pockets in its surface to enclose the particle; this process is *pinocytosis*, as shown in Figure 6.1.2. Both phagocytosis and pinocytosis can be reversed to rid the cell of particulate waste matter.

Section 6.2

The Mathematics of Diffusion

In this section we derive Fick's laws of diffusion by studying a random walk model. Using the normal approximation to the binomial distribution, we obtain the Gaussian solution of the diffusion equation for a point source concentration. It is seen that particles disperse on the average in proportion to the square root of time.

Fick's laws are applied to investigate one aspect of diffusion through biological membranes. It is shown that the rate of mass transport is proportional to the concentration difference across the membrane and inversely proportional to the thickness of the membrane.

Random processes in the biosphere play a major role in life.

In Section 6.1 we described the membrane system that surrounds and pervades a cell. In this section we show how the random motion of substances can carry materials across these membranes and through the bulk of a cell.

Chance plays a major role in the processes of life. On the microscopic scale, molecules are in constant random motion corresponding to their temperature. Consequently chance guides the fundamental chemistry of life. On a larger scale, genes mutate and recombine by random processes. Thus chance is a fundamental component of evolution. Macroscopically, unpredictable events such as intra-species encounters lead to matings or the transmission of disease, while inter-species encounters claims many a prey victim, but not with certainty. The weather can affect living things throughout an entire region and even an entire continent. And on a truly grand scale, astronomical impacts can cause mass extinction.

Diffusion can be modeled as a random walk.

Molecules are in a constant state of motion as a consequence of their temperature. According to the kinetic theory of matter, there is a fundamental relationship between molecular motion and temperature, which is simplified by measuring the latter on an absolute scale, degrees Kelvin. Zero degrees Kelvin, or absolute zero, is -273.15 °C. Moreover, in 1905 Einstein showed that the principle extends to particles of any size, for instance, to pollen grains suspended in water. Einstein's revelation explained an observation made in 1828 by the English botanist Robert Brown, who reported on seeing a jittery, undirected motion of pollen grains in the water of his microscope plate. We now refer to this phenomenon as *Brownian motion*. It is a visible form of diffusion.

The relationship between temperature and particle motion can be precisely stated: *The average kinetic energy of a particle along a given axis is $kT/2$, where T is temperature in degrees Kelvin and k is the universal Boltzmann's constant,* $k = 1.38 \times 10^{-16}$ ergs per degree (Reference [3]). The principle is stated in terms of the *time average* of a single particle, but we will assume that it applies equally well to the average of an ensemble or collection of identical particles taken at the same time, the *ensemble average*.

Now the kinetic energy of an object of mass m and velocity v is $\frac{1}{2}mv^2$, so the average kinetic energy of N particles of the same mass m but possibly different velocities is

$$\frac{\overline{mv^2}}{2} = \frac{\sum_{i=1}^{N} mv_i^2/2}{N}$$
$$= \frac{m}{2N} \sum_{i=1}^{N} v_i^2 = \frac{\overline{mv^2}}{2}.$$

In this we have used an overline to denote average ensemble value.

Therefore, for a collection of particles of mass m, the kinetic theory gives

$$\frac{\overline{mv^2}}{2} = \frac{kT}{2}$$

or

$$\overline{v^2} = \frac{kT}{m}. \tag{6.2.1}$$

Table 6.2.1 gives the average thermal velocity of some biological molecules at body temperature predicted by this equation.

A particle does not go very far at these speeds before undergoing a collision with another particle or the walls of its container and careening off in some new direction. With a new direction and velocity the process begins anew, destined to suffer the same fate. With all the collisions and rebounds, the particle executes what can be described as a random walk through space. To analyze this process,

Table 6.2.1 RMS Velocities at Body Temperature

Molecule	Molecular Weight	Mass (g)	r.m.s. speed at 36°C (m/sec)
H_2O	18	3×10^{-26}	652
O_2	32	5.4×10^{-26}	487
Glucose	180	3×10^{-25}	200
Lysozyme	1,400	2.4×10^{-23}	23
Hemoglobin	65,000	1×10^{-22}	11
Bacteriophage	6.2×10^6	1×10^{-20}	1.1
E. coli	$\approx 2.9 \times 10^{11}$	2×10^{-15}	.0025

we model it by stripping away as much unnecessary complication as possible while still retaining the essence of the phenomenon, see References [4,5].

For simplicity, assume time is divided into discrete periods Δt and in each such period a particle moves one step Δx to the left or right along a line, the choice being random. After n time periods the particle lies somewhere in the interval from $-n(\Delta x)$ to $n(\Delta x)$ relative to its starting point, taken as the origin 0.

For example, suppose $n = 4$. If all four choices are to the left, the particle will be at -4; if three are to the left and one is to the right, it will be at -2. The other possible outcomes are 0, 2, and 4. Notice that the outcomes are separated by two steps. Also notice that there are several ways most of the outcomes can arise, the outcome 2, for instance. We can see this as follows. Let R denote a step to the right and L a step to the left. Then a path of four steps can be coded as a string of four choices of the symbols R or L. For example, $LRRR$ means the first step is to the left and the next three are to the right. For an outcome of four steps to be a net two to the right, three steps must be taken to the right and one to the left, but the order doesn't matter. There are four possibilities that do it, namely $LRRR$, $RLRR$, $RRLR$, and $RRRL$.

In general, let $p(m, n)$ denote the probability that the particle is at position $x = m(\Delta x)$, m steps right of the origin, after n time periods, $t = n(\Delta t)$. We wish to calculate $p(m, n)$. It will help to recognize that our random walk with n steps is something like tossing n coins. For every coin that lands heads we step right and for tails we step left. Let r be the number of steps taken to the right and l the number to the left; then to be at position $m(\Delta x)$ it must be that their difference is m,

$$m = r - l \qquad \text{where} \qquad n = r + l.$$

Thus r can be given in terms of m and n by adding these two equations, and l is given by subtracting:

$$r = \frac{1}{2}(n + m) \qquad \text{and} \qquad l = \frac{1}{2}(n - m). \qquad (6.2.2)$$

As in a coin toss experiment, the number of ways of selecting r moves to the right out of n possibilities is the problem of counting combinations and is given by (see Section 2.7)

$$C(n, r) = \frac{n!}{r! \, (n - r)!}.$$

For example, three moves to the right out of four possible moves can happen in $4!/(3! \, 1!) = 4$ ways in agreement with the explicitly written out $L \, R$ possibilities noted above. Therefore, if the probabilities of going left or right are equally likely, then

$$p(m, n) = \text{probability of } r \text{ steps right} = \frac{C(n, r)}{2^n}, \qquad r = \frac{1}{2}(n + m). \quad (6.2.3)$$

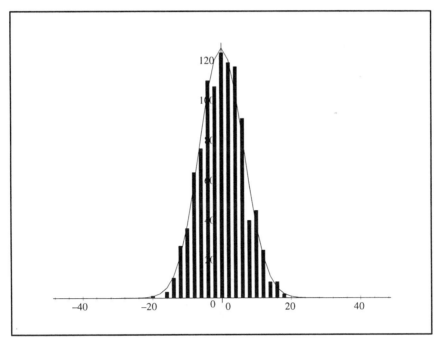

Figure 6.2.1   Graph of $p(m, 40)$

This is the *binomial* distribution with $p = q = 1/2$. The solid curve in Figure 6.2.1 is a graph of $p(m, 40)$. If the random walk experiment with $n = 40$ steps were conducted a large number of times, then a bar graph of the resulting particle postions will closely approximate this figure. Such a bar graph is also shown in Figure 6.2.1. Equivalently, the same picture applies to a large number of particles randomly walking at the *same* time, each taking 40 steps, provided they may slide right past each other without collisions.

Particles disperse in proportion to the square root of time.

The average, or *mean*, position, $\overline{m}$, of a large number of particles after a random walk of n steps with equal probabilities of stepping left or right is 0. To show this, start with equation (6.2.2) to get $m = 2r - n$. Then, since r has a binomial distribution, we can write down its mean and variance from Chapter 2. Equations (2.7.10) and (2.7.11) with $p = 1/2$ and $q = 1 - p = 1/2$ give

$$\overline{r} = np = \frac{n}{2}$$
$$\mathrm{var}(r) = \overline{(r - \overline{r})^2} = npq = \frac{n}{4}. \tag{6.2.4}$$

Hence

$$\bar{m} = \overline{2r - n} = 2\bar{r} - \bar{n} = 2\frac{n}{2} - n = 0,$$

since the average value of the constant n is n. Unfortunately, knowing that the average displacement of particles is 0 does not help in expressing how quickly particles are moving away from the origin. The negative values of those that have moved to the left cancel the positive values of those that have gone right.

We can avoid the left versus right cancellation by using the squares of displacements; we will get thereby the *mean square* displacement, $\overline{m^2}$. Since the mean displacement is 0, the mean square displacement is equal to the variance here; hence from equation (6.2.4),

$$\overline{m^2} = \overline{(m - \bar{m})^2}$$
$$= \overline{(2r - n)^2} = 4\overline{(r - \frac{n}{2})^2} = 4\frac{n}{4} = n.$$

Since $m = x/\Delta x$ and $n = t/\Delta t$, we can convert this into statements about x and t:

$$\overline{x^2} = \frac{\Delta x^2}{\Delta t}t. \qquad (6.2.5)$$

But mean square displacement is not a distance, it is measured in square units, cm^2, for example. To rectify this, the square root of the mean square displacement, or *root mean square (rms)* distance is used to quantify dispersion:

$$\sqrt{\overline{m^2}} = \sqrt{n}$$

and

$$\sqrt{\overline{x^2}} = \sqrt{\frac{\Delta x^2}{\Delta t}}\sqrt{t}. \qquad (6.2.6)$$

Hence particles disperse in proportion to the square root of time. Thus there is no concept of velocity for diffusion. To traverse a distance twice as far requires four times as much time.

The exact equation for $p(m, n)$, equation (6.2.3), has a simple approximation. There is a real need for such an approximation because it is difficult to compute the combinatorial factor $C(n, r)$ for large values of n. Moreover, the approximation improves with an error that tends to 0 as $n \to \infty$. The binomial distribution (see equation (6.2.3) and Figure 6.2.1) looks very much like that of a normal distribution (see Chapter 2). Although Stirling's formula for approximating $n!$,

$$n! \approx \sqrt{2\pi n}n^n e^{-n},$$

may be used to prove it, we will not do this. Instead we will match the means and standard deviations of the two distributions. First recall that the probability that a normally distributed observation will fall within an interval of width dm centered at m is, approximately

$$\frac{1}{\sqrt{2\pi\sigma^2}}e^{-(m-\mu)^2/2\sigma^2}\,dm$$

(see Section 2.7), where μ is the mean and σ is the standard deviation of the distribution. On the other hand, $p(m,n)$ is the probability the walk will end between $m-1$ and $m+1$, an interval of width 2, and, from above, its mean is 0 and its standard deviation is $\sqrt{n}$. Hence

$$p(m,n) \approx \frac{1}{\sqrt{2\pi n}}e^{-m^2/2n}(2)$$

$$\approx \sqrt{\frac{2}{\pi n}}e^{-m^2/2n}. \tag{6.2.7}$$

Our last refinement is to let Δx and Δt tend to 0 to obtain a continuous version of $p(m,n)$. Of course, without care, $p(m,n)$ will approach zero, too, because the probability will spread out over more and more values of m. But since each value of m corresponds to a probability over a width of $2\Delta x$, we take the quotient of $p(m,n)$ by this width. That is, let $u(x,t)$ denote the probability the particle lies in an interval of width $2(\Delta x)$ centered at x at time t. Then

$$u(x,t) = \frac{P(\frac{x}{\Delta x}, \frac{t}{\Delta t})}{2(\Delta x)}$$

$$= \frac{1}{2(\Delta x)}\sqrt{\frac{2}{\pi(\frac{t}{\Delta t})}}e^{-(x/\Delta x)^2/2(t/\Delta t)}$$

And, upon simplification,

$$u(x,t) = \frac{e^{-(x^2/4(\Delta x^2/2\Delta t)t)}}{\sqrt{4\pi\left(\frac{\Delta x^2}{2(\Delta t)}\right)t}}.$$

Now keeping the ratio

$$D = \frac{\Delta x^2}{2(\Delta t)} \tag{6.2.8}$$

fixed as Δx and Δt tend to 0, we obtain the Gaussian distribution

$$u(x,t) = \frac{e^{-x^2/4Dt}}{\sqrt{4\pi Dt}}. \tag{6.2.9}$$

Table 6.2.2 Diffusion Coefficients in Solution

Molecule	Solvent	T, °C	$D\,(10^{-6}\,cm^2/sec)$	seconds to cross	
				.01 mm	1mm
O_2	blood	20	10.0	0.05	500
Acetic acid	water	25	12.9	0.04	387
Ethanol	water	25	12.4	0.04	403
Glucose	water	25	6.7	0.07	746
Glycine	water	25	10.5	0.05	476
Sucrose	water	25	5.2	0.10	961
Urea	water	25	13.8	0.04	362
Ribonuclease	water	20	1.07	0.46	4672
Fibrinogen	water	20	2.0	0.25	2500
Myosin	water	20	1.1	0.45	4545

The parameter D is called the *diffusion coefficient* or *diffusivity* and has units of area divided by time. Diffusivity depends on the solute, the solvent, and the temperature, among other things. See Table 6.2.2 for some pertinent values.

Diffusivity quantifies how rapidly particles diffuse through a medium. In fact, from equation (6.2.6), the rate at which particles wander through the medium in terms of root mean square distance is

$$\text{r.m.s. distance} = \sqrt{\frac{\overline{\Delta x^2}}{\Delta t}}\,t = \sqrt{2Dt}. \tag{6.2.10}$$

In Table 6.2.2 we give some times required for particles to diffuse the given distances. As seen, the times involved become prohibitively long for distances over 1 mm. This explains why organisms whose oxygen transport is limited to diffusion cannot grow very large in size.

The function u has been derived as the probability for the ending point, after time t, of the random walk for a single particle. But, as noted above, it applies equally well to an ensemble of particles if we assume they "walk" independently of each other. In terms of a large number of particles, u describes their concentration as a function of time and position. Starting them all at the origin corresponds to an infinite concentration at that point, for which equation (6.2.9) does not apply. However, for any positive time, $u(x, t)$ describes the concentration profile (in number of particles per unit length). See Figure 6.2.2 for the times 1, 2, 4. Evidently, diffusion transports particles from regions of high concentration to places of low concentration. Fick's First Law, derived below, makes this precise.

To treat diffusion in three dimensions, it is postulated that the random walk proceeds independently in each dimension. The mean transport in the x, y, and z directions are each given by equation (6.2.10), $\overline{x^2} = 2Dt$, $\overline{y^2} = 2Dt$, and $\overline{z^2} = 2Dt$. Hence in two dimensions, if $r^2 = x^2 + y^2$,

$$\overline{r^2} = 4Dt,$$

```
> plot({exp(-x^2/4)/sqrt(4*Pi), exp(-x^2/(4*2))/sqrt(4*Pi*2),
  exp(-x^2/(4*4))/sqrt(4*Pi*4)},x=-10..10,color=BLACK);
```

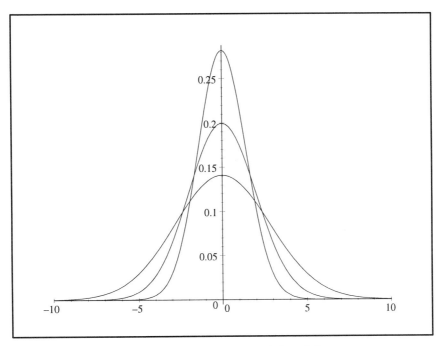

Figure 6.2.2   Dispersion of a unit mass after time 1, 2, and 4

and in three dimensions, if $r^2 = x^2 + y^2 + z^2$,

$$\overline{r^2} = 6Dt.$$

Fick's laws describe diffusion quantitatively.

Again consider a one-dimensional random walk, but now in three dimensional space, for example, along a channel of cross-sectional area A, as in Figure 6.2.3.

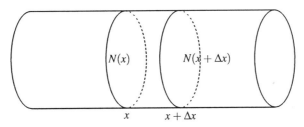

Figure 6.2.3   One-dimensional diffusion along a channel

Let $N(x)$ denote the number of particles at position x. We calculate the net movement of particles across an imaginary plane perpendicular to the channel between x and $x + \Delta x$. In fact, half the particles at x will step to the right and cross the plane, and half the particles at $x + \Delta x$ will step to the left and cross the plane in the reverse direction. The net movement of particles from left to right is

$$-\frac{1}{2}(N(x + \Delta x) - N(x)).$$

The number of particles crossing a unit area in a unit time is the *flux* of particles, denoted by J, and is measured in unit of moles per square centimeter per second, for instance. Dividing by $A\,\Delta t$ gives the flux in the x direction,

$$J_x = -\frac{1}{2A\,\Delta t}(N(x + \Delta x) - N(x)).$$

Let $c(x)$ denote the concentration of particles at x in units of number of particles per unit volume such as moles per liter. Since $c(x) = N(x)/A\,\Delta x$, the previous equation becomes

$$J_x = -\frac{\Delta x}{2\,\Delta t}(c(x + \Delta x) - c(x)) = -\frac{\Delta x^2}{2\,\Delta t}\frac{c(x + \Delta x) - c(x)}{\Delta x}.$$

Now let $\Delta x \to 0$ and recall the definition of diffusivity, equation (6.2.8); we get Fick's First Law,

$$J = -D\frac{\partial c}{\partial x}. \qquad (6.2.11)$$

A partial derivative is used here because c can vary with time as well as location.

To obtain Fick's Second Law, consider the channel again. The number of particles $N(x)$ in the section running from x to $x + \Delta x$ is $c(x,t)A\Delta x$. If concentration is not constant, then particles will diffuse into (or out of) this section according to Fick's First Law:

the decrease in the number of particles in the section $=$
(the flux out at $x + \Delta x$ $-$ the flux in at x)A.

More precisely,

$$-\frac{\partial}{\partial t}(c(x,t)A\Delta x) = (J(x + \Delta x, t) - J(x,t))A.$$

Dividing by Δx and letting $\Delta x \to 0$ gives

$$-A\frac{\partial c}{\partial t} = A\frac{\partial J}{\partial x}.$$

Cancelling the A on each side we obtain the *continuity equation*,

$$\frac{\partial c}{\partial t} = -\frac{\partial J}{\partial x}. \tag{6.2.12}$$

Differentiating J in Fick's First Law and substituting into this gives Fick's Second Law of diffusion, also known as the diffusion equation,

$$\frac{\partial c}{\partial t} = D\frac{\partial^2 c}{\partial x^2}. \tag{6.2.13}$$

Direct substitution shows that the Gaussian distribution u, equation (6.2.9), satisfies the diffusion equation.

Oxygen transfer by diffusion alone limits organisms in size to about one-half millimeter.

As an application of Fick's laws, we may calculate how large an organism can be if it has no circulatory system. Measurements taken for many organisms show that the rate of oxygen consumption by biological tissues is on the order of $R_{O_2} = 0.3$ microliters of O_2 per gram of tissue per second. Also note that the concentration of oxygen in water at physiological temperatures is 7 microliters of O_2 per cm^3 of water. Assuming an organism of spherical shape, balance the rate of oxygen diffused through the surface with that consummed by interior tissue; we get, using Fick's First Law (6.2.11), ($V = \frac{4}{3}\pi r^3$ is the volume of the organism)

$$AJ = DA\frac{dc}{dr} = VR_{O_2}$$

$$D(4\pi r^2)\frac{dc}{dr} = \frac{4}{3}\pi r^3 R_{O_2}.$$

Isolate dc/dr and integrate; use the boundary condition that at the center of the sphere the oxygen concentration is zero, and at the surface of the sphere, where $r = r_m$, the concentration is C_{O_2}. We get

$$\frac{dc}{dr} = \frac{R_{O_2}}{3D}r$$

$$\int_0^{C_{O_2}} dc = \int_0^{r_m} \frac{R_{O_2}}{3D}r\,dr$$

$$C_{O_2} = \frac{R_{O_2}}{6D}r_m^2.$$

Using the values for C_{O_2} and R_{O_2} above and the value 2×10^{-5} cm^2/sec for a maximal value of D, we get

$$r_m^2 = \frac{6DC_{O_2}}{R_{O_2}}$$
$$= \frac{6 \times 2 \times 10^{-5}(\text{cm}^2/\text{sec}) \times 7(\mu l/\text{cm}^3)}{0.3(\mu l/(gm \times \text{sec}))}$$
$$= 0.0028(\text{cm}^2),$$

where we have assumed that one gram of tissue is about 1 cm^3 water. Taking the square root gives the result

$$r_m = 0.53 \text{ mm.}$$

We will use this value in Chapter 7 as a limitation on the size of certain organisms.

Resistance to fluid flow is inversely proportional to the fourth power of the radius of the vessel.

From the discussion above, it is clear that large organisms must actively move oxygen to the site of its usage, possibly dissolved in a fluid. In this section we derive the equation governing resistance to flow imposed by the walls of the vessel through which the fluid passes. In Chapter 7 we will discuss the anatomical and physiological consequences of this resistance to flow.

As a fluid flows through a circular vessel, say of radius a, resistance to the flow originates at the walls of the vessel. In fact, at the wall itself, the fluid velocity $u(a)$ is barely perceptible, thus $u(a) = 0$. A little further in from the wall the velocity picks up and at the center of the vessel the velocity is largest. By radial symmetry, we only need one parameter to describe the velocity profile, namely the radius r from the center of the vessel. The fluid travels downstream in the form of concentric cylinders, the cylinders nearer the center moving fastest (see Figure 6.2.5). This results in a shearing effect between the fluid at radius r and the fluid just a little farther out, at radius $r + \delta r$. Shear stress, τ, is defined as the force required to make two sheets of fluid slide past each other, divided by their contact area. It is easy to imagine that the shear stress depends on the difference in velocity of the two sheets, and, in fact, the two quantities are proportional, thus

$$\tau = \mu \frac{du}{dr}$$

where μ is the constant of proportionality.

Consider a portion of the vessel of length ℓ and let Δp denote the difference in fluid pressure over this length. This pressure, acting on the cylinder of fluid of radius r, is opposed by the shear stress mentioned above. The force on the cylinder is being applied by the difference in pressure acting on its end, thus

$$\text{force} = \Delta p \pi r^2,$$

while the equal opposing force is due to the shear stress acting on the circular side of the cylinder,

$$\text{force} = \tau(2\pi r\ell) = \mu \frac{du}{dr}(2\pi r\ell).$$

Since these forces are equal and opposite,

$$\Delta p \pi r^2 = -\mu \frac{du}{dr}(2\pi r\ell).$$

This simple differential equation can be solved by integrating to obtain

$$u = -\frac{r^2 \Delta p}{4\ell\mu} + C$$

where C is the constant of integration. Using the zero velocity condition at the vessel walls gives the value of C to be

$$C = \frac{a^2 \Delta p}{4\ell\mu}.$$

Hence, for any radius, the velocity is given by

$$u(r) = \frac{\Delta p}{4\ell\mu}(a^2 - r^2).$$

Thus the velocity profile is parabolic (see Figure 6.2.4).

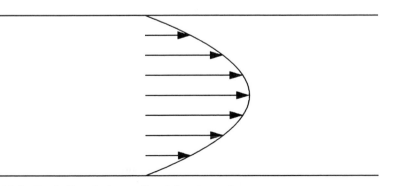

Figure 6.2.4   Parabolic velocity profile of flow in a tube

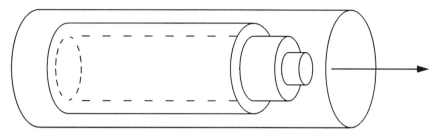

Figure 6.2.5 Circular sheet of fluid moving together

Now we can calculate the total flow rate, Q, of a volume of fluid through a cross-section of the vessel per unit time. A thin ring of fluid contains molecules that are at the same distance r from the axis of the vessel and that move together, see Figure 6.2.5. The volume of fluid per unit time which passes through a given cross-section of the vessel arising from such a ring, dQ, is given as the product of its velocity, $u(r)$, times its area, dS:

$$dQ = u(r)dS = u(r)(2\pi r)dr.$$

Substituting the velocity profile from above and integrating gives

$$Q = \int_0^a \frac{\Delta p}{4\ell\mu}(a^2 - r^2)2\pi r\,dr$$
$$= \frac{\pi\Delta p}{2\ell\mu}\left[\frac{a^2 r^2}{2} - \frac{r^4}{4}\right]_0^a$$
$$= \frac{\pi\Delta p a^4}{8\mu\ell}.$$

This is known as *Poiseuille's* equation, see Reference [6]. It shows that the flow rate increases as the fourth power of the vessel radius, which means that a vessel of twice the radius can carry 16 times the fluid volume for the same pressure drop per unit length.

It is natural to think of the shear stress in the moving fluid due to its contact with the walls as a *resistance* to flow. Fluid resistance R is defined as $R = \frac{\Delta p}{Q}$ and is given by

$$R = \frac{8\mu\ell}{\pi a^4}.$$

It is seen that the fluid resistance is inversely proportional to the fourth power of the radius. We will note in Chapter 7 that this dependence affects the size of vertebrates' hearts.

Startup effects decay exponentially as exemplified by diffusion across a slab.

A biological membrane is a complicated structure, as explained in Section 6.1, and we will take account of some of the details of the structure in the next section. In this section we want to illustrate the decay of transient phenomena, in the form of startup effects, in diffusion. Further, while crude, this slab approximation shares with real membrane diffusion its dependence on the concentration difference to the first power as the driving force behind the transport of solute particles.

By a slab we mean a solid homogenous material throughout which the diffusivity of solute particles is D. The slab has thickness h but is infinite in its other two dimensions, so diffusion through it takes place one-dimensionally.

To complete the statement of the problem, additional information, referred to as *boundary conditions* or *initial conditions*, must be specified, see Reference [7]. We will assume the concentrations of solute on the sides of the slab are maintained at the constant values of C_0 at $x = 0$ and C_h at $x = h$; assume $C_0 > C_h$,

$$c(0,t) = C_0, \qquad c(h,t) = C_h \qquad \text{for all } t \geq 0.$$

This could happen if the solvent reservoirs on either side of the slab were so large that the transport of solute is negligible. Or it could happen if solute particles are wisked away as soon as they appear at $x = h$ or are immediately replenished at $x = 0$ as they plunge into the slab. Or, of course, any combination of these. Further, we assume the startup condition that the concentration in the slab is 0,

$$c(x,0) = 0, \qquad 0 \leq x \leq h.$$

We begin by assuming the solution, $c(x,t)$, can be written in the form of a product of a function of x only with a function of t only,

$$c(x,t) = X(x)T(t).$$

Then $\partial c/\partial t = X(x)T'(t)$ and $\partial^2 c/\partial x^2 = X''(x)T(t)$. Substituting into the diffusion equation, equation (6.2.13), and dividing, we get

$$\frac{1}{T}\frac{dT}{dt} = \frac{D}{X}\frac{d^2X}{dx^2}. \qquad (6.2.14)$$

Now the left-hand side of this equation is a function of t only and the right-hand side is a function of x only and the equality is maintained over all values of x and t. This is only possible if both are equal to a constant, which we may write as $-\lambda^2 D$. Then equation (6.2.14) yields the two equations

$$\frac{1}{T}\frac{dT}{dt} = -\lambda^2 D, \qquad (6.2.15)$$

and

$$\frac{1}{X}\frac{d^2X}{dx^2} = -\lambda^2. \tag{6.2.16}$$

It is easy to verify that the solution of equation (6.2.15) is

$$T = Ae^{-\lambda^2 Dt}$$

and the solution of equation (6.2.16) is

$$X = \begin{cases} ax + b, & \text{if } \lambda = 0; \\ c \sin \lambda x + d \cos \lambda x, & \text{if } \lambda \neq 0, \end{cases}$$

where A, a, b, c, and d are constants.

The solution so far is

$$c(x,t) = \begin{cases} ax + b, & \text{if } \lambda = 0, \text{ and} \\ (c \sin \lambda x + d \cos \lambda x)e^{-\lambda^2 Dt}, & \text{if } \lambda \neq 0. \end{cases}$$

The constant A has been absorbed into the other constants, all of which have yet to be determined using the boundary conditions. For $\lambda = 0$, the conditions on either side of the slab give

$$a0 + b = C_0, \qquad \text{and} \qquad ah + b = C_h.$$

Hence it must be that $b = C_0$ and $a = -(C_0 - C_h)/h$. But this solution cannot satisfy the initial condition; we will use the $\lambda \neq 0$ case for that.

First note that if two functions $c_1(x,t)$ and $c_2(x,t)$ both satisfy the diffusion equation, then so does their sum, as follows:

$$\frac{\partial(c_1 + c_2)}{\partial t} = \frac{\partial c_1}{\partial t} + \frac{\partial c_2}{\partial t},$$

and

$$-D\frac{\partial^2(c_1 + c_2)}{\partial x^2} = -D\frac{\partial^2 c_1}{\partial x^2} - D\frac{\partial^2 c_2}{\partial x^2}.$$

And so

$$\frac{\partial(c_1 + c_2)}{\partial t} = -D\frac{\partial^2(c_1 + c_2)}{\partial x^2}.$$

In particular the sum of the $\lambda = 0$ and $\lambda \neq 0$ solutions

$$c(x,t) = -\frac{C_0 - C_h}{h}x + C_0 + (c \sin \lambda x + d \cos \lambda x)e^{-\lambda^2 Dt}, \tag{6.2.17}$$

will satisfy the diffusion equation. It remains to satisfy the boundary conditions.

At $x = 0$ in equation (6.2.17),

$$C_0 = -\frac{C_0 - C_h}{h}0 + C_0 + (c\sin\lambda 0 + d\cos\lambda 0)e^{-\lambda^2 Dt}.$$

Upon simplifying this becomes

$$0 = de^{-\lambda^2 Dt}$$

which must be valid for all t; thus $d = 0$. Continuing with the $x = h$ boundary condition in equation (6.2.17), we have

$$C_h = -\frac{C_0 - C_h}{h}h + C_0 + (c\sin\lambda h)e^{-\lambda^2 Dt},$$

or

$$0 = (c\sin\lambda h)e^{-\lambda^2 Dt}.$$

As before, this must hold for all $t \geq 0$. We cannot take $c = 0$ for then the initial condition cannot be satisfied. Instead we solve $\sin\lambda h = 0$ for λ, which is as yet unspecified. The permissible values of λ are

$$\lambda_n = \frac{n\pi}{h}, \qquad n = \pm 1, \pm 2, \ldots, \qquad (6.2.18)$$

known as the *eigenvalues* of the problem. The negative values of n may be absorbed into the positive ones since $\sin\lambda_{-n}x = -\sin\lambda_n x$. Remembering that solutions of the diffusion equation may be added, we can form an infinite series solution with a term for each eigenvalue, and, possibly, each with a different coefficient, c_n,

$$c(x,t) = -\frac{C_0 - C_h}{h}x + C_0 + \sum_{n=1}^{\infty}(c_n\sin\lambda_n x)e^{-\lambda_n^2 Dt}. \qquad (6.2.19)$$

Finally, in order to fulfill the initial condition, the coefficients c_n must be chosen to satisfy the initial condition

$$c(x,0) = 0 = -\frac{C_0 - C_h}{h}x + C_0 + \sum_{n=1}^{\infty}(c_n\sin\lambda_n x)e^{-\lambda_n D0},$$

or, upon simplifying,

$$\sum_{n=1}^{\infty}c_n\sin\lambda_n x = \frac{C_0 - C_h}{h}x - C_0.$$

We will not show how to calculate the c_n's; we only note that it can be done (Reference [7]). The infinite series is referred to as the *Fourier series* representation of the function on the right.

Thus the solution occurs in two parts; in one part, every term contains the decaying exponential $e^{-\lambda_n D t}$ for constants λ_n given above. These terms tend to zero and, in time, become negligible. That leaves the *steady state* part of the solution,

$$c(x,t) = -\frac{C_0 - C_h}{h}x + C_0,$$

a linear concentration gradient. The amount of solute delivered in the steady state is the flux given by Fick's First Law

$$J = -D\frac{\partial c}{\partial x} = \frac{D}{h}(C_0 - C_h). \tag{6.2.20}$$

Membrane diffusion is proportional to the concentration difference across the membrane.

The structure of cell membranes was described in Section 6.1. It consists of a double layer of lipid molecules studded with proteins. Some of the latter penetrate entirely through the lipid bilayer and serve to mediate the movement of various substances into and out of the cell's interior. This section is not about the transport of such substances. Rather, we describe the transport of those molecules that pass through the lipid bilayer itself by diffusion. These are mainly lipid soluble molecules.[4]

We use the results of the previous section to model the diffusion through the lipid part of the membrane. However, the membrane molecules themselves have two ends: a hydrophlic head and a lipid tail. Functionally, the head end of one layer faces outward in contact with the aqueous environment of the cell while the head end of the other layer faces inward in contact with the aqueous interior of the cell. The lipid tails of both layers face together and constitute the interior of the membrane. Thus, the concentration of solute just under the head of the membrane molecule is not necessarily the same as in the aqueous phase. Denote by C' the solute concentration just inside the membrane on the environmental side, and by c' the concentration just inside the membrane on the cell interior side. Then, according to the slab equation (6.2.20), the flux of solute through the lipid part of the membrane is given by

$$J = \frac{D}{h}(C' - c').$$

---

[4]In Section 6.1 we noted that water and carbon dioxide, although polar molecules, can move through the lipid part of a membrane.

We next assume a linear relation between the concentrations across the molecular head of the membrane molecule; thus

$$C' = \Gamma C, \quad \text{and} \quad c' = \Gamma c$$

where C is the environmental concentration of the solute and c is the concentration inside the cell. The constant Γ is called the *partition coefficient*. With this model, the partition coefficient acts as a diffusivity divided by thickness ratio for the diffusion of solute across the head of the membrane molecule. The partition coefficient is less than 1 for most substances, but can be greater than 1 if the solute is more soluble in lipid than in water.

Combining the equations above we calculate flux in terms of exterior and interior concentrations as

$$J = \frac{\Gamma D}{h}(C - c) \tag{6.2.21}$$

this is in moles/cm^2/sec for instance.

As solute molecules accumulate inside the cell, the concentration difference in equation (6.2.21) diminishes, eventually shutting off the transport. Denote the volume of the cell by V and the surface area by S. The quantity SJ is the rate of mass transport across the membrane in moles/sec, that is

$$SJ = \frac{dm}{dt} = V\frac{dc}{dt}$$

since concentration is mass per unit volume. Therefore, multiplying equation (6.2.21) by S and using this relation we get

$$V\frac{dc}{dt} = \frac{S\Gamma D}{h}(C - c)$$

or

$$\frac{dc}{dt} = k(C - c)$$

where $k = S\Gamma D/(Vh)$ is a constant. The solution is given by

$$c = C - c_0 e^{-kt} \tag{6.2.22}$$

where c_0 is the initial concentration inside the membrane. That is, the interior concentration becomes exponentially asymptotic to that of the environment.

Exercises

1. Instead of presenting a theoretical distribution of the position of particles after 40 steps, we simulate the random movement using the random number

generator of *Maple*. First, initialize a counter to record how many points end up at each site.

```
> with(stats): with(plots):
> for i from 1 to 100 do
    count.i:=0
    od:
```

Choose integers 0 or 1 randomly.

```
> N:=rand(0..1):
```

Decide how many particles to follow.

```
> particles:=500;
```

Choose how many steps each particle will take. Make this an even integer.

```
> steps:=40;
```

Count how many particles end at each place.

```
> for m from 1 to particles do
    place:=sum('2*N(p)-1','p'=1..steps)+steps:
  count.place:=count.place + 1:
  od:
```

Finally, set up a histogram plot of the data.

```
> ranges:=[seq(-steps/2+2*(i-1)..-steps/2+2*i,i=1..steps/2)];
> movement:=[seq(count.(20+2*j),j=1..20)];
> diffusion:=[seq(Weight(ranges[i],movement[i]),i=1..20)];
> statplots[histogram](diffusion);
```

2.

    a. Plot the Gaussian distribution of equation (6.2.9) with $D = 1$ and for $t = 1, 2$, and 3.

```
> plot({exp(-x^2/4)/sqrt(4*Pi),exp(-x^2 /(4*2))/sqrt(4*Pi*2),
   exp(-x^2/(4*3))/sqrt(4*Pi*3)},x=-10..10);
```

    b. Verify that equation (6.2.9) satisfies the partial differential equation (6.2.13) with $D = 1$.

```
> u:=(t,x)->exp(-x^2/(4*t))/sqrt(4*Pi*t);
> diff(u(t,x),t)-diff(u(t,x),x,x);
> simplify(" );
```

c. The analogue for equation (6.2.13) for diffusion in a plane is

```
> diff(U(t,x,y),t) = diff(U(t,x,y),x,x)+diff(U(t,x,y),y,y);
```

Show that the function U given below satisfies this two dimensional diffusion equation:

```
> U:=(t,x,y)->exp(-(x^2+y^2)/(4*t))/t;
> diff(U(t,x,y),t) - diff(U(t,x,y),x,x)-diff(U(t,x,y),y,y);
> simplify(" );
```

d. Give visualization to these two diffusions by animation of (6.2.9) and of the two-dimensional diffusion.

```
> plot({ exp(-x^2/4)/sqrt(4*Pi),exp(-x^2 /(4*2))/sqrt(4*Pi*2),
    exp(-x^2/(4*3))/sqrt(4*Pi*3)},x=-10..10);
> with(plots):
> animate(exp(-x^2/(4*t))/sqrt(4*Pi*t),x=-10..10,t=0.1..5);
> animate3d(exp(-(x^2+y^2)/ (4*t))/t,x=-1..1,y=-1..1,t=0.1..0.5);
```

3. A moment's reflection on the form of equation (6.2.13) suggests a geometric understanding. The right side is the rate of change in time of $c(t,x)$. The equation asserts that this rate of change is proportional to the curvature of the function $c(t,x)$ as a graph in x and as measured by the second derivative. That is, if the second derivative in x is positive and the curve is concave up, expect $c(t,x)$ to increase in time. If the second derivative is negative and the curve is concave down, expect $c(t,x)$ to decrease in time. We illustrate this with a single function. Note that the function $\sin(x)$ is concave down on $[0, \pi]$ and concave up on $[\pi, 2\pi]$. We produce a function so that with $t = 0$, $c(0,x) = \sin(x)$, and for arbitrary t, $c(t,x)$ satisfies equation (6.2.13).

```
> c:=(t,x)->exp(-t)*sin(x);
```

Here, we verify this is a solution of (6.2.13).

```
> diff(c(t,x),t)-diff(c(t,x),x,x);
```

Now we animate the graph. Observe where $c(t,x)$ is increasing and where it is decreasing.

```
> with(plots):
    animate(c(t,x),x=0..2*Pi,t=0..2);
```

Section 6.3

Interplacental Transfer of Oxygen: Biological and Biochemical Considerations

A fetus must obtain oxygen from its mother. Oxygen in the mother's blood is attached to a blood pigment called hemoglobin and is carried to the placenta, where it diffuses across a system of membranes to the fetus' hemoglobin. A number of physical factors cause the fetal hemoglobin to bind the O_2 more securely than does the maternal, or adult, hemoglobin, thus assuring a net O_2 movement from mother to fetus.

The blood of a mother and her unborn child do not normally mix.

The circulatory systems of a mother and her unborn child face one another across a plate-like organ called the *placenta*. The placenta has a maternal side and a fetal side, each side being fed by an artery and drained by a large vein, the two vessels being connected in the placenta by a dense network of fine capillaries. The two sides of the placenta are separated by membranes and the blood of the mother and that of the child do not mix. All material exchange between mother and child is regulated by these placental membranes, which can pass ions and small-to-medium biochemical molecules. Large molecules, however, do not usually transit the placental membranes.

Hemoglobin carries oxygen in blood.

The chemical hemoglobin is found in anucleate cells called red blood cells or *erythrocytes*. Hemoglobin picks up O_2 at the mother's lungs and takes it to the placenta, where the O_2 crosses the placenta to the hemoglobin of fetal red blood cells for distribution to metabolizing fetal tissues.

Oxygen-affinity measures the strength with which hemoglobin binds oxygen.

A fixed amount of hemoglobin can hold a fixed maximum amount of oxygen, at which point the hemoglobin is said to be saturated. Figure 6.3.1 is an *oxygen dissociation curve*; it shows that the extent to which saturation is approached is determined by the *partial pressure* of the oxygen.[5] The partial pressure of O_2 at which the hemoglobin is half-saturated is a measure of the *oxygen-affinity* of the hemoglobin. Thus, hemoglobin that reaches half-saturation at low O_2 partial pressure has a high oxygen-affinity (see References [1] and [8] for further discussion).

---

[5]The *partial pressure* of a gas is the pressure exerted by that specific gas in a mixture of gases. The partial pressure is proportional to the concentration of the gas. The total pressure exerted by the gaseous mixture is the sum of the partial pressures of the various constituent gases.

The reversible attachment of O_2 to hemoglobin is represented by

$$Hb + O_2 \quad \underset{\longleftarrow}{\longrightarrow} \quad O_2\text{-}Hb$$

$$\underset{\text{hemoglobin}}{} \qquad\qquad \underset{\text{oxyhemoglobin}}{}$$

At equilibrium the relative amounts of hemoglobin and oxyhemoglobin are fixed, the reaction going to the right as often as it goes to the left. The relative amounts of oxyhemoglobin, hemoglobin and oxygen at equilibrium are determined by the oxygen affinity of the hemoglobin. The greater the oxygen affinity, the more oxyhemoglobin there will be relative to the amounts of hemoglobin and oxygen, i.e., the more the equilibrium will move toward the right in the above reaction scheme.

Oxygen affinity depends on a variety of factors.

In practice, oxygen affinity is determined by multiple factors: First, we would surely expect that the structure of hemoglobin would be important, and that will be discussed below. Second, oxygen affinity is affected by the extent to which oxygen molecules are already attached. Hemoglobin can bind to as many as four O_2 molecules. The second, third and fourth are progressively easier to attach because the oxygen affinity of the hemoglobin increases as more O_2 molecules are added. Third, blood pH affects its oxygen affinity. The pH of the blood and the presence of CO_2 are related; this will be discussed in Section 7.5. Finally, a chemical constituent of red blood cells, called *D*-2, 3-*biphosphoglycerate (BPG)*, plays an important role in the oxygen-binding properties of hemoglobin by binding to it and thereby decreasing its O_2 affinity. The role of BPG is a crucial one because

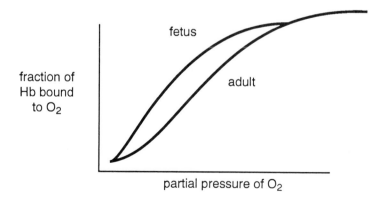

Figure 6.3.1 Oxygen dissociation curves for adult and fetal hemoglobin. Note that, for a given partial pressure (concentration) of oxygen, the fetal hemoglobin has a greater fraction of its hemoglobin bound to oxygen than does the adult hemoglobin.

the more BPG is bound to hemoglobin the less tightly the hemoglobin binds oxygen. Therefore, the oxygen will be released more easily and will be provided to metabolizing tissues in higher concentration. In terms of the chemical reaction above, BPG moves the equilibrium toward the left.

Fetal hemoglobin has a greater affinity for oxygen than does adult hemoglobin.

Adult and fetal hemoglobins have somewhat different structures. The result is that fetal hemoglobin binds less BPG than does adult hemoglobin, and therefore fetal hemoglobin has the higher oxygen affinity of the two. Figure 6.3.1 shows oxygen dissociation curves for the hemoglobin of an adult and for that of a fetus. Note that, at a given partial pressure of O_2, the fetal hemoglobin has a greater O_2 affinity than does maternal hemoglobin. Thus, there is a net movement of oxygen from the mother to the fetus.

We must be very careful here: We must not think that the fetal hemoglobin somehow drags O_2 away from that of the mother. This would require some sort of "magnetism" on the part of the fetal hemoglobin, and such magnetism does not exist. What does happen is represented by

$$O_2\text{-}Hb_{\text{adult}} \;\; \overset{\longrightarrow}{\longleftarrow} \;\; O_2 + Hb_{\text{adult}}$$

$$\cdots\cdots\cdots\cdots\cdots\cdots\cdots\uparrow \Big\downarrow \cdots\cdots\cdots\overset{\text{placenta}}{\cdots\cdots}$$

$$O_2 + Hb_{\text{fetal}} \;\; \overset{\longrightarrow}{\longleftarrow} \;\; O_2\text{-}Hb_{\text{fetal}}$$

Both kinds of hemoglobin are constantly attaching to, and detaching from, oxygen consistent with their oxygen affinities. The mother's breathing gives her blood a high concentration of oxyhemoglobin, and that leads to a high concentration of *free* oxygen on her side of the placenta. On the fetal side of the placenta, the fetus, which does not breathe, has a low O_2 concentration. Therefore, O_2, once released from maternal oxyhemoglobin, moves by simple diffusion across the placenta, in response to the concentration gradient. On the fetal side, fetal hemoglobin attaches to the oxygen and holds it tightly because of its high oxygen affinity. Some oxygen will dissociate from the fetal hemoglobin but little will diffuse back to the maternal side because the concentration gradient of the oxygen across the placenta is in the other direction.[6] In summary, oxygen diffuses across the placenta from mother to fetus, where it tends to stay because of its concentration gradient and the high oxygen affinity of fetal hemoglobin, compared to that of adult hemoglobin.

---

[6]In Chapter 7 we will see that the concentration of CO_2 in the blood also affects the oxygen affinity of hemoglobin.

Section 6.4

Oxygen Diffusion Across the Placenta: Physical Considerations

The delivery of fetal oxygen typifies the function of the placenta. In this organ, fetal blood flow approaches maternal blood flow but the two are separated by membranes. Possible mechanisms for oxygen transfer are simple diffusion, diffusion facilitated by some carrier substance, or active transport requiring metabolic energy. No evidence for facilitated diffusion or active transport has been found. We will see that simple diffusion can account for the required fetal oxygen consumption. In this section we draw on References [9-15].

The oxygen dissociation curve is sigmoid in shape.

Oxygen in blood exists in one of two forms, either dissolved in the plasma or bound to hemoglobin as oxyhemoglobin. Only the dissolved oxygen diffuses; oxyhemoglobin is carried by the moving red blood cells. The binding of oxygen to hemoglobin depends mostly on O_2 partial pressure but also on blood acidity. The relationship, given by a dissociation curve, posseses a characteristic sigmoid shape as a function of partial pressure (see Figure 6.4.1 or Table 6.4.1).

    The affect of increasing acidity is to shift the curve rightward. (The dissociation curves can be constructed using the data in Table 6.4.1.)

```
> with(plots):
> fetal74:=[0,0, 10, 3.5, 20,10.5, 30,15.2, 40,17.4, 50,18.6,60,19.2,
> 70,19.5, 80,19.7, 90,19.8, 100,19.9 ];
> f74:=plot(fetal74);
> fetal72:=[0,0, 10, 2.2, 20,7.3, 30,12.0, 40,15.2, 50,16.9,60,18.0,
> 70,18.6, 80,19.1, 90,19.5, 100,19.9 ];
> f72:=plot(fetal72);
> maternal74:=[0,0, 10, 1.3, 20,4.6, 30,8.7, 40,11.5, 50,13.2,60,14.2,
> 70,14.7, 80,14.9, 90,15.0, 100,15.1 ];
> m74:=plot(maternal74);
> maternal72:=[0,0, 10, 1.0, 20,4.0, 30,7.8, 40,10.6, 50,12.5,60,13.7,
> 70,14.4, 80,14.7, 90,14.9, 100,15.1 ];
> m72:=plot(maternal72);
> display({f74,f72,m74,m72});
```

Table 6.4.1 O_2 Concentration in ml per 100 ml Blood, References [9,10]

pO_2mm Hg $\longrightarrow$	10	20	30	40	50	60	70	80	90	100
fetal (pH 7.4)	3.5	10.5	15.2	17.4	18.6	19.2	19.5	19.7	19.8	19.9
fetal (pH 7.2)	2.2	7.3	12.0	15.2	16.9	18.0	18.6	19.1	19.5	19.8
maternal (pH 7.4)	1.3	4.6	8.7	11.5	13.2	14.2	14.7	14.9	15.0	15.1
maternal (pH 7.2)	1.0	4.0	7.8	10.6	12.5	13.7	14.4	14.7	14.9	15.1

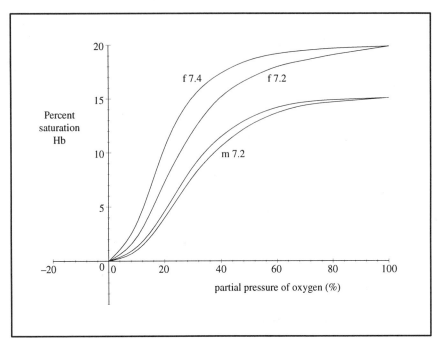

Figure 6.4.1 O_2 Concentration in ml per 100 ml blood

When maximally saturated, hemoglobin (Hb) holds about 1.34 ml O_2 per gm. Fetal blood contains about 15 gm Hb per 100 ml while maternal blood has 12 gm per 100 ml.

Although only the dissolved oxygen diffuses, hemoglobin acts like a moving reservoir on both maternal and fetal sides of the placenta. On the maternal side, O_2 diffuses across the placental membrane from the maternal blood plasma causing a decrease in the partial pressure of O_2, written pO_2. But a lower oxygen partial pressure dissociates oxygen out of the hemoglobin to replace what was lost. This chemical reaction is very fast. Consequently hemoglobin acts to preserve the partial pressure while its oxygen, in effect, is delivered to the fetal side. Of course as more and more oxygen dissociates, pO_2 gradually decreases.

On the fetal side the opposite occurs. The incoming oxygen raises the partial pressure with the result that oxygen associates with fetal hemoglobin with gradual increase of pO_2.

Fetal oxygen consumption rate at term is 23 ml per minute.

The first step in showing that simple diffusion suffices for oxygen delivery is to determine how much oxygen is consumed by the fetus. By direct measurement, oxygen partial pressure and blood *pH* at the umbilical cord are as shown in Table 6.4.2.

Table 6.4.2 Placental Oxygen and Flow Rate Data, from References [9,10]

umbilical artery	pO_2: 15 mm Hg, pH: 7.24
umbilical vein	pO_2: 28 mm Hg, pH: 7.32
umbilical flowrate	250 ml per minute
maternal artery	pO_2: 40 mm Hg
maternal vein	pO_2: 33 mm Hg
maternal flowrate	400 ml per minute
placental membrane surface	12 square meters
placental membrane thickness	3.5×10^{-4} cm
pO_2 diffusivity (see text)	3.09×10^{-8} cm^2/min/mm Hg

It follows from Figure 6.4.1 that each 100 ml of venous blood in the fetus contains approximately 13.5 ml O_2 while arterial blood contains about 4.5 ml.

Evidently an O_2 balance for fetal circulation measured at the umbilical cord is given by

$$O_2 \text{ in } - O_2 \text{ out } = O_2 \text{ consumed.}$$

For each minute this gives

$$\text{rate } O_2 \text{ consumed} = 250 \frac{\text{ml blood}}{\text{min}} \times (13.5 - 4.5) \frac{\text{ml } O_2}{100 \text{ ml blood}}$$
$$= 22.5 \text{ ml } O_2/\text{min.}$$

Maximal oxygen diffusion rate is 160 ml per minute.

Next we estimate the maximum diffusion possible across the placenta. Recall the membrane transport equation (6.2.21)

$$J = -\frac{D}{w}\Delta c, \tag{6.4.1}$$

where we have taken the partition coefficient $\Gamma = 1$ and the membrane thickness to be w. This holds for those sections of the membrane that happen to have thickness w and concentration difference Δc. Normally both these attributes will vary throughout the placenta. However, since we are interested in the maximal diffusion rate, we assume them constant for this calculation. Placental membrane thickness has been measured to be about 3.5 microns (3.5×10^{-4} cm). Since flux is the diffusion rate per unit area, we must multiply it by the surface area, S, of the membrane. Careful measurements show this to be about 12 square meters at term [15].

Actually taking a constant average value for w is a reasonable assumption. But taking O_2 concentrations to be constant is somewhat questionable. Mainly, doing so ignores the effect of the blood flow. We will treat this topic below.

Here we assume O_2 dissociates out of maternal blood in response to diffusion maintaining the concentration constant on the maternal side. On the fetal side O_2 associates with fetal blood maintaining a constant concentration there. We will see that in the countercurrent flow model this tends to be realized.

A lesser difficulty in applying Fick's Law has to do with the way an oxygen concentration is normally measured, namely in terms of partial pressure. By Henry's Law (Reference [15], p. 1714), there is a simple relationship between them: The concentration of a dissolved gas is proportional to its partial pressure, in this case pO_2. Hence

$$c = \delta(pO_2),$$

for some constant δ. Incorporating surface area and Henry's Law, equation (6.4.1) takes the form

$$SJ = -\delta D \frac{S}{w} \Delta(pO_2).$$

The product δD has been calculated to be about 3.09×10^{-8} cm^2/min-mm Hg (derived from data in [15], see the problems).

For the constant fetal partial pressure we take the average of the range 15 to 28 mm Hg noted in Table 6.4.2, so about 21.5 mm Hg. Maternal arterial pO_2 is 40 mm Hg while venous pO_2 is 33 mm Hg for an average of 36.5 mm Hg. Hence, using the values in Table 6.4.2,

$$
\begin{aligned}
O_2 \text{ diffusion rate} &= 3.09 \cdot 10^{-8} \frac{\text{cm}^2}{\text{min-mm Hg}} \\
&\quad \cdot \frac{12 \text{ m}^2 \cdot 10^4 \text{ cm}^2/\text{m}^2}{3.5 \cdot 10^{-4} \text{ cm}} (36.5 - 21.5) \text{ mm Hg} \\
&= 159 \frac{\text{cm}^3}{\text{min}}.
\end{aligned}
$$

Recalling that only 22.5 ml of oxygen per minute are required, the placenta, in its role transfering oxygen, need only be about $22.5/159 = 14\%$ efficient.

The fetal flow rate limits placental transport efficiency.

The placenta as an exchanger is not 100% efficient due to (1) maternal and fetal shunts, (2) imperfect mixing, and (3) most importantly, flow of the working material, which we have not taken into consideration. Let F be the maternal flow rate, f fetal the flow rate; C maternal O_2 concentration, and c fetal O_2 concentration. Use the subscript i for in and o for out (of the placenta). Let r denote the transfer rate across the placenta ($r = SJ$). From the mass balance equation,

$$O_2/\text{min in} \pm O_2 \text{ gained/lost per min} = O_2/\text{min out},$$

we get

$$fc_i + r = fc_o; \qquad FC_i - r = FC_o \qquad (6.4.2)$$

since the oxygen rates in or out of the placenta are the product of conentration times flow rate. From the membrane equation (6.2.21),

$$r = K(\Delta\text{concentration across membrane}) \quad \text{where} \quad K = \frac{\Gamma DS}{w}.$$

For the fetal and maternal concentrations we use C_o and c_o. From equations (6.4.2)

$$r = K(C_o - c_o)$$
$$= K(C_i - \frac{r}{F} - \frac{r}{f} - c_i).$$

Solve this for r and get

$$r = \frac{C_i - c_i}{\frac{1}{K} + \frac{1}{F} + \frac{1}{f}}. \qquad (6.4.3)$$

Now consider the magnitude of the three terms in the denominator. If F and f were infinite, then the denominator reduces to $1/K$, and the transfer rate becomes

$$r = K(C_i - c_i)$$

as before. In this case diffusion is the limiting factor.

Since the flow terms are not infinite, their effect is to increase the denominator and consequently reduce the transfer coefficient, that is,

$$\frac{1}{\frac{1}{K} + \frac{1}{F} + \frac{1}{f}} < \frac{1}{\frac{1}{K}} = K.$$

Moreover, depending on the relative size of the three terms in the denominator of equation (6.4.3), diffusion may not be the limiting factor. In particular, the smallest of the quantities K, F, or f, corresponds to the largest of the reciprocals $1/K$, $1/F$, or $1/f$. Using the values of S and w from Table 6.4.2, and taking $\Gamma D = 4 \times 10^{-7}$ cm^2/sec. [15], gives

$$K = \frac{4 \cdot 10^{-7} \cdot 12 \cdot 10^4}{3.5 \cdot 10^{-4}}$$
$$= 137\frac{cm^3}{sec}.$$

Compare this with a maternal flow rate F of 400 ml/min or 6.7 ml/sec and a fetal flow rate f of 250 ml/min or 4.2 ml/sec. Thus fetal flow rate is the smallest term and so is the limiting factor. Furthermore, from equation (6.4.3) we can see that diffusion is a relatively minor factor compared to the maternal and fetal flow rates. That is,

$$\frac{1}{\frac{1}{137} + \frac{1}{6.7} + \frac{1}{4.2}} = 2.53$$

while

$$\frac{1}{\frac{1}{6.7} + \frac{1}{4.2}} = 2.58.$$

Countercurrent flow is more efficient than concurrent flow.

In this section we will compare the diffusion properties of the placenta depending on whether the maternal and fetal blood flow in the same or the opposite directions through the placenta. For this we assume the placenta to be a channel separated by the placental membrane. Assume first that fetal and maternal blood flow in the same direction. As shown in Figure 6.4.2, on the maternal side we take the channel height to be H and the velocity to be v_m while these will be h and v_f respectively on the fetal side. Let the channel width be b. Assume steady state has been reached and take the oxygen concentrations at position x along the maternal and fetal sides to be $C(x)$ and $c(x)$ respectively.

On the fetal side, a block ΔV of blood at position x gains in concentration in moving distance Δx due to the flux $J(x)$ at x. Let ΔS denote the area of contact

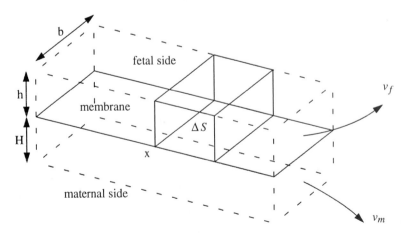

Figure 6.4.2 Diffusion of oxygen into a moving incremental volume

of the block with the placental membrane. Since the time required to move this distance is $\Delta t = \Delta x / v_f$, we have

$$c(x + \Delta x) = \frac{c(x)\Delta V + J(x)\Delta S(\Delta x / v_f)}{\Delta V}.$$

But from the membrane equation (6.2.21) (with $\Gamma = 1$)

$$J(x) = (D/w)(C(x) - c(x))$$

and so

$$\frac{c(x + \Delta x) - c(x)}{\Delta x} = \frac{D}{w v_f} \frac{\Delta S}{\Delta V}(C(x) - c(x)).$$

Now $\Delta S / \Delta V = 1/h$, so in the limit we have

$$\frac{dc}{dx} = \frac{(D/w)}{h v_f}(C - c). \tag{6.4.4}$$

A similar calculation on the maternal side gives

$$\frac{dC}{dx} = -\frac{(D/w)}{H v_m}(C - c). \tag{6.4.5}$$

Denote by T the *tension* or concentration difference $C(x) - c(x)$. By subtracting the first equation from the second we get

$$\frac{dT}{dx} = -\frac{D}{w}\Big(\frac{1}{H v_m} + \frac{1}{h v_f}\Big)T. \tag{6.4.6}$$

For this parallel flow, denote by k_p the constant coefficient,

$$\begin{aligned}
k_p &= \frac{D}{w}\Big(\frac{1}{H v_m} + \frac{1}{h v_f}\Big) \\
&= \frac{D}{w}\Big(\frac{b}{F} + \frac{b}{f}\Big)
\end{aligned} \tag{6.4.7}$$

where we have replaced the velocities v_m and v_f by the maternal and fetal flow rates F and f, respectively, using the fact that a flow rate is the product of velocity with cross-sectional area,

$$F = (bH)v_m \quad \text{and} \quad f = (bh)v_f.$$

Now the solution of the differential equation (6.4.6) is

$$T = T_0 e^{-k_p x}, \qquad (6.4.8)$$

with T_0 the initial tension, $40 - 15 = 25$ mm Hg. Assuming the channel has length L and the final tension is $33 - 28 = 5$ mm Hg, see Table 6.4.2, equation (6.4.8) with $x = L$ becomes

$$5 = 25 e^{-k_p L} \qquad \text{therefore} \quad k_p L = -\log(1/5). \qquad (6.4.9)$$

Next we calculate the average tension $\bar{T}$ over the run of the channel. The average of equation (6.4.8) is given by integral

$$\bar{T} = \frac{1}{L} \int_0^L T_0 e^{-k_p x}\, dx = \left. \frac{-T_0}{k_p L} e^{-k_p x} \right|_0^L$$

$$= \frac{T_0 - T_0 e^{-k_p L}}{k_p L} = \frac{25 - 5}{\log(5)} = 12.4 \text{ mm Hg}$$

where equation (6.4.9) was used to substitute for $k_p L$.

Next consider countercurrent flow. Arguing in the same manner as above, the differential equations for countercurrent flow are similar to equations (6.4.4) and (6.4.5). The sign of the second is reversed because the flow is reversed here:

$$\frac{dc}{dx} = \frac{(D/w)}{h v_f}(C - c)$$

on the fetal side, and

$$\frac{dC}{dx} = \frac{(D/w)}{H v_m}(C - c)$$

on the maternal side. Subtracting, we get

$$\frac{dT}{dx} = -\frac{D}{w}\left(\frac{1}{h v_f} - \frac{1}{H v_m}\right)T.$$

For this countercurrent flow, denote by k_c the coefficient

$$k_c = -\frac{D}{w}\left(\frac{b}{f} - \frac{b}{F}\right). \qquad (6.4.10)$$

As before, the solution is

$$T = T_0 e^{-k_c x}, \qquad (6.4.11)$$

Now however $T_0 = 33 - 15 = 18$ mm Hg. And for $x = L$, $T = 40 - 28 = 12$ mm Hg, therefore $12 = 18e^{-k_cL}$ from which it follows $k_cL = \log(1.5)$. Hence the average tension in this case is

$$\bar{T} = \frac{T_0 - T_0e^{-k_cL}}{k_cL} = \frac{6}{\log(1.5)} = 14.8 \text{ mm Hg}.$$

Therefore countercurrent flow is somewhat more efficient than concurrent flow, by these calculations, $14.8/12.4 = 1.2$ times more efficient. Note that the average tension for countercurrent flow is approximately equal to the numerical average, $(18 + 12)/2 = 15$.

Section 6.5

The Spread of Infectious Diseases

Mathematical models can be created to describe the spread of infectious diseases. We will only describe the qualitative details. One thing is clear at the outset: A model based on chance alone is unrealistic because every disease spreads in its own characteristic way. Thus, there are specific biases in the spread of each disease that must be considered. Among these are the facts that some diseases are not infectious, and that infectious diseases show a wide variety of transmission modes. Further, different hosts react differently to a pathogen, depending on the dose and host's health.

Some diseases are not infectious.

We often think of diseases in terms of "germs," suggesting a causative agent that is specific, isolatable and biological. Actually, many diseases are not even "caught," in the sense of being transmitted from one person to another. Examples are heart disease and headaches, which may result from stress, lead poisoning, which is caused by a metallic element, and scurvy, which results from a vitamin C deficiency. Genetic disorders, like sickle-cell anemia and diabetes, are diseases but are likewise not infectious.

Some diseases are infectious.

Here we are interested in diseases that one person can transmit to another through the action of some intermediate, biological system, say a fungus, bacterium or virus. Even here the notion of infection is complicated because the pathogen may require an intermediate, non-human host between the two human hosts. For example, while smallpox is transmitted directly from one person to another by contact, malaria is transmitted via an intermediate mosquito vector and cannot be given directly from one person to another. Many infectious diseases are very specific: Cat distemper is highly infectious between cats, but humans cannot contract it.

The spread of a disease depends on many factors.

In order to understand the rate and pattern of the spread of a disease we have to answer several questions:

1. How does infection occur?   We must consider the mode of transmission, e.g., body contact, drinking water, insect vector, etc. What is the portal of entry for the pathogen: an open wound, inhalation, ingestion? In particular, can we assume that infection is a random event? This latter question is relevant to any treatment that involves a random walk approach.

2. Once infection occurs, does the disease actually occur?   Here we must know the dose, i.e., the number of microorganisms that caused the infection. Then we must inquire about the condition of the host. A healthy host may never develop the disease whereas, with the same dose, a weakened host may show symptoms quickly. In the latter category are the very young, the aged and people already ill from other diseases. Further, a host who has previously been exposed to the disease may have immunity.

3. Once the disease occurs, is it serious enough to cause problems?   The word "problems" here covers the spectrum of symptoms from minor discomfort to death. If the symptoms are minor, the patient may never report them, in which case epidemiology statistics may be faulty and a mathematical treatment based on these statistics would then be faulty also.

For more information see References [16] and [17].

Section 6.6

Questions for Thought and Discussion

1. What are the functions of the endoplasmic reticulum?
2. Describe the physical and chemical factors that affect the attachment of oxygen to hemoglobin.

References and Suggested Further Reading

1. **Membrane structure:** William S. Beck, Karel F. Liem and George Gaylord Simpson; *Life: An Introduction to Biology*, 3rd ed., Harper Collins Publishers, New York, 1991.
2. **Membrane transport:** Edward K. Yeargers, *Basic Biophysics for Biology*, CRC Press, Inc., Boca Raton, 1992.

3. **Diffusion:** Russell K. Hobbie, *Intermediate Physics for Medicine and Biology,* John Wiley and Sons, 2nd ed, New York, p. 65, 1988.

4. **Diffusion:** H. C. Berg, *Random Walks in Biology*, Princeton University Press, Princeton, NJ, 1993

5. **Diffusion in Biology:** J. D. Murray, *Mathematical Biology*, Springer-Verlag, New York, 1989.

6. **Fluid resistance:** S.I. Rubinow, *Introduction to Mathematical Biology*, John Wiley and Sons, New York, 1975.

7. **Diffusion across a slab:** David L. Powers, *Boundary Value Problems*, Academic Press, New York, 1979.

8. **Oxygen dissociation curves, fetal blood:** William T. Keeton and James L. Gould, *Biological Science*, 5th ed. W. W. Norton and Company, New York, 1993.

9. **Placenta:** H. Bartels, W. Moll, J. Metcalfe, Physiology of gas exchange in the human placenta, *Am. J. Obstet. Gynecol.* **84**, 1714–1730, 1962.

10. **Placenta:** J. Metcalfe, H. Bartels, W. Moll, "Gas Exchange in the Pregnant Uterus," *Physiol. Rev.* **47**, 782–838, 1967.

11. **Placenta:** R. E. Forster II, "Some Principles Governing Maternal–Foetal Transfer in the Placenta," *Foetal and Neonatal Physiology*, Cambridge University Press, Cambridge, 223–237, 1973.

12. **Placenta:** K. S. Comline, K. W. Cross, G. S. Dawes, P. W. Nathanielsz, *Foetal and Neonatal Physiology*, Cambridge University Press, Cambridge, 1973.

13. **Placenta:** F. C. Battaglia, G. Meschia, *An Introduction to Fetal Physiology*, Academic Press, Inc., Harcourt Brace Jovannovich, New York, 1986.

14. **Placenta:** A. Guettouche, et. al., Mathematical Modeling of the Human Fetal Arterial Blood Circulation, *Int. J. Biomed. Comput.* **31**, 127–139, 1992.

15. **Placenta:** A. Costa, M. L. Costantino, R. Fumero, Oxygen exchange mechanisms in the human placenta: mathematical modelling and simulation, *J. Biomed. Eng.* **14**, 85–389, 1992.

16. **Epidemiology:** John P. Fox, Carrie E. Hall and Lila R. Elveback, *Epidemiology: Man and Disease*, The Macmillan Company, New York, 1970.

17. **Epidemiology and disease:** Julius P. Krier and Richard F. Mortenson, *Infection, Resistance and Immunity*, Harper and Row, Publishers, New York, 1990.

Chapter 7

The Biological Disposition of Drugs and Inorganic Toxins

Introduction to this chapter

This chapter is a discussion of how some foreign substances get into the body, how they become distributed, what their effects are and how they are eliminated from the body. Lead is the exemplar in the biological discussion, but the biological concepts can be applied to many other substances. The mathematical discussion focuses on lead poisoning and on pharmaceuticals.

Lead can be eaten, inhaled or absorbed through the skin; it is distributed to other tissues by the blood. Some of it is then removed from the body by excretion and defecation. Any lead that is retained in the body can have unpleasant biological consequences—anemia and mental retardation, for instance. These processes can only be understood at the levels of organ systems and of the tissues that comprise those organs. Thus, this chapter includes discussions of the lungs, the digestive tract, the skin, blood, the circulatory system, bones, and the kidneys, all of which are involved in the effects of lead on humans.

Section 7.1

The Biological Importance of Lead

No biological requirement for lead has ever been demonstrated. Rather, there is much experimental evidence that it is toxic. Lead is sparsely distributed in nature, but mining activities to support the manufacture of batteries, leaded gasoline and other products have concentrated it.

Trace metals play an important role in nutrition.

A number of metallic elements play crucial roles in our nutrition, usually in small amounts. Sodium is necessary to nerve conduction, iron is an essential part of hemoglobin and magnesium is a component of chlorophyll. In many cases, traces of metals are required for the correct functioning of biomolecules; examples are copper, manganese, magnesium, zinc and iron. On the other hand, no human metabolic need for lead has ever been demonstrated, whereas the toxic effects of lead are well-documented.

Most lead enters the biosphere through human-made sources.

Lead is found in the earth's crust in several kinds of ore. Its natural concentration in any one place is almost always quite low and is of little concern to biology. Commercial uses for lead, however, have produced locally high concentrations of the metal in air, water and soil (see References [1] and [2]).

About 40% of the refined lead in the western world is used in the manufacture of batteries and another 10% finds its way into antiknock compounds in gasoline. Cable sheathing and pipes account for about another 15%. Still more is used in paints, glassware, ceramics and plastics. Fortunately, the use of lead in interior paints and gasoline has lately been approaching zero in the USA and some other countries.

A biological compartment model shows the paths of lead into, around and out of an organism.

It often happens that a particular material, introduced into one part of an organism, quickly reaches a common concentration in several other parts of the organism. The various parts in which this equilibration occurs constitute a single *biological compartment*. Note that the parts of a compartment can be organ systems, organs or parts of an organ, and that they need not be near each other if a suitable distribution system is available (see Reference [3]).

Figure 7.1.1 illustrates these concepts in a biological compartment model for lead. One compartment is the blood, in which flow and turbulence cause rapid mixing. The blood carries the lead to "soft tissues," which we take to mean all tissues that are neither bone nor blood. All these soft tissues behave similarly toward lead and take it up to about the same degree, so they can be considered to constitute a second compartment.[1] A third compartment is bone, in which lead has a very long half life. The fourth compartment is implied: It is the environment, from which the lead originates and to which it must ultimately return, either through living biological processes or at the death of the organism. The arrows connecting the compartments show the direction of lead movement between the various compartments.

---

[1]This is a somewhat rough approximation: The aorta, the main artery out of the heart, is a soft tissue that seems not to equilibrate lead quickly with other soft tissues. Further, the behavior of a few other soft tissues toward lead seems to depend on the national origin and age of the cadavers used in the data collection. (See the reference by Schroeder and Tipton [4].)

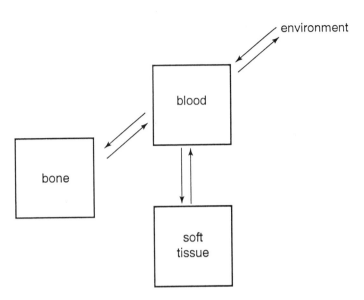

Figure 7.1.1   A compartment model of the three biological tissue types among which lead becomes distributed, and the relationship of the compartments to the environmental compartment

Section 7.2

Early Embryogenesis and Organ Formation

In discussions of the uptake and metabolism of toxins and drugs, the relevant compartments are almost always organs and groups of organs. This section presents some information about how organs develop in embryos. We show that organs and their interdependence on one another develop from the very start. This will allow us to treat organs as a group of interacting compartments for mathematical purposes. Thus, the stage will be set for our discussion of lead poisoning.

Specific organs have evolved in multicellular organisms to perform very specific biological tasks. This functional specialization has a benefit and a cost. The benefit is that a given organ is usually very efficient at performing its assigned biological tasks. The cost is that the organ generally can do little else and must therefore rely on other organs to support it in other functions. As examples, the heart is a reliable blood pump and the kidneys efficiently remove nitrogenous wastes from blood. However, the heart is dependent on the kidneys to remove blood wastes which would be detrimental to the heart, and the kidneys are dependent on the heart to pump blood at a high enough pressure to make the kidneys work.

Early divisions of fertilized eggs result in an unchanged total cell mass.

A fertilized egg, or *zygote*, starts to divide by mitosis soon after fertilization. In the case of humans, this means that division starts while the zygote is still in the oviduct. Early divisions merely result in twice the number of cells, without any increase in total cell mass, and are thus called *cleavage divisions* (see Reference [5]).

A one-celled zygote contains the full complement of genes available to the organism.[2] As mitosis proceeds, many of the new cells *differentiate*, or take up more specialized functions, a process that coincides with the inactivation of un-needed genes. The genes that remain active in a particular cell are those that determine the functional nature of that cell. Thus, cells of the heart and of the liver require different sets of active genes (although the two sets certainly over-lap).[3] The gene inactivation process starts early: Even before the heart organ itself becomes obvious there are heart precursor cells that individually contract. Soon, however, a functioning heart and blood vessels develop in concert with the fetus' need for a continuing blood supply.

Higher organisms have a three-layer body plan.

Embryos of each species have their own unique behavior, but there are several events in embryogeny that are shared among most multicellular species: Early cleavage generates a solid mass of cells, which then hollows out. Next, cell pro-liferation and movement create three basic tissue layers. Finally, the various organ systems develop out of these basic tissue layers. We will follow a human zygote through these steps.

Fertilization of the human egg occurs in the upper third of the oviduct, after which the zygote moves toward the *uterus*, a powerful muscular organ (see Figure 7.2.1). The initial cleavage divisions take place in the oviduct, and result in a solid ball of cells that resembles a mulberry; its name, *morula*, is taken from the Latin word for that fruit. The morula hollows out to form the structure shown in Figure 7.2.1. It is called a *blastocyst* and consists of an outer sphere of cells, with an *inner cell mass* at one end. When the blastocyst reaches the uterus, it imbeds into the uterine wall, a process called *implantation*, at about Day 8.

Figures 7.2.2 and 7.2.3 illustrate embryonic development after implantation. To start, the embryo is little more than a disk of cells, but an elongated structure called the *primitive streak* soon forms along what will be the head-to-foot axis. Cells on the outer margins of the primitive streak migrate toward it, downward through it and into the interior of the disk of cells (Figure 7.2.2 is a perspective view and Figure 7.2.3 is a cross-section through the disk). This migration, called *gastrulation*, establishes three *germ layers* of tissue from which all subsequent organs will develop. These germ layers are the *endoderm*, the *mesoderm* and the *ectoderm*. Next, the tips of the embryo fold down and around, a process

---

[2]In Chapter 10 we will discuss an exception to this statement: Certain viruses can insert genes into cells.

[3]Inactivation of unneeded genes by a cell is a common event. For example, different genes are active in different parts of a single division cycle. Even the inactivation of embryonic genes during early development is not irreversible: Crabs can grow new claws, plant stems can be induced to grow roots and the genes for cell division are reactivated in cancer cells.

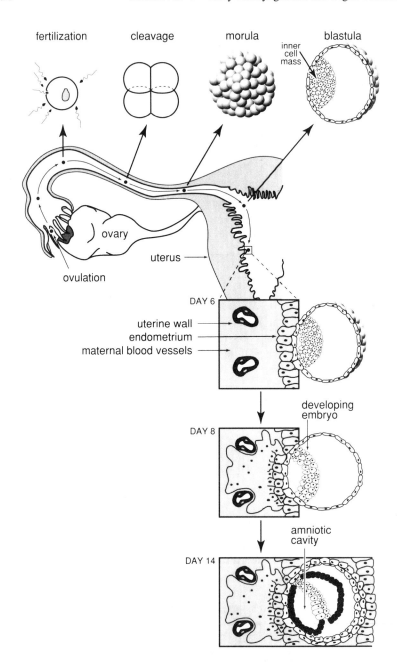

Figure 7.2.1 The stages of development of a mammalian fetus, arranged with respect to where they occur in the female reproductive system. A single cell becomes a mass of cells (called a *morula*) and then forms a hollow ball (a *blastula*), with the inner cell mass at one side. (Redrawn from "Life – The Science of Biology," 4th edition, by William Purves, Gordon Orians and Craig Heller, Sinauer Associates, Inc., Sunderland, MA; 1995. Used with permission.)

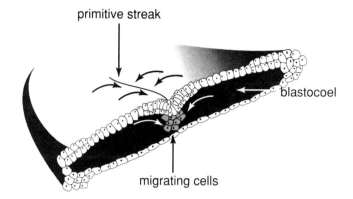

primitive streak

blastocoel

migrating cells

Figure 7.2.2 A perspective view of gastrulation. Cells move from the sides, into the primitive streak and downward into the embryo. (Redrawn from "Life – The Science of Biology," 4th edition, by William Purves, Gordon Orians and Craig Heller, Sinauer Associates, Inc., Sunderland, MA; 1995. Used with permission.)

that establishes the tubular nature of the embryo (part (c) of Figure 7.2.3). The endoderm will form much of the digestive tract and lungs, the ectoderm will form the skin and nervous system and the mesoderm, lying in between the other two layers, will form many of the internal organs. Part (d) of Figure 7.2.3 is a cross section of the fetus, which projects outward from the plane of the diagram.

The placenta is the interface between mother and fetus.

Mammalian fetuses are suspended in a watery *amniotic fluid* in a membranous sac called the *amnion*; this structure cushions the fetus against mechanical injury.[4] As pointed out in Chapter 6, the blood of the mother and her fetus do not mix, a fact that necessitates a special structure for the exchange of maternal and fetal materials: A flat, plate-like organ called the *placenta* develops between mother and child at the point of implantation. All materials exchanged between mother and child cross at the placenta, one side of which is composed of fetal tissues and the other side of which is composed of maternal uterine tissue. The two sides of the placenta have interdigitating projections into one another to increase their area of contact, thus facilitating material exchange. Lead is among the many substances that cross the placenta; thus, a mother can poison her fetus by passing lead to it.

Shortly after it leaves the fetal heart, the fetus' blood is shunted outward into a vessel in the *umbilical cord* and into the placenta. At the placenta the fetal blood takes O_2 and nutrients from the mother's blood and gives up CO_2 and wastes. The fetus' blood, after having been enriched in O_2 and nutrients and cleansed of CO_2 and wastes, returns to the fetal body through the umbilical cord.

---

[4] An amnion and amniotic fluid are also found in the eggs of egg-laying mammals and reptiles and in bird eggs.

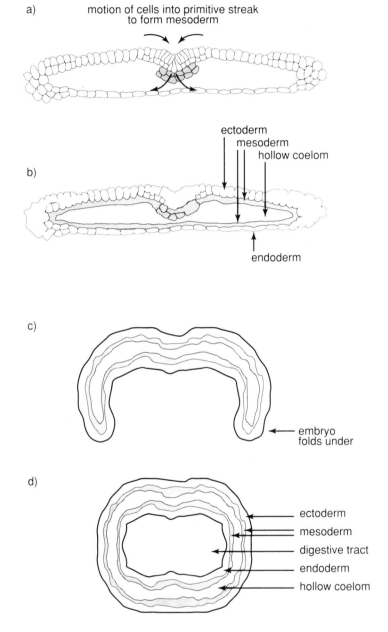

Figure 7.2.3   A cutaway view of gastrulation, showing how the mesodermal layer defines the coelom. All the structures shown project out of the plane of the paper. (a) The cells move toward the primitive streak, then down into the interior of the zygote. (b) These cells form the mesodermal lining of the coelom. (c) This process is followed by a curling of the embryo to form the digestive tract. (d) This is a cross-section of the elongated fetus. It forms a tube, with the digestive tract running down the center.

The evolutionary development of the coelom facilitated the evolution of large animals.

During embryogenesis in the higher animals, a cavity forms in the center of the mesoderm (Figure 7.2.3). This cavity is the *coelom*, and it has played a major role in the evolution of large animals (larger than ~ 1 cm). By definition a body cavity is called a coelom only if it is completely surrounded by mesoderm, the latter being identified by its creation during gastrulation and its role in forming specific internal organs, e.g., bones, muscles, heart, sex organs and kidneys.

A coelom provides room for the seasonal enlargement of the reproductive systems of some animals, notably birds. In addition, a coelom separates the muscles of the digestive tract from those of the body wall, allowing the two to function independently of one another. For purposes of our discussion of lead poisoning, however, a coelom plays two roles (to be described at length below): First, a coelom provides room into which the lungs can expand during breathing. Second, a coelom is important in determining the structural properties of the circulatory system: large animals require a powerful heart, one that can expand and contract appreciably. A coelomic cavity provides space for a beating heart to change its size.

Section 7.3

Gas Exchange

The lungs are gas exchange organs. This means, however, that they can provide an efficient entry route into our bodies for foreign substances such as lead. For example, the air we breathe may contain lead from leaded gasoline and particulate lead, mainly from lead smelters. About 40% of the lead we inhale is absorbed into the blood from the lungs. In this section we discuss the anatomy and functioning of our lungs.

Animals can exchange gas with the outside world.

Gas exchange between an animal and the atmosphere is diffusion-controlled. Thus, two physical factors influence the rate at which an animal gives up CO_2 and takes up O_2 across membranes from its surroundings—these factors are concentration gradients and surface area (see equation (6.2.11)). The concentration gradients are provided naturally by the metabolic usage of O_2 and the subsequent production of CO_2 in respiring tissues. Carbon dioxide-rich/oxygen-poor blood arrives at the animals' gas exchange organs, where passive diffusion causes CO_2 to move outward to the environment and O_2 to move inward from the environment (Figure 7.3.1).

Several organs other than lungs can serve as sites of gas exchange: In insects, tiny *tracheal tubes* carry air directly to and from target tissues. The skin of some animals, notably amphibians, is a gas exchange organ. Many aquatic animals have *gills*, which are feathery structures that contain networks of blood vessels

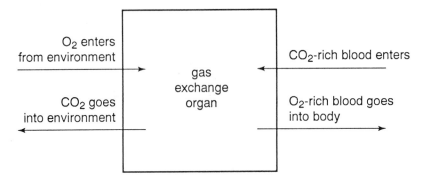

Figure 7.3.1 A gas exchange organ. Blood with an excess of CO_2 and a deficiency of O_2 arrives at the gas exchange organ. Concentration gradients drive CO_2 from the blood into the atmosphere and O_2 from the atmosphere into the blood.

over which water can flow. The water brings in O_2 and carries away CO_2. We are most concerned, however, with humans, who, along with all other mammals and birds, exchange gases in *lungs* (see References [6] and [7]).

Lungs themselves have no muscles, and are therefore incapable of pumping air in and out. Instead, air is pushed in and out of the lungs indirectly: Inhalation occurs when a set of *intercostal* muscles moves the ribs so as to increase the front-to-back size of the chest cavity (i.e., the coelom). At the same time, a dome-shaped sheet of muscle (the *diaphragm*) at the base of the chest cavity contracts and moves downward (Figure 7.3.2). The two actions expand the volume of the

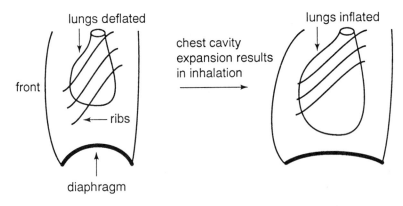

Figure 7.3.2 The process of inhalation. The lungs have no muscles of their own. Their expansion and contraction are driven by the expansion and contraction of the chest cavity. Exhalation can also occur if the person simply relaxes, allowing the chest cavity to become smaller.

chest cavity and thereby draw air into the lungs. On the other hand, exhalation occurs when those intercostal muscles and the diaphragm relax.[5]

Air enters the lungs through tubes that branch out profusely, and which lead to small sacs called *alveoli*. It is the large number of alveoli that gives the lungs their extensive surface exposure to the outside world—about $100m^2$ of area. The alveoli are lined with tiny blood vessels that exchange CO_2 and O_2 with the atmosphere, giving up CO_2 to the environment and then carrying O_2 to tissues.

Section 7.4

The Digestive System

This section is a discussion of the function of our digestive tract, a system of organs uniquely able to take nutrient materials, as well as toxins like lead, from our environment and route them into our blood.

The digestive system is an important path for lead intake. Rain washes atmospheric lead into municipal water supplies and people drink it. Lead in the pipes and solder of home water distribution systems leaches into drinking water.[6] Wine may have a high lead content. Plants absorb lead through their roots or bear it on their surfaces; the plants may then be consumed by humans. Organ meats, particularly kidneys, may contain high lead concentrations. Children may eat lead-based paint from old furniture. The fraction of ingested lead that is absorbed by the digestive tract is usually about 10–15%, but may approach 45% during fasting (perhaps because it does not compete with food for absorption).

Digestion is the splitting of biopolymers into smaller pieces.

The word "digestion" has a very restricted meaning in physiology: It is a particular way of splitting the linkage between the components in a macromolecular polymer. The process is modeled:

macromolecular polymer:

$\downarrow$ digestion

macromolecular subunits: $6\cdot$

We will discuss macromolecular structure in more detail in Chapter 9, but for now it is sufficient to note that when we eat a large molecule the process

---

[5]Forcible exhalation of air does not result from the reverse action of the muscles mentioned for inhalation. Muscles can only exert force in contraction and thus a second set of intercostal muscles and certain abdominal muscles act to push air forcibly from the lungs by moving the front of the ribs *downward* and decreasing the volume of the chest cavity.

[6]Lead is poorly soluble in water unless oxygen is present, a condition unfortunately met in most drinking water.

of digestion breaks it into smaller molecules. These smaller molecules are then either metabolized further to extract energy or are used to make building blocks for our own macromolecules (see References [6] and [7]).

The mammalian digestive tract is a series of specialized organs.

At a fairly early stage in the evolution of animals, different parts of the digestive tract assumed different roles. In particular, the various organs of the digestive system reflect the animal's lifestyle. Figure 7.4.1 is a diagram of a mammalian digestive system; we will use it for the ensuing discussion.

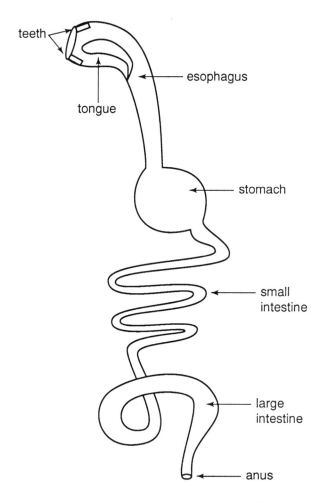

Figure 7.4.1   The digestive tract is a convoluted tube, with different portions performing different specialized functions. See text for details.

a. *Dentition.* Teeth are used for cutting, piercing and grinding. Extreme development of the piercing teeth is typical of predators; cutting and grinding teeth are prominent in herbivores. Human dentition is more characteristic of herbivores than of carnivores.

b. *Tongue.* This organ pushes food back into the esophagus, which leads to the stomach. The act of swallowing pushes a small flap over the opening to the windpipe *(trachea)*, which minimizes the possibility of choking on food.

c. *Esophagus.* This is the tube leading from the back of the mouth cavity to the stomach.

d. *Stomach.* This is an organ of mixing and, to a lesser extent, digestion; except for small molecules like water and ethanol, very little absorption takes place from the stomach into the blood. The presence of food in the stomach triggers the process of digestion: Glands in the wall of the stomach generate hydrochloric acid, which contributes directly to the chemical breakup of food and creates the acidic environment required by other digestive agents (these agents are called *enzymes*; they will be discussed in detail in Chapter 9).

   Stomach muscles churn the food and the stomach secretions to mix them. In our earlier discussion, it was pointed out that the coelom separates the voluntary muscles of the outer body from the involuntary muscles of the digestive tract, and allows them to work independently of each other. Thus, the involuntary movements of the digestive tract (called *peristalsis*) relieve us of the need to walk around waving our arms and wiggling in order to move food along our digestive tract.

e. *Small intestine.* Peristalsis moves the food along and mixes it with a large variety of digestive enzymes from the pancreas and from the small intestine itself. Surprisingly, these enzymes function only near neutral pH, which is attained by the secretion of alkali ions from the pancreas to neutralize stomach acid.

   The lining of the small intestine has a very large surface area, generated in several ways. First, the small intestine is very long, especially in herbivores. (Plant matter contains a great deal of a carbohydrate called *cellulose*, which is very difficult to digest, so plant-eaters need a long intestine.) Second, there are the numerous intestinal folds, called *villi*. Third, each cell in the intestinal lining has hundreds of small projections, called *microvilli*. The total absorptive surface of the human small intestine is several hundred square meters!

   Most absorption of digested food takes place in the small intestine. Molecular subunits of carbohydrates (e.g., starch and sugars), proteins (e.g., meat) and fats are absorbed throughout the highly convoluted intestinal surface and into the blood.

f. *The large intestine.:* Undigested matter and water collect in the large intestine, also called the *colon*. Most of the water is pumped out into the blood, leaving the waste, called *fecal matter*, which is expelled through the *anus*. Ingested lead, if not absorbed in the digestive tract, is removed from the body in fecal matter.

Section 7.5

The Skin

The third pathway by which chemicals can enter our bodies is through the skin. We will examine the unique properties of our skin and the conditions under which lead can cross it.

Skin is a sensor of, and a waterproof barrier to, the outside world.

Skin is uniquely constructed to be the interface to our environment. Working from the inside to the outside of our skin, there is first a layer rich in small blood vessels, or *capillaries*. Capillaries bring nutrients and oxygen to the skin to support the needs of the many nearby nerve endings and other specialized cells that detect stimuli like pain, pressure and heat.

The next skin layer, also requiring materials brought by the blood, is a group of rapidly dividing, pancake-shaped cells. As these cells divide they push toward the outside and die. The final, outer skin layer, called the *stratum corneum*, consists of these dead cells. Thus we are surrounded by a layer of dead cells, of which the membranes are a principal remnant. It is this layer that we need to examine more closely in our discussion of lead poisoning.

Cellular membranes were described in Chapter 6; we can review that discussion by pointing out that the principal structural components of cell membranes are closely related to fats. Thus, cell membranes are waterproof, a property that makes good sense. After all, most biological chemistry is water-based and therefore we need to protect our interior aqueous environment from our exterior environment, which is also mostly aqueous.

There are, however, chemicals that can penetrate the cell membranes of the stratum corneum and other cells and thereby get into our blood streams. Because membranes contain a lot of fat-like molecules we should not be surprised that some fat-soluble compounds may move across membranes via a temporary state in which the substances become transiently dissolved in the membrane. Absorption of compounds across the skin is said to be *percutaneous*.[7] Examples of such compounds are tetramethyl lead and tetraethyl lead, "antiknock" compounds found in leaded gasoline. It was common in days past to see people hand-washing machine parts in leaded gasoline, an activity virtually guaranteed to cause lead absorption.

---

[7]The compound dimethylsulfoxide (DMSO) rapidly penetrates the skin and is sold as a remedy for certain bone joint disorders. It is possible to taste DMSO by sticking one's finger into it: The compound crosses the skin of the finger, gets into the bloodstream and goes to the tongue, where it generates the sensation of taste—said variously to be like garlic or oysters.

Section 7.6

The Circulatory System

Our circulatory system partitions chemicals throughout the body. It picks them up at the lungs, the digestive tract and the skin and distributes them to other body tissues. In the case of lead poisoning, the circulatory system plays another key role: One of the most important toxic effects of lead is to interfere with the synthesis of the oxygen-carrying pigment hemoglobin, found in red blood cells. We will describe the circulatory system and show how its anatomy promotes the rapid distribution of materials to all other body tissues.

The discussion of oxygen transfer across the placenta in Section 6.3 was a short introduction to some of the material in this section.

Circulatory systems move a variety of materials around an animal's body.

Living organisms are open thermodynamic systems, which means that they are constantly exchanging energy and matter with their surroundings. The exchange of materials between the cells of a multicellular animal and the animal's environment is mediated by a circulatory system. This system picks up O_2 at the lungs, gills, tracheal tubes or skin, and delivers it to metabolizing cells. The CO_2 that results from the metabolic production of energy is returned to the gas exchange organ and thus to the organism's surroundings. The circulatory system also picks up nutrients at the digestive tract and delivers them to the body's cells; there it picks up wastes which it takes to the kidneys or related organs for excretion. Further, the blood carries minerals, proteins and chemical communicators, or *hormones*, from one part of the body to the other. Hormones regulate such activities as growth, digestion, mineral balances and metabolic rate (see References [6] and [7]).

Open circulatory systems are convenient, but inefficient.

The blood of most small invertebrate animals spends most of its time circulating leisurely around the animal's internal organs. An example is shown in Figure 7.6.1. Note that the blood is not necessarily confined to vessels at all; rather, it merely bathes the internal organs. This kind of circulatory system is said to be *open*.

The correct functioning of an open circulatory system relies on a very important physical property: The materials carried by the blood, and listed in the previous section, can diffuse effectively over distances of no more than about one millimeter (see Section 6.2 and the reference by Dusenbery [8]). Thus, an upper limit is set for the size of internal organs; no part of any organ can be more than about 1 mm. from the blood. Of course, this also sets a limit on the size of the entire organism; animals with open circulation seldom exceed a few centimeters in size. Typical examples are snails and houseflies; atypical examples are giant clams and lobsters but, being aquatic, they have the advantage of being bathed

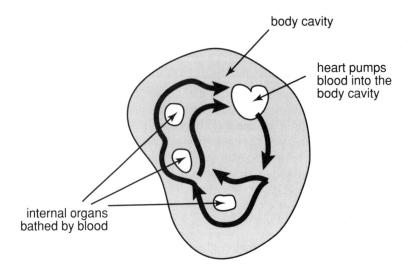

Figure 7.6.1 An open circulatory system. The heart pumps blood into an open body cavity. Organs in the body cavity are bathed by the blood, exchanging gases, nutrients and wastes with it. The blood eventually returns to the heart.

in water, which provides for easy waste and CO_2 removal. Further, these large animals have low specific metabolic rates, which minimizes their O_2 and nutrient needs, as well as their production of wastes.

Open circulatory systems are structurally simple, but the size restriction they place on organisms is a major shortcoming. There are plenty of ecological niches into which large animals could fit, especially on land, but to do so has required the evolution of a different kind of circulatory system. We examine that next.

Closed circulatory systems are efficient, but require a powerful heart.

An open circulatory system can be likened to a large fan at the front of a classroom. Air gets moved from the fan toward the rear of the room and it eventually returns to the back of the fan. Most people in the room feel a small breeze from time to time. On the other hand, a *closed circulatory system* can be likened to an enclosed air pump, with a network of conduits that take air directly to each person individually and return the air directly to the pump.

As pointed out above, the problem with open circulation is that most materials can diffuse distances of no more than about one millimeter during biologically realistic times, thus limiting the size of the organism. Closed circulatory systems remove this restriction by taking blood, via tiny vessels, directly to the immediate vicinity of all metabolizing cells. This blood distribution is independent of the size of the organism, how far the cells are from the heart, and how deep the cells are inside an organ.

A vertebrate closed circulatory system contains a heart that pumps blood into a thick-walled artery, called the *aorta*. The latter then progressively branches

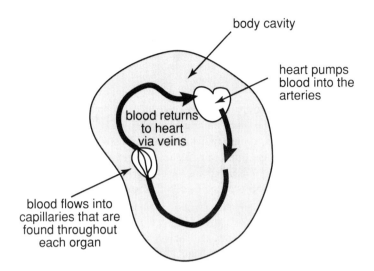

Figure 7.6.2 A closed circulatory system. The heart pumps blood into a system of arteries, which deliver the blood directly to capillaries in organs. The blood in these capillaries exchanges gases, nutrients and wastes with tissues in their immediate vicinity. The blood returns to the heart via a system of veins.

into smaller arteries and then into capillaries, whose walls are only one cell thick, across which material exchange between blood and other tissues must take place (Figure 7.6.2). Capillaries have such narrow lumens that many blood cells must be bent in order to get through. Thus, capillaries have the large surface-to-volume ratio necessary for their material exchange function. Eventually, capillaries join together in groups to form small veins that combine into a few large veins and return blood to the heart.

A closed circulatory system is very efficient because it delivers blood right to the doorstep of metabolizing tissues; no cell is very far from a capillary. An important problem is built into closed circulatory systems, however: The frictional resistance to blood flow in the capillaries is much greater than that in arteries. This is true in spite of the fact that the total cross-sectional area of the artery leading from the heart (the aorta) is much less than the total cross-sectional area of all the body's capillaries.

The reason for this apparent contradiction is well-known: Blood is viscous and thus adheres to vessel walls as it passes. Only in the center of the vessel lumen is this friction reduced. The difficulty is that capillaries have a very small lumen and a lot of wall area (to increase their material-exchange properties). Therefore, the frictional resistance of capillaries to blood flow is very high. A general rule is that the friction encountered by a viscous material passing through a tube is inversely proportional to the fourth power of the radius of the tube (see Section 6-2 and the reference by Vogel [9]). Thus, if the radius is halved, the frictional force goes up by 16 times. The end result is that a closed circulatory system requires a very powerful heart to force blood through the many tiny capillaries, in spite of the large total cross-sectional area of the latter.

The cellular fraction of blood serves a variety of functions.

Whole blood has both a liquid fraction, called *plasma*, and a cellular fraction. Plasma is mostly water, but it also contains many inorganic ions, hormones, bio-chemicals like sugars, fats, amino acids and proteins. The levels of all these plasma solutes are critical and thus are maintained at relatively constant levels.

     Blood cells are usually classified on the basis of their appearance. All of them originate from common precursor cells in the bone marrow and become specialized later.

a.   Red blood cells, or *erythrocytes*, will be of special importance to us in our later discussion of lead poisoning. Mammalian erythrocytes lose their nuclei during formation; the remaining cytoplasm contains large quantities of a bio-chemical called *hemoglobin*, which has a high affinity for O_2. We introduced the biological role of hemoglobin in Section 6.3; we now elaborate on it.

Erythrocytes pick up O_2 at the lungs:

Reaction #1:[8]
$$Hb + O_2 \longrightarrow O_2\text{-}Hb$$
(hemoglobin)   (oxyhemoglobin)

The erythrocytes are then carried to the sites of metabolism, where the oxygen is needed, and reaction #1 is reversed to free the oxygen for use in metabolic processes:

Reaction #2:
$$Hb + O_2 \longleftarrow O_2\text{-}Hb$$

Interestingly, the affinity of hemoglobin for CO_2 is fairly low, therefore most CO_2 is carried from the sites of respiration back to the lungs in the form of carbonic acid or the bicarbonate ion, dissolved in the water of the plasma:

Reaction #3: $CO_2 + H_2O \longrightarrow \quad H_2CO_3 \quad \longrightarrow H^+ + HCO_3^-$
(carbonic acid)   (bicarbonate ion)

In the lungs the CO_2 is reconstituted from the bicarbonate ion and carbonic acid, and then exhaled:

Reaction #4: $CO_2 + H_2O \longleftarrow H_2CO_3 \longleftarrow H^+ + HCO_3^-$

---

[8]Recall from Chapter 6 that one hemoglobin molecule can bind up to four O_2 molecules.

Reactions #1 through #4 are related: Reaction #4 takes place at the lungs and raises the blood pH by removing carbonic acid. The rate of reaction #1 is pH-dependent; conveniently, it goes faster at higher pH. Thus the removal of CO_2 at the lungs promotes the attachment of hemoglobin to O_2. The opposite occurs in metabolizing tissues: The production of CO_2 lowers the blood pH there via reaction #3, and the lower pH promotes reaction #2, the release of oxygen from oxyhemoglobin.[9]

While the affinity of hemoglobin for CO_2 is low, its affinity for carbon monoxide (CO) is very high. As a result, the presence of even small amounts of CO can prevent the attachment of O_2 to hemoglobin, accounting for the lethal effect of CO.

The best-studied toxic effect of lead is its role in causing *anemia*, a reduction in erythrocyte concentration. This in turn reduces the oxygen-carrying ability of the blood. The anemia is evidently the result of two processes: First, lead interferes with hemoglobin synthesis and, second, lead causes the lifetimes of mature erythrocytes to be reduced from the usual four months.

b. *Platelets* are blood cells involved in clotting; they are actually cell fragments that lack nuclei. Platelets collect at the site of an injury and disintegrate. This releases platelet proteins that generate a cascade of reactions, finally resulting in the formation of a clot consisting of a plasma protein called *fibrin*.

c. *Leukocytes* are nucleated and are often called white blood cells; they are used to fight off infections. One class of leukocytes, the lymphocytes, will be discussed in Chapter 10 in the context of HIV infections. A second group, the granulocytes, functions in certain general responses to infections and allergens.

There are four overlapping functional paths in our circulation: systemic, pulmonary, lymphatic and fetal.

Blood cells and plasma can take several routes around the human body, the differences between the routes being both anatomical and functional.

Look at Figure 7.6.3. The mammalian heart has four chambers: the two upper ones are atria and the two lower ones are the muscular *ventricles*. Blood, rich in CO_2 from metabolizing tissues, enters the heart at the right atrium and goes to the right ventricle. The powerful ventricle pushes the blood to capillaries in the lungs; these capillaries surround the many small air sacs (*alveoli*), which are filled with air when we inhale. Here CO_2 is moved from the plasma into the alveoli for exhalation, and the erythrocytes pick up O_2 from the alveoli; the chemistry for these two processes was outlined in Reactions #1 and #4 in the last section. The blood then returns to the heart at the left atrium. The path of the

---

[9] At least *some* CO_2 does bind to hemoglobin. It has the useful effect that it decreases the oxygen affinity of the hemoglobin (in the vicinity of metabolizing tissues).

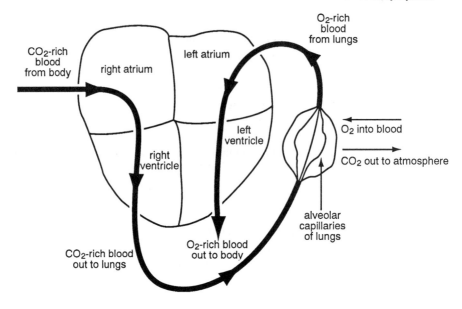

Figure 7.6.3 The flow of blood through a mammalian heart and lungs. Note that the blood enters and leaves the heart twice.

blood just described—from the heart to the lungs and back to the heart—is called *pulmonary circulation.*

> Pulmonary circulation in outline form:
>   right ventricle $\longrightarrow$ pulmonary artery $\longrightarrow$ lung capillaries $\longrightarrow$
> $\longrightarrow$ pulmonary vein $\longrightarrow$ left atrium

    After moving from the left atrium to the left ventricle, blood exits the heart and goes to respiring tissues all over the body via arteries and then capillaries. At the capillaries, O_2 is given up to the respiring cells and CO_2 is taken from them into the plasma.[10] These reactions were described earlier as reactions #2 and #3. The blood now returns to the heart by way of veins. The path of blood from the heart to respiring cells and back to the heart is called *systemic circulation.*

> Systemic circulation in outline form:
>   left ventricle $\longrightarrow$ aorta $\longrightarrow$ other arteries $\longrightarrow$
> $\longrightarrow$ capillaries of respiring tissues $\longrightarrow$ veins $\longrightarrow$ right atrium

    Note the importance of the powerful ventricles—they must overcome the high frictional resistance that blood meets in the narrow lumens of the capillaries.

---

[10]The phrase "respiring cells" here means cells that are breaking down sugar to get energy, and which are therefore giving off carbon dioxide.

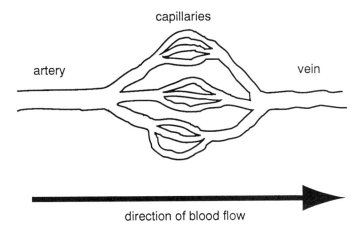

Figure 7.6.4 A capillary bed. As blood moves from arteries to capillaries, the overall cross-section of the circulatory system increases, but overall frictional resistance increases also.

Figure 7.6.4 shows a connected group, or *bed*, of capillaries. The lumen of the artery narrows down as the blood approaches the capillaries and, as a result, the frictional resistance increases dramatically. For much of the blood, the effect is almost like hitting a dead end. Consequently, the hydrostatic pressure in the blood at the front (upstream) end of the capillary bed rises sharply. The high hydrostatic pressure pushes some of the liquid fraction of the plasma through gaps in the vessel walls (Figure 7.6.5). The only part of the plasma that cannot be pushed out is the plasma protein fraction, because these molecules are too big. Thus, this *interstitial fluid* forced out of capillaries lacks the large plasma proteins.

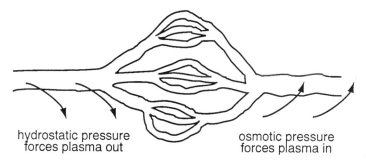

Figure 7.6.5 The increase in frictional resistance to blood flow imposed by capillaries generates a high hydrostatic pressure upstream from the capillaries. This hydrostatic pressure forces some of the liquid fraction of the blood out of the upstream end of the capillaries and into the surrounding tissues. On the downstream end of the capillaries the blood, now with a high concentration of dissolved substances, draws some of the liquid from the surrounding tissues back into the circulatory system by osmotic pressure.

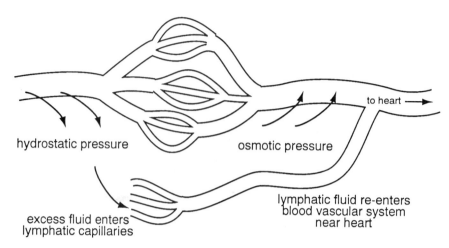

hydrostatic pressure

osmotic pressure

to heart →

excess fluid enters
lymphatic capillaries

lymphatic fluid re-enters
blood vascular system
near heart

Figure 7.6.6 Some of the liquid fraction of the blood, forced out of capillaries by hydrostatic pressure, does not return to the blood at the downstream end of the capillaries. This excess is picked up by lymphatic capillaries and eventually returns to the bloodstream near the heart.

At the far (downstream) end of the capillary bed the capillaries join together, friction decreases, and the hydrostatic pressure also decreases. The blood plasma at this point contains everything that the interstitial fluid contains, plus plasma proteins. Thus, the plasma has more dissolved solute than does the interstitial fluid. As a result, most of the interstitial fluid is osmotically drawn back into the vessels to dilute the plasma proteins. Not all the interstitial fluid makes it back to the blood, however; there is a positive pressure differential of several millimeters of mercury between the hydrostatic pressure on the upstream end of the capillary bed and the osmotic pressure on the downstream end. This would cause a build-up of fluid in the tissues if it were not for the fact that there is another path of circulation to collect the excess fluid.

The extra fluid is collected in a set of capillaries of the *lymphatic circulation* and brought by lymphatic veins to the upper body, where they empty into blood veins near the heart (see Figure 7.6.6). During the journey to the heart, the flow of the *lymphatic fluid*, as the interstitial fluid is called at this point, is pushed by contractions of the nearby skeletal (voluntary) muscles. This movement should be bi-directional, but lymphatic veins have valves that allow flow only toward the

Lymphatic circulation in outline form:
   heart ⟶ arteries ⟶ systemic capillaries ⟶
   ⟶ intercellular spaces ⟶ lymphatic capillaries ⟶
   ⟶ lymphatic veins ⟶ blood veins ⟶ heart

heart. Along the way the lymphatic veins pass through *lymph nodes*, packets of lymphatic tissue that filter out pathogens.

In addition to the interstitial fluid, some white blood cells can also move between the blood circulation and the lymphatic circulation. They do so by squeezing between the endothelial cells that constitute the walls of blood capillaries.

The fourth path of the circulatory system is found in fetuses and is called *fetal circulation*. A fetus does not breathe, so its pulmonary circulation is minimized via two shunts: There is an opening in the fetal heart (the *foramen ovale*), between the right and left atria, that directs blood away from the pulmonary circulation (Figure 7.6.7). Secondly, a special vessel, called the *ductus arteriosus*, carries blood from the pulmonary artery directly to the aorta. Finally, the umbilical artery

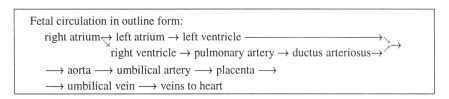

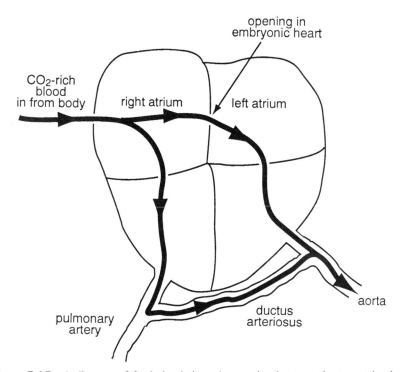

Figure 7.6.7 A diagram of fetal circulation. An opening between the two atria, the *foramen ovale,* and a vessel called the *ductus arteriosus*, shunt fetal blood away from the lungs.

and vein carry the fetus' blood to and from the placenta, respectively. All the above fetal structures close off permanently in the first few minutes after birth.

Section 7.7

Bones

Lead has a strong tendency to localize, or sequester, in bones; its half-life there is about 20 years. Thus, the skeleton can serve as a chronic systemic source of lead long after the original exogenous lead exposure.

Bones are not static. Besides producing blood cells, they are being "remodeled" in response to our physical activities throughout our lives. In this section we will examine the anatomy, function and growth of bones.

Bones support, protect and move.

The set of our bones is called our *skeleton*. It supports the rest of our body structure. All large terrestrial animals require an internal skeleton because the non-skeletal tissue would collapse under its own weight and a hard external skeleton (like an insect's) would weigh too much. Our skeleton surrounds our internal organs, protecting them from mechanical injury. In the cases where organs are not protected by bone, e.g., eyeballs and testicles, reflexive reactions and very low pain thresholds are necessary for protection. With the aid of our voluntary muscles we can use our skeleton to project effects at a distance—walking, reaching and hugging, for instance.

Bone marrow is the source of blood cells.

In the earlier section on the circulatory system a number of blood cells were described. All of these cells originate from a single variety of cell in the core, or *marrow*, of bones. These cells are called *stem cells*, and they divide rapidly, generating large numbers of daughter cells. The subsequent fate of a daughter cell depends on the conditions of its maturation environment. Some lymphocytes, for instance, mature in the thymus gland, just behind the breast-bone of children. Red blood cells mature in the marrow, synthesize hemoglobin and then (in mammals) lose their nuclei. We will have more to say about blood cell origins in Chapter 10.

Bony tissue is replaced throughout life.

There are special cells, called *osteoblasts*, in bone that secrete a protein about which the compound *hydroxyapatite* (mainly calcium phosphate) crystallizes; thus, hard bone is formed. When the muscles to which a bone is attached become stronger, the stresses cause the bone to thicken to adapt to the new need. This is initiated by another group of cells, called *osteoclasts*, which secrete chemicals to dissolve hard bone tissue. Osteoblasts then reconstitute the bone in a new, thicker form.[11] It has been estimated that our bones are replaced up to ten times in our lives.

Virtually all bone growth after adolescence affects the thickness of a bone. Bone growth before that time may be in thickness, but may also be in the length of the bone, accounting for the dramatic rate of change in a child's height.

Lead tends to be deposited in the region of bones near their ends, as revealed by X-ray pictures. The lead is then very slowly released over a period of many years. This long half-life means that lead tends to accumulate in bones. Indeed, about 95% of a person's total-body lead can be found in the bones [2].

Section 7.8

The Kidneys

Our kidneys provide a mechanism for ridding the body of water-soluble substances. They are very selective, however: They can maintain a constant chemical composition in our bodies by removing materials in excess and retaining materials in short supply. Thus, we should correctly expect that the kidneys would help to excrete lead. The problem is that much lead becomes sequestered in bone and is therefore not solubilized in blood where the kidneys can get at it. Nevertheless, if absorbed lead is to be removed from the body, kidney excretion will be the major route out.

The kidneys remove nitrogenous wastes and help regulate the concentration of materials in the blood.

When we eat protein, e.g., the muscle from a cow's leg, the process of digestion breaks the protein down into its component compounds, called *amino acids.* In our bodies amino acids have two possible fates. First, they can be incorporated into our own proteins. We can synthesize most, but not all, of the amino acids we need from related precursors we get in our diet. Second, ingested amino acids can be broken down to extract some of their energy. In this section we will be concerned with one aspect of the latter of these two fates—the removal of a certain chemical group ($-NH_2$) from an amino acid prior to extraction of the amino acid's energy.

Unless it is to be used in the synthesis of other nitrogenous compounds in our bodies, the amino group ($-NH_2$) of ingested amino acids must be removed from our body as waste. The problem is that the amino group quickly forms *ammonia* (NH_3), which is toxic. Aquatic animals can get rid of ammonia by releasing it directly into the surrounding water, in which the ammonia is highly soluble. Many terrestrial animals, humans included, convert the ammonia to urea (H_2NCOH_2),

---

[11]It should now be evident how archeologists can tell so much about an animal by examining a few bones. The pattern of bumps and thickenings on a bone constitute a graphic history of the animal's life.

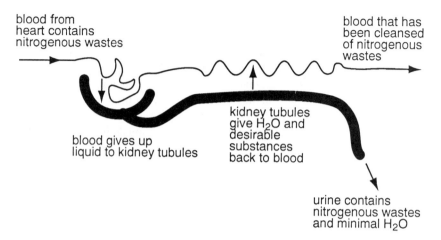

blood from
heart contains
nitrogenous wastes

blood that has
been cleansed
of nitrogenous
wastes

kidney tubules
give H$_2$O and
desirable
substances
back to blood

blood gives up
liquid to kidney tubules

urine contains
nitrogenous wastes
and minimal H$_2$O

Figure 7.8.1 A simplified model of the mammalian kidney. Blood pressure at the kidneys forces some of the liquid fraction of the blood into kidney tubules. This liquid fraction contains wastes, but also contains useful solutes, like sugars. Later, the kidney takes back the useful substances and most of the water, leaving urine with the concentrated wastes (to be voided).

which is moderately water-soluble and much less toxic than ammonia.[12] The urea is then removed from our bodies by our kidneys in urine.

At the kidneys, blood pressure forces some of the liquid fraction of blood into small kidney tubules (see Figure 7.8.1). This liquid contains water, ions, many essential biological molecules and, of course, urea. As the liquid moves along these convoluted tubes, most of the water and most of the desirable dissolved substances are resorbed back into the blood. The aqueous liquid left behind in the kidneys is *urine*, which contains a high concentration of urea. Urine also contains other dissolved substances that the blood has in excess of normal needs. Thus, the kidneys serve the *homeostatic function* of maintaining the concentration of dissolved substances in the blood and other tissues at normal levels.[13]

Lead is absorbed into bones about 100 times faster than it is released. Thus an important therapeutic approach to lead poisoning is to try to prevent the lead from becoming sequestered in bone, from which it would be slowly released back into the blood over a period of decades. The trick is to make the lead very soluble shortly after exposure so the kidneys can excrete it efficiently. There is a class of chemical compounds, called *metal chelates*, that react rapidly with lead and many other metals to form very soluble compounds that the kidneys can excrete. *Penicillamine*, for example, can be administered orally and the com-

---

[12] Egg-laying animals go one step further and convert the urea to uric acid, which is water-insoluble. Thus, uric acid can accumulate inside an avian or reptilian egg and not harm the fetus prior to hatching.

[13] Patients with *diabetes mellitus* produce insufficient quantities of the protein *insulin*. This leads to an excess of the sugar glucose in their blood. The kidneys remove some of the excess *glucose* in urine—thus providing a means of medical diagnosis.

$$
\begin{array}{c}
\overset{\displaystyle OH}{\underset{|}{}} \quad \overset{\displaystyle H}{\underset{|}{}} \\
O = \overset{C}{}\diagdown \quad \diagup \overset{N}{}\diagdown \\
\quad \overset{|}{C}\diagup \qquad Pb \\
H\diagup \quad \diagdown \qquad \diagup \\
\qquad \overset{|}{C} - S \\
H_3C \diagup \quad \underset{|}{} \\
\qquad CH_3
\end{array}
$$

pound it forms with lead, penicillamine-lead, is highly soluble in water. The structure of penicillamine-lead is shown above.[14]

In summary, kidneys first remove most dissolved materials, both useful and waste, from the blood and later put the useful materials back into the blood. We could easily imagine a simpler system, in which only a small amount of water, all the urea and any other material in excess is removed directly, without benefit of resorption, but that is not how our kidneys work. The forces of evolution do not necessarily yield the simplest system but, instead, yield a modification of a pre-existing system. This frequently generates a more complicated system, but if the new system provides a selective advantage it can become fixed in the population.

Lead has been shown to have a direct, pathological effect on kidney function. Acute exposure in children leads to malfunctioning of the resorption process, re-sulting in a high concentration of glucose and other desirable compounds in the urine. Chronic lead exposure eventually results in general kidney failure.

Section 7.9

Clinical Effects of Lead

Lead poisoning is indicated by a variety of clinical symptoms, including gastroin-testinal and mental disorders.

Lead poisoning causes a wide variety of general symptoms.

We conclude with a short description of symptoms of lead poisoning that a physi-cian might see in a patient. In adults there are gastrointestinal disorders like vomiting and pain. In children there are central nervous system disorders, e.g., drowsiness, irritability and speech disturbances, as well as gastrointestinal symp-toms. An interesting symptom in some cases is a blue line along the gums, formed when lead reacts with sulfur, the accumulation of the latter being associated with poor dental hygiene.

---

[14]Chelating agents are not without risks: A chelating agent that picks up lead may also pick up other divalent metals, e.g. calcium. Loss of blood calcium can lead to uncontrollable muscle tremors and even death.

The effect of lead on IQ is an interesting one. As mentioned above, lead-induced neurophysiological disorders are especially noted in children. This is not unexpected because lead seems to affect the velocity of nerve impulse conduction. Evidence suggests that lead poisoning can reduce a child's IQ by about five points.

Section 7.10

A Mathematical Model for Lead in Mammals

While lead interacts differently with the various tissues of the body, as a first approximation we need only distinguish three tissue types: bone, blood, and the other soft tissue of the body. Bone tends to take up lead slowly but retain it for very long periods of time in contrast to soft tissue, other than blood, in which the turnover of lead is much quicker. Blood is the transport agent of the metal. The disposition of lead in the body can be followed as a three compartment system by tracking its movement into and out of these three tissue types. In this section we analyze such a model proposed by Rabinowitz, Wetherill, and Kopple.

The uptake and movement of lead can be modeled by the use of compartments.

The activity of lead in the body depends on the tissue in which it is located (recall the end of Section 7.1). To construct a mathematical model for the movement of lead, at least three distinct tissue types must be recognized: bone, blood, and soft tissue (other than blood).[15] These will be our mathematical compartments. Lead enters the system by ingestion and through the skin and lungs. These intake paths usher the substance to the blood. From the blood, lead is taken up by bone and by soft tissue. This uptake is reversible; lead is also released by these organic reservoirs back to the blood. However, especially for bone, lead's half-life in the tissue is very long. Lead can be shed from the body via the kidneys from blood and, to a lesser extent, through hair. Thus blood is the main conduit through which lead moves among our compartments.

To begin the model, let Compartment 1 be the totality of the victim's blood, Compartment 2 the soft tissue, and Compartment 3 the skeletal system. We must also treat the environment as another compartment to account for lead intake and elimination; we designate it as Compartment 0. Let x_i, $i = 1, \ldots, 3$, denote the amount of lead in compartment i and let a_{ij}, $i = 0, \ldots, 3$, $j = 1, \ldots, 3$, denote the rate of movement of lead *to* Compartment i *from* Compartment j. The product $a_{ij}x_j$ is the rate at which the amount of lead increases in Compartment i due to lead in Compartment j. There is no reason that a_{ij} should equal a_{ji} and, as noted above, the rate of movement from blood to bone is very different than the reverse rate. The units of the a's are per day.

Because we will not keep track of the amount of lead in the environment, this is an *open compartment* model. Instead, we account for environmental intake

---

[15] As in the discussion ending Section 7.1, throughout this section, by soft tissue, we mean soft tissue other than blood.

by including a separate term, $I_L(t)$, applied to Compartment 1, blood. From the discussion above, some of the rates are zero; namely $a_{03} = a_{23} = a_{32} = 0$, signifying no direct elimination to the environment from bone and no direct exchange between bone and soft tissue. In addition, all rates $a_{i0} = 0$ since there is no x_0 term. Finally, there is no need for terms of the form a_{ii} since a compartment is our finest unit of resolution.

With these preparations, we may now present the model, which derives from the simple fact that the rate of change of lead in a compartment is equal to the difference between the rate of lead entering and the rate leaving.

$$\frac{dx_1}{dt} = -(a_{01} + a_{21} + a_{31})x_1 + a_{12}x_2 + a_{13}x_3 + I_L(t)$$

$$\frac{dx_2}{dt} = a_{21}x_1 - (a_{02} + a_{12})x_2 \qquad (7.10.1)$$

$$\frac{dx_3}{dt} = a_{31}x_1 - a_{13}x_3$$

In words, the first equation, for example, says that lead leaves blood for the environment, soft tissue and bone at a rate in proportion to the amount in the blood; lead enters blood from soft tissue and bone in proportion to their respective amounts; and lead enters blood from the environment according to $I_L(t)$. The algebraic sum of these effects is the rate of change of lead in blood. In line with our discussion of Section 2.4, this system can be written in matrix form as

$$\mathbf{X}' = A\mathbf{X} + \mathbf{f}. \qquad (7.10.2)$$

Here $\mathbf{X}$ is the vector of x's, $\mathbf{f}$ is the vector

$$\mathbf{f} = \begin{bmatrix} I_L(t) \\ 0 \\ 0 \end{bmatrix}$$

and A is the 3×3 matrix

$$A = \begin{bmatrix} -(a_{01} + a_{21} + a_{31}) & a_{12} & a_{13} \\ a_{21} & -(a_{02} + a_{12}) & 0 \\ a_{31} & 0 & -a_{13} \end{bmatrix}.$$

From equation (2.4.12), the solution is

$$\mathbf{X} = e^{At}\mathbf{X}_0 + e^{At} \int_0^t e^{-As}\mathbf{f}(s)\,ds. \qquad (7.10.3)$$

We will now suppose that the intake of lead, $I_L(t)$, is constant; this is a reasonable assumption if the environmental load remains constant. Then $\mathbf{f}$ is also constant and we may carry out the integration on the right-hand side of equation (7.10.3). In keeping with the result that $-a^{-1}e^{-at}$ is the integral of the ordinary exponential function e^{-at}, we get

$$e^{At} \int_0^t e^{-As} \mathbf{f}(s) \, ds = e^{At} \left[-A^{-1} e^{-As} \right] \Big|_0^t \mathbf{f}$$
$$= -e^{At} \left[e^{-At} - I \right] A^{-1} \mathbf{f}$$
$$= - \left[I - e^{At} \right] A^{-1} \mathbf{f}.$$

Substitution of this result into equation (7.10.3) gives

$$\mathbf{X} = e^{At} \mathbf{X}_0 - \left[I - e^{At} \right] A^{-1} \mathbf{f}$$
$$= e^{At} \left[\mathbf{X}_0 + A^{-1} \mathbf{f} \right] - A^{-1} \mathbf{f}. \qquad (7.10.4)$$

This is the solution of system (7.10.1) provided A^{-1} exists and that the exponential e^{At} is computable.[16]

The long-term predictions of the model

Recall from the discussion of Section 2.7 that the long-term behavior of the solution is predicted by knowledge of the eigenvalues of the matrix A. But it is easily seen that this is a compartment matrix (cf. Section 2.7). The diagonal terms are all negative, the first column sum is $-a_{01}$, the second column sum is $-a_{02}$ and the third column sum is 0. Therefore by the Gershgorin Circle Theorem, the eigenvalues of A have negative, or zero, real parts. In the case that they are all strictly negative, then the exponential e^{At} tends to the zero matrix as $t \to \infty$. As a result, the long-term fate of lead in the body is given by the term $A^{-1} \mathbf{f}$,

$$\mathbf{X} \to -A^{-1} \mathbf{f} \qquad \text{as } t \to \infty. \qquad (7.10.5)$$

A study on human subjects

Rabinowitz, Wetherill, and Kopple studied the lead intake and excretion of a healthy volunteer living in an area of heavy smog. Their work is reported in Reference [3], and extended by Batschelet, Brand, and Steiner in Reference [11]. (See also Reference [12].) The data from this study were used to estimate the rate constants for the compartment model equation (7.10.1). Lead is measured in micrograms and time in days. For example, the rate 49.3 is given below as the

---

[16] See *Nineteen Dubious Ways to Compute the Exponential of a Matrix* by Reference [10] for a delightful discussion of the problems involved.

Table 7.10.1   Lead Exchange Rates

Coefficients	a_{01}	a_{12}	a_{13}	a_{21}	a_{02}	a_{31}	I_L
Value	.0211	.0124	.000035	.0111	.0162	.0039	49.3

ingestion rate of lead in micrograms per day and the other coefficients are as given in Table 7.10.1.

This model has significant biological implications. The output of the computation of the exponential of this matrix does not seem to merit printing. More important for the purposes of understanding this model is the graph of the solutions. These graphs are shown in Figure 7.10.15. This figure shows graphs of the total lead in Compartments 1, 2, and 3 over a period of 365 days. The horizontal axis is days and the vertical axis is in units of micrograms of lead.

Our calculation of the solution for equation (7.10.1) follows the procedure of equation (7.10.4) exactly. As we will see, the eigenvalues for this matrix are negative. Further, since the trend of the solution is independent of the starting condition—recall equation (7.10.5)—we take the initial value to be

$$X_0 = \begin{pmatrix} 0 \\ 0 \\ 0 \end{pmatrix}.$$

The solution is computed and graphed with

```
> with(linalg):
> A:=matrix(3,3,[-0.0361, 0.0124, 0.000035, 0.0111, -0.0286, 0.0, 0.0039,
       0.0, -0.000035]);
> etA:=exponential(A,t);
> AinvF:= evalm(inverse(A) * vector([493/10,0,0]));
> u:= evalm(-AinvF + etA * AinvF);
> plot({u[1](t),u[2](t),u[3](t)},t = 0..365);
```

One observation is that the level of lead in blood and soft tissue approaches a steady state quickly. Lead achieves a steady state in blood of about 1800 units and of about 700 in soft tissue. The level of lead in the bones continues to rise after a period of one year. In this model, the bones continue to absorb lead because of the constant rate of input. On the other hand, the bones release lead slowly and steadily. As we have already seen in the discussion in Section 7.8, a high level of lead in the bones has implications for severe and chronic biological problems.

The levels of lead in the steady state is the next subject for discussion. For this lead model, the long-term behavior of solutions can be computed as follows:

```
> lambda:=eigenvals(A):
```

This computation yields

$$-0.0447, \qquad -0.02, \quad \text{and} \quad -0.00003.$$

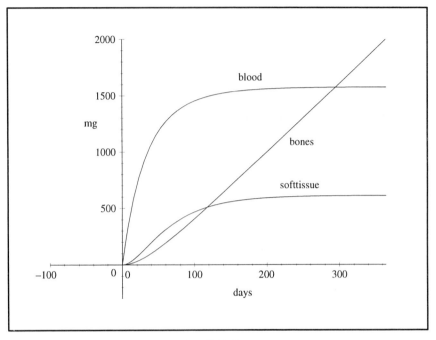

Figure 7.10.15   Solutions for equation (7.10.1)

We find that the eigenvalues for the matrix A associated with the lead problem are all negative. Further, a computation of $-A^{-1}\mathbf{f}$

```
> v3:=vector([493/10,0,0]);
> leadlim:=evalm(-inverse(A)*v3);
```

yields

$$-A^{-1}\mathbf{f} = (1800, 698, 200582) \quad \text{where } \mathbf{f} = (439/10, 0, 0).$$

Hence, this model predicts the long-range forecast summarized as follows: The level of lead in the blood will rise to about 1800 micrograms, the level of lead in other soft tissue will rise to about 698 micrograms, and the level of lead in the bones will rise to about 200582 micrograms.

It should be recognized that the coefficients for our absorption of lead are not constants. By way of data for long-range forecasts, Ewers and Schlipkoter point out that after age 20, the lead content of most soft tissue does not show age-related changes [1]. The lead content of bones and teeth increases throughout life, since lead becomes localized and accumulates in human calcified tissues. This accumulation begins with fetal development and continues to approximately the age of 60 years. Various studies show that approximately 95% of the total body lead is lodged in the bones of human adults.

Exercises

1. This exercise is an investigation of the lead model. The exercise is broken
    into a set of questions that can be answered by modifications of the syntax in
    this chapter.
    a. What is the long-range forecast for lead in each of the compartments
        using model (7.10.1)?
    b. Approximately what is the lead level achieved in each of the compart-
        ments in "one year"?
    c. Ewers and Schlipkoter state that 95% of the total body lead is lodged in
        the bones of human adults [1]. Schroeder and Tipton state that "Bones
        contain 91% of the total body lead" [4]. What percentage of the total
        body lead does this model place in the bones?
    d. Re-do Questions (a) and (b) by doubling or halving the ingestion rate.
        What is the revised long-range forecast for each of the compartments?
        Approximately what is the revised lead level achieved in each of the
        compartments in "1 year"?
2. The remaining questions are concerned with a person who has lived in a lead-
    contaminated environment so long that a level of 2500 micrograms of lead
    has accumulated in the bones. You may continue to assume that in the con-
    taminated environment 49.3 micrograms per day are absorbed. In their 1968
    paper, Schroeder and Tipton stated that an average of 17 micrograms of lead
    are retained per day. We take this absorption rate to be a "clean" environment.
    a. What level of lead do you expect in the tissue and blood for a person
        living in a contaminated environment long enough that 2500 micrograms
        of lead has accumulated in the bones?
    b. Suppose that the person described is moved to a relatively lead-free en-
        vironment. What is the approximate level of lead in the bones, tissue,
        and blood at the end of one year after living in the new environment?
    c. We have seen that there are drugs that alleviate the effects of lead in the
        bones by increasing the rate of removal from the bones. What should
        that rate be to cut in half the amount of lead in the bones at the end of
        one year in the cleaner environment?
    d. Suppose that the person takes the drug you have designed, but is not
        moved to the cleaner environment. What are the levels of lead in the
        bones, tissue, and blood after one year of taking the drug while living in
        the contaminated environment?
    e. Ewers and Shlipkoter give the half-life of lead in blood as 19 days, in
        soft tissue as 21 days, and in bones as 10 to 20 years [1]. What is the
        half-life as assumed in our model?
    f. According to this model, what percentage of the lead ingested into the
        body is returned to the environment during the 100th day in the initial
        situation?

Section 7.11

Pharmacokinetics

The routes for dispersion of drugs through the body follow the same pattern as those of lead. The previous section followed lead through the body. The model of this section examines how the body handles the ingestion of a decongestant. We keep track of this drug in two compartments of the body: the gastrointestinal tract and the circulatory system. The mathematical importance of this model is that the limit for the system is a periodic function.

A two-compartment pharmacokinetic model is used to construct a drug utilization scenario.

Among all the means for the delivery of therapeutic drugs to the blood stream, oral ingestion/gastro-intestinal absorption is by far the most popular. In this section we study this delivery mechanism following closely the work of Spitznagel [13]. The working hypothesis of the study is the following series of events. The drug is taken orally on a periodic basis resulting in a *pulse* of dosage delivered to the GI tract. From there, the drug moves into the blood stream, without delay, at a rate proportional to its concentration in the GI tract and independent of its concentration in the blood. Finally the drug is metabolized and cleared from the blood at a rate proportional to its concentration there.

Evidently the model should have two compartments; let $x(t)$ denote the concentration of drug in the GI tract and $y(t)$ denote its concentration in the blood. In addition, we need the drug intake regimen; denote by $D(t)$ the drug dosage, as seen by the GI tract, as a function of time t.[17] The governing equations are

$$\frac{dx}{dt} = -ax + D,$$
$$\frac{dy}{dt} = ax - by. \tag{7.11.1}$$

Since equations (7.11.1) constitute a linear system with forcing function $D(t)$, its solution, in matrix form, is given by equation (2.4.12):

$$\mathbf{Y} = e^{Mt}\mathbf{Y}_0 + e^{Mt}\int_0^t e^{-Ms}\mathbf{P}(s)ds \tag{7.11.2}$$

where $\mathbf{Y}$ and $\mathbf{P}$ are the vectors

$$\mathbf{Y} = \begin{bmatrix} x \\ y \end{bmatrix} \quad \text{and} \quad \mathbf{P} = \begin{bmatrix} D(s) \\ 0 \end{bmatrix},$$

---

[17]With the use of time-release capsules, a drug may not be immediately available to the GI tract even though the medication has been ingested.

and M is the coefficient matrix

$$\begin{bmatrix} -a & 0 \\ a & -b \end{bmatrix}.$$

Note that, as a compartment model, the diagonal terms of this matrix are negative and the column sums are negative or zero. Consequently, the first term of the solution, $e^{Mt}Y_0$, is transient, that is, it tends to 0 with time. Therefore, asymptotically the solution tends to the second term

$$Y \to e^{Mt} \int_0^t e^{-Ms} \begin{bmatrix} D(s) \\ 0 \end{bmatrix} ds \qquad (7.11.3)$$

which is periodically driven by D.

Periodic solutions predict serum concentration cycles.

In conjunction with specific absorption and metabolism rates for a given drug, system (7.11.1) and its solution, equation (7.11.2), may be used to predict cycles of drug concentration in the blood. Fortunately the required data are available for a variety of drugs such as PPA and CPM, as reported and defined in Reference [13]. As mentioned above, the exact shape of the dosage profile, $D(t)$, depends

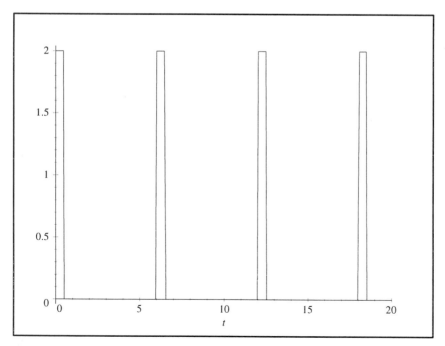

Figure 7.11.1 Graph of drug dosage $D(t)$

```
> a:=ln(2)*2; b:=ln(2)/5;
> Dose:=t->sum((Heaviside(t-n*6)-Heaviside(t-(n*6+1/2)))*2,n=0..10);
> J:=DEplot2([diff(x(t),t)=Dose(t)-a*x,diff(y(t),t)=a*x-b*y],[x,y],
   0..50,{[0,0,0]},stepsize=0.5,scene=[t,x]):
> K:=DEplot2([diff(x(t),t)=Dose(t)-a*x,diff(y(t),t)=a*x-b*y],[x,y],
   0..50,{[0,0,0]},stepsize=0.5,scene=[t,y]):
> plots[display]({J,K});
```

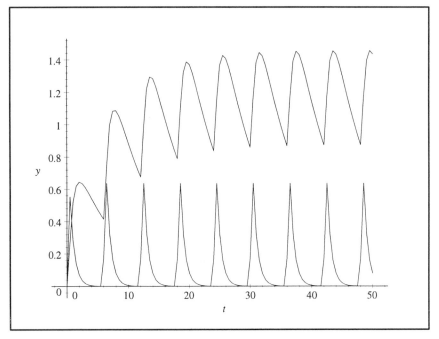

Figure 7.11.2   Loading of the blood stream and the GI tract from a dosage regime

on how the producer, the pharmaceutical company, has buffered the product. We assume the drug is taken every six hours (four times a day) and dissolves within about half an hour providing a unit-pulse dosage with height 2 and pulse width 1/2 on the interval [0,6], see Figure 7.11.1. The rate parameters a and b are typically given as half-lives, cf Section 3.4. For PPA we use a 1/2-hour half-life in the GI tract, so $a = 2\ln(2)$, and a five-hour half-life in the blood, $b = \ln(2)/5$.

For a numerical solution, we take zero initial conditions, $x(0) = y(0) = 0$, that is, initially no drug is present in the GI tract or circulatory system, and use a Runge–Kutta method to obtain Figure 7.11.2. The behavior of $x(t)$ is the lower graph and predicts an oscillating increase and decrease of concentration in the GI tract. On the other hand, the concentration in the circulatory system, $y(t)$, is predicted to be an oscillation superimposed on a gradually increasing level of concentration.

In Figure 7.11.3 we show the phase–plane plot, x vs y, of this solution. It shows that, asymptotically, the solution tends to a periodic (but non-sinusoidal) oscillation; this is called a *limit cycle*.

```
> phaseportrait([diff(x(t),t)=Dose(t)-a*x,diff(y(t),t)=a*x-b*y],[x,y],
  0..50,{[0,0,0]},stepsize=0.5);
```

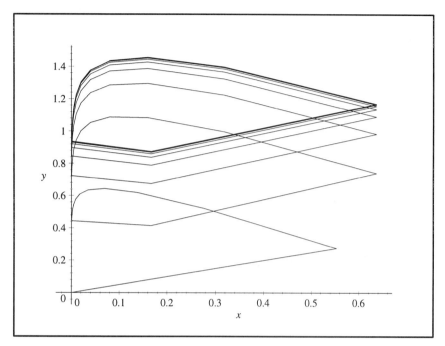

Figure 7.11.3 $\{x(t), y(t)\}$ with limit cycle

From the figure, the high concentration level in the blood is about 1.5 while the low is about 0.9 on the limit cycle. In designing a drug, it is desirable to keep the concentration as uniform as possible and to come up to the limit cycle as quickly as possible. Toward that end, the parameter a can be more easily adjusted, for example by a "time release" mechanism. The parameter b tends to be characteristic of the drug itself.

The asymptotic periodic solution may be found from equation (7.11.3) or by the following manual scheme.

We can solve for $x(t)$ explicitly. There are two parts: one part for $0 < t < 1/2$ where the dosage function has value 2, and the other part for $1/2 < t < 6$ where the dosage function has value 0. We call the first part $x_1(t)$ and find the solution here. In this case, the input from $D(t) = 2$ and the initial value for the periodic solution is yet unknown. Call the initial value x_0. We have

$$x(t) = -ax(t) + 2$$

with $x(0) = x_0$.

```
> dsolve({diff(x(t),t)+a*x(t)=2,x(0)=x0},x(t));
> x1:=unapply(op(2,"),t);
```

This equation has solution

$$x_1(t) = 1/\ln(2) + (x_0 - 1/\ln(2))2^{-2t}.$$

Follow this part of the solution from $t = 0$ to $t = 1/2$. Next we compute the solution $x(t)$ for (7.11.1) with $1/2 < t < 6$. In this case, the input from $D(t) = 0$ and the initial value for the continuation of the periodic solution starts where the first part left off. Thus,

$$x(t) = -ax(t)$$

with

$$x(1/2) = x_1(1/2)$$

```
> dsolve({diff(x(t),t)+a*x(t)=0,x(1/2) = x1(1/2)},x(t));
> x2:=unapply(op(2,"),t);
```

This differential equation has solution

$$x_2(t) = 2^{-2t}(x_0 + 1/\ln(2)).$$

In order for $x(t)$ to be a periodic solution for (7.11.1) with period 6, it should be true that $x_0 = x(0) = x(6)$. We find x_0 by setting $x_1(0) - x_2(6)$ equal to zero and solving for x_0.

```
> x0:=solve(x2(6)-x0=0,x0);
```

The solution is

$$x_0 = 1/(\ln(2)4095).$$

It remains to find the periodic solution y for the second equation. Equation (7.11.2) can be rewritten now that we have a formula for $x(t)$:

$$y(t) + by(t) = ax(t).$$

The function $y_1(t)$ will be the solution for $0 < t < 1/2$ and $y_2(t)$ is the solution for $1/2 < t < 6$.

Now continue the solution for $1/2 < t < 6$ and for $y(1/2) = y_1(1/2)$.

```
> dsolve({diff(y(t),t) = a *x1(t) - b*y(t), y(0) = y0},y(t));
> simplify(rhs(" )); y1:=unapply(" ,t);
```

```
> dsolve({diff(y(t),t)=a*x2(t)-b*y(t),y(1/2)=y1(1/2)},y(t));
> y2:=unapply(op(2, " ),t);
```

The final requirement is that $y_2(6) = y_1(0)$.

```
> solve(y2(6)=y0,y0); y0:=evalf(" );
```

The result is that y_0 is about .8864. To visualize the result, we make two graphs: a plot of $x(t)$ and $y(t)$ superimposed on the same graph and a phase plane plot $\{x(t), y(t)\}$.

The upper graph is the graph of $y(t)$. The two functions have periodic extensions.

```
> plot({[t,x1(t),t=0..1/2],[t,x2(t),t=1/2..6],
   [t,y1(t),t=0..1/2],[t,y2(t),t=1/2..6]},color=BLACK);
```

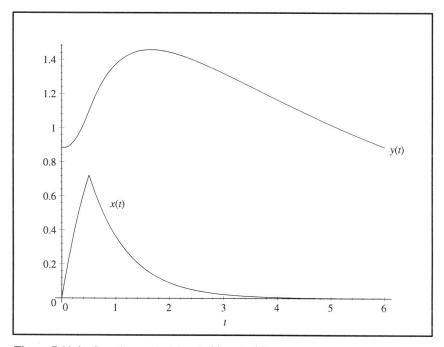

Figure 7.11.4 Superimposed plots of $x(t)$ and $y(t)$

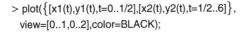

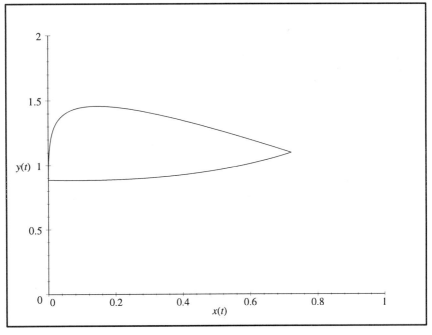

Figure 7.11.5   Parametric plots of $\{x(t), y(t)\}$

This model continues to raise many questions. Because $y(t)$ represents the level of the drug in the circulatory system, we note that the level should be high enough for the drug to be effective, and but not so high as to cause side effects. It is up to the drug companies to adjust the constants a and b so that these levels for the periodic solutions are maintained. The job of the pharmaceutical company is to determine the appropriate level that y should have, and to adjust a and b to maintain that level.

Exercises

1. Part of the interest in [13] was to contrast the behaviors of PPA and CPM. CPM has a half life of one hour in the GI tract and 30 hours in the blood system. This contrast can be made by modifying the code suggested in Section 7.11.
2. With the model of this section, suppose that a and b are as specified for the two drugs. What could be an initial intraveneous injection level, x_0, with subsequent equal dose levels as specified and taken orally, so that the periodic steady state is achieved in the first six hours, and maintained thereafter?

Section 7.12

Questions for Thought and Discussion

1. For what anatomical and molecular reasons do we expect the lead in leaded gasoline to move percutaneously into our bodies?
2. Describe the path of ingested lead from mouth to bone marrow.
3. For what reasons might we expect a person who has ingested lead to become anemic?
4. Discuss why the evolution of small animals into large animals required the evolution of a closed circulatory system and a concomitant coelom.
5. Starting with the number "2," number the following events in the order in which they occur.

blood enters right atrium	1
fluid from blood enters lymphatic system	
blood gives up CO_2 at alveoli	
blood enters systemic capillaries	
blood enters pulmonary artery	
blood enters aorta	

References and Suggested Further Reading

1. **Lead poisoning:** Ulrich Ewers and Hans-Werner Schlipkoter, "Lead," pp. 971–1014, in *Metals and Their Compounds in the Environment*, ed. Ernest Merian, VCH Publishers, Inc. New York, 1991.
2. **Lead poisoning:** *The Chemical Environment*, pp. 28–92, eds. John Lerihan and William W. Fletcher, Academic Press, 1977.
3. **Lead poisoning:** Michael B. Rabinowitz, George W. Wetherill and Joel D. Kopple, *Science* **182**, 725–727, 1973.
4. **Lead poisoning:** Henry A. Schroeder and Isabel H. Tipton, *Arch. Environ. Health* **17**, 956, 1968.
5. **Embryology:** Leland Johnson, *Biology*, William C. Brown, Publishers, Dubuque, Iowa, Second edition, 1987.
6. **Biological organ systems:** William S. Beck, Karel F. Liem and George Gaylord Simpson, *Life – An Introduction To Biology*, 3rd ed., Harper-Collins Publishers, New York, 1991.
7. **Biological organ systems:** Eldra Pearl Solomon, Linda R. Berg, Diana W. Martin and Claude Villee, *Biology*, 3rd ed., Saunders College Publishing, Fort Worth, 1993.
8. **Diffusion distances in cells:** David Dusenbery, *Sensory Ecology*, Chapter 4, W. H. Freeman and Co., San Francisco, 1992.
9. **Resistance to fluid flow:** Steven Vogel, *Life in Moving Fluids: The Physical Biology of Flow*, Princeton University Press, Princeton, NJ, pp. 165–169, 1989.
10. **Exponential of a matrix:** "Nineteen dubious ways to compute the exponential of a matrix," Moler and Van Loan, *SIAM Review* **20**, no. 4, 1978.
11. **Smog:** E. Batschelet, L. Brand, and A. Steiner, *J. Math. Biol.* **8**, 15–23, 1979.
12. **Computer modeling:** Robert L. Borrelli, Courtney S. Coleman, William E. Boyce, *Differential Equations Laboratory Workbook,* John Wiley & Sons, Inc., New York, 1992.
13. **Pharmacokinetics:** Edward Spitznagel, *Two-Compartment Pharmacokinetic Models C-ODE-E*, Fall, 1992 (published by Harvey Mudd College, Claremont, CA).

Chapter 8

Neurophysiology

Introduction to this chapter

This chapter presents a discussion of the means, primarily electrical, by which the parts of an organism communicate with each other. We will see that this communication is not like that of a conducting wire; rather, it involves a self-propagating change in the ionic conductance of the cell membrane.

   The nerve cell, or neuron, has an energy-requiring, steady-state condition in which the interior of the cell is at a negative potential relative to the exterior. Information transfer takes the form of a disruption of this steady-state condition, in which the polarity of a local region of the membrane is transiently reversed. This reversal is self-propagating and is called an action potential. It is an all-or-none phenomenon: Either it occurs in full form, or it doesn't occur at all.

   Neurons are separated by a synaptic cleft, and interneuronal transmission of information is chemically mediated. An action potential in a presynaptic neuron triggers the release of a neurotransmitter chemical that diffuses to the postsynaptic cell. The sum of all the excitatory and inhibitory neurotransmitters that reach a postsynaptic cell in a short period of time determines whether or not a new action potential is generated.

Section 8.1

Communication Between Parts of an Organism

Specialization of structure and function in all organisms necessitates some means of communication among the various parts. Diffusive and convective flows of chemicals provide methods of communication, but they are very slow compared to the speed with which many intraorganismal needs must be conveyed. The high-speed alternative is electrical communication, for which complex nervous systems have evolved.

Communication is necessary at all levels of biological organization.

The wide variety of molecular structures available in living systems is necessitated by the wide variety of physicochemical tasks required. Each kind of molecule, supramolecular structure, organelle, cell, tissue and organ is usually suited to just one or a few tasks. This kind of specialization ensures that each task is performed by the structure best adapted to it, one that Darwinian selection has favored above all others. The down side to such specialization is that the resultant structures are often localized into one region, well-isolated from all others. If important parts are separated, there must be some means of communication between them to allow the entire organism to behave as a single integrated unit.

We find such specialization at all biological levels of organization. For example, there are many microscopic organisms that seem to be single cells, but close examination reveals that they possess special anatomical structures dedicated to quite different functions. Because of these differentiated structures these organisms are often said to be "acellular," which is a simple admission that they do not fit into classical descriptive categories. Examples are found in the organisms called *protozoans*: Many have light-sensitive spots to help them orient, while others have elementary digestive tracts, with an opening to the outside and a tube leading into the body of the "cell." Of special interest are the primitive neural structures of the protozoan *ciliates*. These one-celled (or acellular) organisms move from place to place via the rapid beating of many small hair-like *cilia*. If these cilia were to beat at random there would be as many pushing in one direction as in the other, and the ciliate would not move. Observation of the cilia shows that they beat in synchronized waves, pushing the organism in a particular direction. If a small needle is inserted into the ciliate and then moved around to cause mechanical damage, the cilia will continue to beat—but not synchronously. Evidently there is some kind of primitive neural system to coordinate the movements of the cilia, and the needle damages that system, thus desynchronizing the cilia.

In multicellular animals the need for a quick coordination between the perception of stimuli and consequent responses has led to the evolution of an endocrine system and a specialized nervous system. The endocrine system, facilitated by blood flow, provides chemical communication. The nervous system, the most complicated system in our bodies, provides the high-speed network that allows the other organs to work in harmony with each other.

Communication in multicellular organisms can be chemical or electrical.

Hormones are chemicals secreted by one kind of tissue in a multicellular organism and transported by the circulatory system to *target organs*. At its target organ a hormone exerts powerful chemical effects that change the basic physiology of cells. For example, sex hormones manufactured in the reproductive system cause changes in the skeletal and muscular structures of mammals, preparing them for the physical processes of mate attraction and reproduction. Insulin, a pancreatic hormone, affects the way cells in various tissues metabolize sugars.

The target organs of hormones are specifically prepared for hormone recognition. Protein receptors on the surfaces of cells in those organs can recognize certain hormones and not others. For instance, both men and women produce the hormone *follicle stimulating hormone* (FSH), but it stimulates sperm production in males and egg production in females. The different effects are attributable to the target organs, not to the hormone, which is the same in both sexes. On the other hand, FSH has no effect on the voluntary muscles, which evidently have no FSH receptors. In the unusual phenomenon called *testicular feminization* a person with both a man's chromosomal complement and *testosterone* levels has a woman's body, including external genitalia. The reason is that the person has no testosterone receptors on their target organ cells; thus, the testosterone in the blood is not recognized by the organs that manifest secondary sexual characteristics. In humans the default gender is evidently female, the generation of male characteristics requiring the interaction of testosterone and its appropriate receptors.

A second form of communication in an animal is electrical. We have a highly developed nervous system that provides us with a means of perceiving the outside world and then reacting to it. Indeed, the elaborate nervous system of primates and the complex behaviors it supports are defining characteristics.

Stimuli such as light, salt, pressure and heat generate electrical signals in receptors. These signals are relayed to the *central nervous system*, consisting of the brain and spinal cord, where they are processed and where an appropriate response is formulated. The information for the response is then sent out to muscles or other organs where the response actually takes place. Examples of responses are muscular recoil, glandular secretion and sensations of pleasure. There are special cells along which the electrical signals are passed. They are called *neurons*.

Section 8.2

The Neuron

Neurons, and other cells as well, are electrically polarized, the interior being negative with respect to the exterior. This polarization is due to the differential permeability of the neuron's plasma membrane to various ions, of which potassium is the most important. Active transport by sodium/potassium pumps maintains the interior concentration of sodium low and potassium high.

Neurons are highly specialized cells for conveying electrical information.

Figure 8.2.1 shows a model of a typical mammalian neuron. The cell's long, narrow shape suggests its role: It is specifically adapted to the task of conveying information from one location in the body to another. The direction of information flow is from the dendrites, through the cell body, to the axon and on to other neurons, muscles and glands. Neurons also exhibit unusual electrical behavior, variations of which permit these cells to pass information from dendrite to axon.

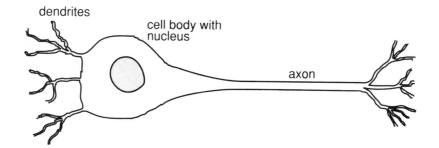

dendrites

cell body with nucleus

axon

Figure 8.2.1   A model of a neuron. The direction of information transfer is from left to right.

a) Membrane completely permeable to K^+ and Q^-.

| $[K^+_L]>0$ | $[K^+_R]>0$ |

| $[Q^-_L]>0$ | $[Q^-_R]>0$ |

Therefore, $[K^+_L] = [K^+_R]$
$[Q^-_L] = [Q^-_R]$

L = left
R = right
K^+ = positive ion
Q^- = negative ion

| = membrane

b) Membrane completely impermeable to K^+ and Q^-.

| $[K^+_L]>0$ | $[K^+_R]=0$ |

| $[Q^-_L]>0$ | $[Q^-_R]=0$ |

Therefore, $[K^+_L] > [K^+_R]$
$[Q^-_L] > [Q^-_R]$
$[K^+_L] = [Q^-_L]$

c) Membrane permeable to K^+ and impermeable to Q^-.

| $[K^+_L]>0$ | $[K^+_R]>0$ |

| $[Q^-_L]>0$ | $[Q^-_R]=0$ |

Therefore, $[K^+_L] > [K^+_R] \neq 0$
$[Q^-_L] > 0$ and $[Q^-_R] = 0$

Figure 8.2.2   This figure shows how a potential difference can be generated across a selectively permeable membrane without the expenditure of energy. See the text for a detailed discussion. Parts (a) and (b) depict extreme situations, in which no potential difference is generated across the membrane: In Part (a) the membrane is equally permeable to both the positive and the negative ions. In Part (b) the membrane is impermeable to both ions. In Part (c), however, the membrane is permeable to the positive ion and impermeable to the negative ion. This generates an equilibrium that is a compromise between electrical and mechanical diffusion properties, and results in an imbalance of charges across the membrane.

These two properties of neurons, shape and electrical behavior, suggest an analogy with the conduction of electric signals by a copper wire. That analogy is
incorrect, however, and we will spend the part of this chapter discussing how that
is so.

A membrane can achieve electrical polarization passively.

Under certain conditions a voltage, or *potential difference*, can be maintained
across a membrane without the expenditure of energy.[1] We will use Figures 8.2.2(a), (b) and (c) to show how this can happen in a model system of
two water-filled compartments that are separated by a membrane. In each case
we begin by dissolving a compound KQ in the water on the left of the membrane, where K is a positive ion, perhaps potassium, and Q is any large organic
group. This compound will immediately dissociate into K^+ and Q^-, such that
$[K^+] = [Q^-]$, the braces indicating concentrations. (We assume that KQ dissociates completely.)

In Figure 8.2.2(a) we imagine that the membrane is completely permeable
to both ions. They will both move to the right by passive diffusion, eventually
reaching an equilibrium state in which the concentrations of each of the two ions
will be the same on both sides of the membrane. The net change will be zero on
each side of the membrane and therefore, no *potential difference* will exist across
the membrane.

In Figure 8.2.2(b) we imagine that the membrane is completely impermeable
to both ions. Neither will therefore move across the membrane, and the electrical
charges on each side of the membrane will total zero. Thus, there will again be
no transmembrane potential difference.

In Figure 8.2.2(c) we will finally see how a potential difference can be generated. We imagine that the membrane is permeable to K^+, but is impermeable
to Q^-, perhaps because Q^- is too big to pass through the membrane. K^+ will
then move across the membrane in response to its own *concentration* gradient.
Very quickly, however, the movement of K^+ away from the Q^- will establish a
transmembrane *electrical charge* gradient that opposes the concentration gradient. The concentration gradient pushes potassium to the right and the electrical
gradient pushes it to the left. The system will then quickly reach *electrochemical
equilibrium*, because there will be no further net change in $[K^+]$ on either side of
the membrane. Note that all of Q^- is on the left, but the K^+ is divided between
the right and left sides. The right side thus will be at a positive electrical potential
with respect to the left.

Note that the potential difference in the system of Figure 8.2.2(c) is achieved
spontaneously, i.e., without the expenditure of energy. Real biological systems
can establish potential differences in much the same way: They possess mem-

---

[1]This voltage difference across the neuron's plasma membrane is called a potential difference
because it represents a form of potential energy, or energy conferred by virtue of the positions of
electrical charges.

branes that are permeable to some materials and not to others, a feature that Figures 8.2.2(a), (b) and (c) show to be essential to the establishment of a potential difference. Our model system, however, lacks some realistic features: For example, real systems have many other ions, like Cl^- and Na^+, that must be considered. Further, consideration of Figure 8.2.2 suggests that the *degree* of permeability will be important in the establishment of transmembrane potentials (see References [1] and [2]).

The membrane of a "resting" neuron is electrically polarized.

If one probe of a voltmeter is inserted into a neuron and the other probe is placed on the outside, the voltmeter will show a potential difference of about −70 millivolts.[2] In other words, the interior of the cell is 70mv more negative than the outside, and a positive charge would tend to be attracted from the outside to the inside.

In the absence of strong stimuli the neuronal transmembrane potential difference does not change over time, and it is therefore referred to as the cell's *resting potential*. This term has wide usage, but can be misleading because it implies an equilibrium situation. The problem is that real equilibria are maintained without the expenditure of energy, and maintenance of the resting potential requires a great deal of energy (see next paragraph). Thus, we should expect that the brain, being rich in nervous tissue, would require considerable energy. This suspicion is confirmed by the observation that the blood supply to the brain is far out of proportion to the brain's volume.

The resting potential across a neuronal membrane is maintained by two competing processes, the principal one of which is passive, and the other of which is active, or energy-requiring. The passive process is the leakage diffusion of K^+ from the inside to the outside, leaving behind large organic ions, to which the membrane is impermeable. This process was shown earlier (in Figure 8.2.2) to generate a transmembrane potential difference.[3]

The problem with leakage diffusion is that many other ions, of both charges, also leak across real biological membranes. After a while this would lead to the destruction of the electrochemical equilibrium. The cell, however, has a means of re-establishing the various gradients. This active, energy-requiring process is under the control of molecular *sodium-potassium pumps*, which repeatedly push three Na^+ ions out of the cell, against a concentration gradient, while pushing two K^+ ions into the cell, also against a concentration gradient.[4] Thus, the Na/K pump and diffusion work against each other in a non-equilibrium, steady-state way. Such non-equilibrium, steady-state processes are very common in living cells.

---

[2]For various kinds of cells, the potential may vary from about −40mv to more than −100mv.
[3]The fraction of potassium ions that must leak out of a neuron to establish a potential difference of 100mv is estimated to be only about 10^{-5}.
[4]The molecular basis for the Na/K pump is not known. What is known is its effect—pumping Na^+ outward and K^+ inward.

Consideration of Figure 8.2.2 suggests that the low permeability of a membrane to sodium eliminates that ion as a contributor to the resting potential, and that the high permeability to potassium implicates that ion in the resting potential. This suspicion is confirmed by experiment: The neuronal resting potential is relatively insensitive to changes in the extracellular concentration of sodium, but highly sensitive to changes in the extracellular potassium concentration.

The concentrations of Na^+, K^+ and Cl^- inside and outside a typical neuron are given in Table 8.5.1. The asymmetric ionic concentrations are maintained by the two factors mentioned above: The Na/K pump and the differential permeabilities of the cell's plasma membrane to the different ions. In particular, we note that Na^+, to which the membrane is poorly permeable, might rush across the membrane if that permeability increased.

Section 8.3

The Action Potential

The neuronal plasma membrane contains voltage-controlled gates for sodium and for potassium. The gates are closed at the usual resting potential. When a stimulus depolarizes the membrane beyond a threshold value, the sodium gates open and sodium rushes into the cell. Shortly therafter, potassium rushes out. The Na/K pump then restores the resting concentrations and potential. The change in potential associated with the stimulus is called an action potential, and it propagates itself without attenuation down the neuronal axon.

The resting potential can be changed.

The usual resting potential of a neuron is about $-70mv$, but this value can be changed either by changing the permeability of the membrane to any ion or by changing the external concentration of various ions. If the potential is increased, say to $-100mv$, the neuron is said to be *hyperpolarized*. If the potential is decreased, say to $-40mv$, the neuron is said to be *depolarized* (see References [1–4]).

The membranes of neurons contain gated ion channels.

In Sections 6.1 and 8.2 we described several ways that materials could move across membranes. They were

a. Simple passive diffusion: Material moves directly through the bulk part of the membrane, including the hydrophobic layer. Water, carbon dioxide and hydrocarbons move in this fashion.

b. Facilitated passive diffusion: Ions and neutral materials move through special selective channels in the membrane. The selectivity of these channels resides in the recognition of the moving material by transport proteins, or permeases, in the membrane. Nevertheless, the process is spontaneous in that the material moves from regions of high concentration to regions of low concentration, and no energy is expended.

c.   Facilitated active transport: Ions and neutral materials move through selective channels (determined by transport proteins) but they move from regions of low concentration to regions of high concentration. Thus, energy is expended in the process.

We are interested in certain aspects of facilitated passive diffusion. Some channels through which facilitated, passive diffusion occurs seem to function all the time. Others, however, are controlled by the electrical or chemical properties of the membrane in the area near the channel. Such channels can open and close, analogous to fence gates, and are therefore called *voltage-gated channels* and *chemically gated channels*, respectively.

We now look more closely at voltage-gated channels; we will return to chemically gated channels a bit later. The following narrative corresponds to Figure 8.3.1. Voltage-gated sodium channels and potassium channels are closed when the potential across the membrane is -70 mv, i.e., the usual resting potential. If a small region of the axonal membrane is depolarized by an electrode (or other external stimulus) to about -50mv the voltage-gated sodium channels in that area will open. Sodium ions will then rush into the cell (because the Na^+ concentration outside the cell is so much higher than it is inside the cell, as will be shown in Table 8.5.1).

The inward rush of sodium will *further* depolarize the membrane, opening even more sodium channels, suggesting an avalanche. This is prevented at the site of the *original* depolarization, however, by the spontaneous closing of the sodium gates about a millisecond after they open. During that millisecond the membrane potential difference not only goes to zero, but becomes positive. At that time the potassium gates open, and potassium rushes out of the neuron in response to the positive charges that have accumulated in the interior. This makes the transmembrane potential drop rapidly, even hyperpolarizing to below -70 mv. The Na/K pump now restores Na^+ and K^- to resting levels inside and outside the neuron.

The entire process of action potential generation, from initiation to complete recovery, takes about two milliseconds. The neuron can be re-stimulated before it has recovered completely, i.e., during the period of hyperpolarization, but such stimulation naturally takes a greater depolarizing stimulus than the original one. Finally, note the *"all-or-none"* nature of action potential generation: No action potential is generated until the threshold degree of depolarization is reached, and then the same size action potential is generated, no matter how far past the threshold the stimulatory depolarization is carried.

The action potential propagates itself down the neuron.

We saw just above how sufficient depolarization of the axonal membrane at a localized point can cause the voltage-controlled sodium gates to open, permitting sodium ions to rush into the cell at that point. These sodium ions further depolar-

a) unstimulated neuron

$[Na^+]_o$ high

$[K^+]_o$ low

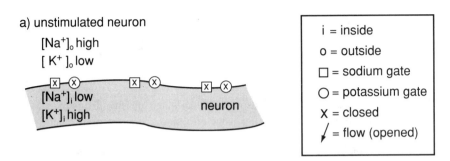

b) part of membrane depolarized; depolarization causes Na^+ gate to open

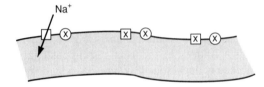

c) original Na^+ gate closes; K^+ gate opens, then nearby Na^+ gate opens

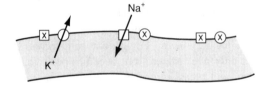

d) disturbance (action potential) propagated down axon

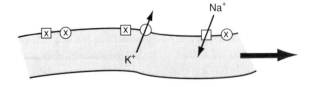

e) Na^+ pumped back out and K^+ pumped back in after action potential passes

Figure 8.3.1 The generation and movement of an action potential. This figure shows how an action potential is generated by an initial, localized depolarization of a neuron. This depolarization causes a depolarization at a neighboring site. Thus the disturbance, or action potential, propagates itself down the neuron.

ize the membrane at that point and permit even more sodium to enter the cell. The inrushing sodium ions now spread laterally along the inside of the membrane, attracting negative ions from the extracellular fluid to the outside of the membrane. As the numbers of these ions increase, they will eventually depolarize the *neighborhood* of the site of the original stimulus and thus open the sodium gates there. Meanwhile, the site of the original stimulus cannot be restimulated during a several millisecond *latent period*. The temporary latency at the site of the original stimulus and the depolarization of the membrane in the neighborhood of that site combine to cause the action potential to *spread away* from the site of the original stimulus.[5]

The decay of the action potential at its original site and its lateral spread from that site causes the disturbance to move along the neuron. The movement of the action potential is often compared to the burning of a fuse: Action at one site triggers action at the adjacent site and then the original action is extinguished. The size of the disturbance is the same as it passes any point on the fuse.

In closing this section we bring up a point made at the beginning of this section: The electrical propagation of an action potential is not like the propagation of an electrical current down a copper wire. The latter involves the lengthwise motion of electrons in the wire; the former involves radial and lengthwise motions of atomic ions.

Section 8.4

Synapses—Interneuronal Connections

When an action potential reaches the junction, or synaptic gap, between two neurons, it triggers the release of a neurotransmitter chemical from the presynaptic cell. The neurotransmitter then diffuses across the synaptic gap and, combining with other incoming signals, may depolarize the postsynaptic cell enough to trigger a new action potential there.

Besides the excitatory neurotransmitters, there are inhibitory neurotransmitters. The latter hyperpolarize the *postsynaptic neuron*, making it more difficult for a new action potential to be generated. Thus, whether or not an action potential is generated is determined by the sum of all incoming signals, both excitatory and inhibitory, over a short time period. The nervous system can use this summation to integrate the signals induced by a complex environment, and thereby generate complex responses.

In some cases, accessory cells wrap around neuronal axons to increase the electrical resistance between the cell's interior and its exterior. The action potential seems to jump between gaps in this sheath, thus greatly increasing the velocity with which the action potential is propagated.

---

[5]The effect of some anaesthetics, e.g., ether and ethyl alcohol, is to reduce the electrical resistance of the neuronal membrane. This causes the action potential to be extinguished. The effect of membrane resistance on action potential velocity is discussed in Section 8.5.

Most communication between neurons is chemical.

The simplest "loop" from stimulus to response involves three neurons: one to detect the stimulus and carry the message to the central nervous system, one in the central nervous system and one to carry the message to the responding organ. Most neural processing is much more complicated than that, however. In any case, some means of cell-to-cell communication is necessary. It surprises many biology students to learn that the mode of communication between such neurons is almost always chemical, not electrical.

Figure 8.4.1 shows the junction of two typical neurons, There is a gap of about 30nm between the axon of the neuron carrying the incoming action potential and the dendrite of the neuron in which a new action potential will be generated.[6]

---

[6] 1nm. = 1 nanometer = 10^{-9} meter.

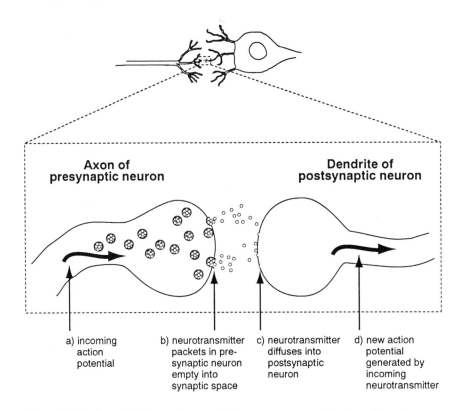

| a) incoming action potential | b) neurotransmitter packets in pre-synaptic neuron empty into synaptic space | c) neurotransmitter diffuses into postsynaptic neuron | d) new action potential generated by incoming neurotransmitter |

Figure 8.4.1   Synaptic information transfer between neurons. The incoming action potential causes the presynaptic neuron to release a neurotransmitter chemical, which diffuses across the synaptic space, or cleft. At the postsynaptic neuron the neurotransmitter causes a new action potential to be generated.

The gap is called a *synapse*. The arriving action potential opens certain voltage gated ion channels, which causes small packets of a *neurotransmitter chemical* to be released from the *presynaptic membrane* of the *presynaptic neuron*. This neurotransmitter then diffuses to the *postsynaptic membrane* of the postsynaptic neuron, where it opens chemically gated ion channels. The opening of these chemically gated channels causes a local depolarization of the dendrites and cell body of the postsynaptic neuron. If the depolarization is intense enough a new action potential will be created in the area between the postsynaptic cell body and its axon.

More than a hundred neurotransmitters have been identified, but *acetyl-choline* is a very common one, especially in the part of our nervous system that controls voluntary movement. If acetylcholine were to remain at the postsynaptic membrane it would continue to trigger action potentials, resulting in neuronal chaos. There is, however, a protein catalyst called *acetylcholine esterase* that breaks down acetylcholine soon after it performs its work. Most *nerve gasses*, including many insecticides, work by inactivating acetylcholine esterase, leading to uncontrolled noise in the animal's nervous system, and therefore death.

Note that synaptic transmission causes information flow to be one-way, because only the end of an axon can release a neurotransmitter and only the end of a dendrite can be stimulated by a neurotransmitter.

Occasionally, synaptic transmission is electrical.

At times chemical transmission by synapses is too slow, because diffusion may result in a delay of more than a millisecond. For these situations there are direct cytoplasmic connections that allow electrical communication between cells. For example, parts of the central nervous system controlling jerky eye movements have electric synapses, as do parts of the heart muscle (which, of course, is not nervous tissue).

Summation of incoming signals occurs at the postsynaptic neuron.

Whether or not a new action potential is generated depends on the degree to which the postsynaptic neuron is depolarized. Generally, no one presynaptic neuron will cause enough depolarization to trigger a new action potential. Rather, many presynaptic neurons working together have an additive effect, and the totality of their effects may be enough to depolarize the postsynaptic neuron. The various signals are simply summed together.

Summation can be spacial or temporal. *Spacial summation* occurs when many presynaptic neurons simultaneously stimulate a single postsynaptic neuron. *Temporal summation* is a little more complicated. To understand it we need to recall the all-or-none nature of the action potential—every stimulus that depolarizes a neuron past its threshold generates the same size action potential. So, we ask, how does a neuron code for stimulus *intensity*, if all action potentials are the same size? The answer is that stimulus intensity is coded by the *frequency* of action potentials, more intense stimuli generating more frequent spikes. Temporal summation occurs when many signals from one, or a few, presynaptic neurons arrive at a postsynaptic neuron in a short time period. Before one depolarizing pulse of neurotransmitter can decay away, many more pulses arrive, finally summing sufficiently to generate a new action potential.

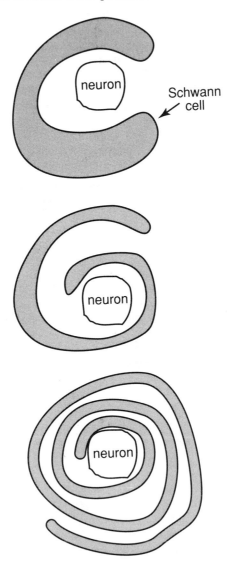

Figure 8.4.2 This picture shows how a Schwann cell wraps around a neuron, resulting in many layers of membrane between the neuron and the outside world. This many-layered structure has a very high electrical resistance, which radically alters the action potential-transmitting properties of the neuron.

Synaptic transmission may be excitatory or inhibitory.

As pointed out earlier, there are many known neurotransmitters. Some depolarize the postsynaptic neuron, leading to the generation of a new action potential. Other neurotransmitters, however, can hyperpolarize the postsynaptic neuron, making the subsequent generation of a new action potential harder. These latter neurotransmitters are therefore *inhibitory*. In general, therefore, the information reaching a postsynaptic cell will be a mixture of excitatory and inhibitory signals. These signals are summed and the resultant degree of depolarization determines whether or not a new action potential is generated.

We can now see an important way that the nervous system integrates signals. The actual response to a stimulus will depend on the pattern of excitatory and inhibitory signals that pass through the network of neurons. At each synapse, spacial and temporal summation of excitatory and inhibitory signals determine the path of information flow.

Myelinated neurons transmit information very rapidly.

As will be demonstrated in the next section, the velocity of conduction of an action potential down an axon depends on the diameter of the axon and on the electrical resistance of its outer membrane. Figure 8.4.2 shows how special accessory cells, called *Schwann cells*, repeatedly wrap around neuronal axons, thus greatly increasing the electrical resistance between the axon and the extracellular fluid. This resistance is so high, and the density of voltage-gated sodium channels so low, that no ions enter or leave the cell in these regions. On the other hand, there is a very high sodium channel density in the unwrapped regions, or *nodes* between neighboring Schwann cells. The action potential exists only at the nodes, internodal information flow occuring by ion flow *under* the myelinated sections. Thus, the action potential seems to skip from one node to another, in a process called *saltatory conduction*. Saltatory conduction in a myelinated nerve is about 100 times faster than the movement of an action potential in a nonmyelinated neuron. We would (correctly) expect to see it in neurons that control rapid-response behavior.

Section 8.5
A Model for the Conduction of Action Potentials

Through a series of meticulously conceived and executed experiments, Hodgkin and Huxley elucidated the physiology underlying the generation and conduction of nervous impulses. They discovered and charted the key roles played by sodium and potassium ions and the change in membrane conductance to those ions as a function of membrane potential. By empirically fitting mathematical equations to their graphs, they formulated a four-variable system of differential equations that accurately models action potentials and their propagation. The equations show that propagation velocity is proportional to the square root of axon diameter for an unmylinated nerve.

The Hodgkin–Huxley model is a triumph in neurophysiology.

We present here the mathematical model of nerve physiology as reported by Hodgkin and Huxley, see References [5–10]. Their experiments were carried out on the giant axon of a squid, the largest axon known in the animal kingdom, which achieves a size sufficient for the implantation of electrodes. Early experiments in the series determined that results at different temperatures could be mathematically transformed to any other (physiological) temperature. Consequently most of their results are reported for a temperature of 6.3°C. Temperatures from 5°C to 11°C are environmental for the animal, help maintain the nerve fiber in good condition, and 6.3°C is approximately 300° Kelvin.

In the resting state a (non-mylinated) axon is bathed in an interstitial fluid containing, among other things, sodium, potassium, and chloride ions. The interior material of the axon, the *axoplasm*, is separated from the external fluid by a thin lipid membrane, and the concentration of these ions differ across it. The concentration of sodium ions, Na^+, is 10 times greater outside than inside; the concentration of potassium ions, K^+, is 30 times greater inside than out; and chloride ions, Cl^-, are 15 times more prevalent outside than in (see Table 8.5.1).

Diffusion normally takes place down a concentration gradient, but when the particles are electrically charged, electrical potential becomes a factor as well. In particular, if the interior of the axon were electrically negative relative to the exterior, then the electrically negative chloride ions will tend to be driven out against its own concentration gradient until a balance between the electrical pressure and concentration gradient is reached (see Section 8.2). The balancing voltage difference, V, is given by the Nernst equation,

$$V = \frac{RT}{F} \ln \frac{C_i}{C_o}, \qquad (8.5.1)$$

where R is the gas constant, $R = 8.314$ Joules/°C/mole, T is absolute temperature, and F is Faraday's constant, 96,485 coulombs, the product of Avagadro's number with the charge on an electron. The concentrations inside and out are denoted by C_i and C_o respectively.

For Cl^- this potential difference is -68 mv. But, in the resting state (see Section 8.2), the interior of an axon is at -70 mv relative to the exterior (see Figure 8.5.1). Thus there is little tendency for chloride ions to migrate. The same

Table 8.5.1   Intra- and Extracellular Ion Concentrations

Inside Axon	Extracellular Fluid	C_i/C_o	Nernst Equivalent
Na^+ 15	Na^+ 145	.10	-55 mv
K^+ 150	K^+ 5	30	82 mv
Cl^- 7.5	Cl^- 110	.068	-68 mv

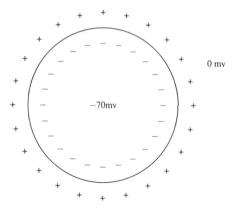

Figure 8.5.1   Charges inside and outside an axon

holds for potassium ions which are electrically positive. Their Nernst equivalent is 82 mv, that is, the outside needs to be 82 mv more positive than the inside for a balance of the two effects. Since the outside is 70 mv more positive, the tendency for K^+ to migrate outward, expressed in potential, is only 12 mv.

The situation is completely different for sodium ions, however. Their Nernst equivalent is -55 mv, but since sodium ions are positive, the interior would have to be 55 mv electrically positive to balance the inward flow due to the concentration gradient. Since it is -70 mv instead, there is an equivalent of 125 mv $(= 70 + 55)$ of electrical potential for sodium ions to migrate inward. But in fact there is no such inward flow; we must therefore conclude that the membrane is relatively impermeable to sodium ions. In electrical terms, the conductance of the membrane to sodium ions, g_{Na}, is small.

Electrical conductance is defined as the reciprocal of electrical resistance,

$$ g = \frac{1}{R}. $$

The familiar Ohm's Law relating voltage V, current i, and resistance R,

$$ V = iR, $$

then takes the form

$$ i = gV. \tag{8.5.2} $$

In discrete component electronics, resistance is measured in ohms and conductance in mhos. However for an axon, conductance and current will depend on surface area. Therefore conductance here is measured in mhos per square centimeter. Letting $E_{Na} = 55$ mv denote the equilibrium potential given by the Nernst equation for sodium ions, the current per square centimeter of those ions can be calculated as

$$i_{Na} = g_{Na}(V - E_{Na}), \qquad E_{Na} = 55 \text{ mv} \qquad (8.5.3)$$

where V is the voltage inside the axon. A negative current corresponds to inward flow. Similarly for potassium ions,

$$i_K = g_K(V - E_K), \qquad E_K = -82 \text{ mv}. \qquad (8.5.4)$$

And, grouping chlorine and all other ionic currents together as *leakage currents*,

$$i_l = g_l(V - E_l), \qquad E_l = -59 \text{ mv}, \qquad (8.5.5)$$

One of the landmark discoveries of the Hodgkin–Huxley study is that membrane permeability to Na^+ and K^+ ions varies with voltage and with time as an action potential occurs.[7] Figure 8.5.2(a) plots potassium conductance against time with the interior axon voltage "clamped" at -45mv. Hodgkin and Huxley's apparatus maintained a fixed voltage in these conductance experiments by controlling the current fed to the implanted electrodes. The large increase in membrane conductance shown is referred to as *depolarization*. By contrast, Figure 8.5.2(b)

---

[7]The biological aspects of this variation were discussed in Section 8.3.

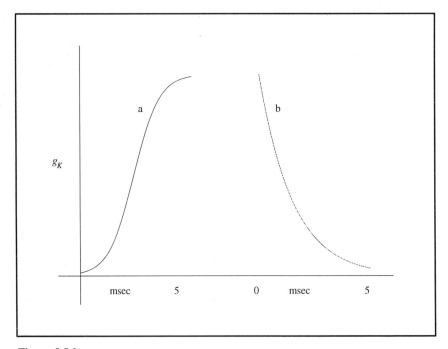

Figure 8.5.2

depicts membrane *repolarization* with the voltage now clamped at its resting value. In these *voltage clamp* experiments, the electrodes run the entire length of the axon, so the entire axon membrane acts in unison; there are no spacial effects. Figure 8.5.3 shows a Na$^+$ conduction response. Somewhat significant in the potassium depolarization curve is its sigmoid shape. On the other hand, the sodium curve shows an immediate rise in conductance. We conclude that the first event in a depolarization is the inflow of Na$^+$ ions. An outflow of K$^+$ ions follows shortly thereafter. The leakage conductance is 0.3 m-mhos/cm^2 and, unlike the the sodium and potassium conductances, is constant.

Potassium conductance is interpolated by a quartic.

Hodgkin and Huxley went beyond just measuring these conductance variations; they modeled them mathematically. To capture the inflexion in the potassium conductance curve, it was empirically modeled as the fourth power of an exponential rise, thus

$$g_K = \bar{g}_K n^4 \tag{8.5.6}$$

where $\bar{g}_K$ is a constant with the same units as g_K and n is a dimensionless exponential variable,

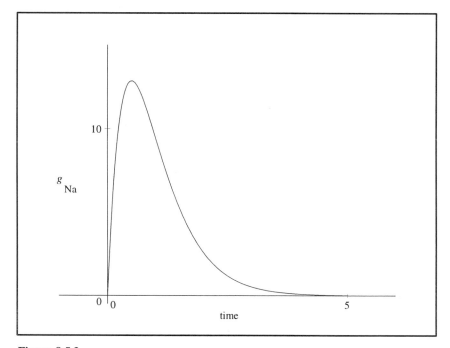

Figure 8.5.3

$$n = n_\infty - (n_\infty - n_0)e^{-t/\tau_n},$$

increasing from n_0 to the asymptote n_∞, see Figure 8.5.4. If $n_0 = 0$, then, for small t, n behaves like t^4 as seen from a Taylor expansion,

$$n^4 = (n_\infty - n_\infty e^{-t/\tau_n})^4$$
$$= (\frac{n_\infty}{\tau_n})^4 t^4 - 4(\frac{n_\infty}{\tau_n})^3 (\frac{n_\infty}{2!\tau_n^2})t^5 + \cdots.$$

Therefore, the n^4 curve has a horizontal tangent at 0, as desired.

In order that n become no larger than 1 in the fit to the experimental data, $\bar{g}_K$ is taken as the maximum conductance attainable over all depolarization voltages. This value is

$$\bar{g}_K = 24.34 \text{ m-mhos/cm}^2.$$

The function n may also be cast as the solution of a differential equation, thus

$$\frac{dn}{dt} = \alpha_n(1 - n) - \beta_n n, \qquad (8.5.7)$$

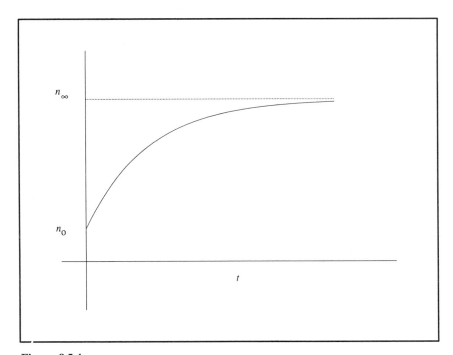

Figure 8.5.4

where the coefficients α_n and β_n are related to the rising and falling slopes of n respectively, see Figure 8.5.5. Their relationship with the parameters of the functional form is

$$\tau_n = \frac{1}{\alpha_n + \beta_n} \quad \text{and} \quad n_\infty = \frac{\alpha_n}{\alpha_n + \beta_n}.$$

Experimentally, these coefficients vary with voltage. Hodgkin and Huxley fit the experimental relationship by the functions

$$\alpha_n = \frac{0.01(10 - (V - V_r))}{e^{1 - 0.1(V - V_r)} - 1} \tag{8.5.8}$$

and

$$\beta_n = 0.125 e^{-(V - V_r)/80}, \tag{8.5.9}$$

where $V_r = -70$ mv is resting potential. Note that equation (8.5.8) is singular at $V = -60$; however the singularity is removable with limiting value 0.1.

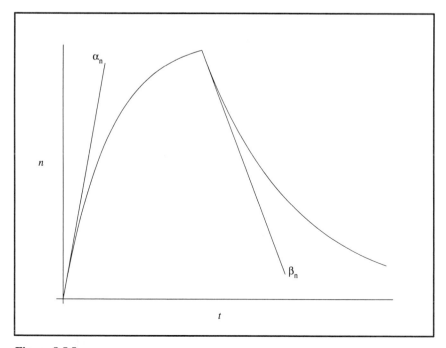

Figure 8.5.5

Substituting equation (8.5.6) into (8.5.4) gives

$$i_K = \bar{g}_K n^4 (V - E_K). \tag{8.5.10}$$

Sodium conductance is interpolated by a cubic.

Sodium conductance is modeled as

$$g_{Na} = \bar{g}_{Na} m^3 h \tag{8.5.11}$$

where $\bar{g}_{Na} = 70.7$ m-mhos/cm^2 is a fixed parameter carrying the units and m and h are dimensionless. The use of two dimensionless variables helps smooth the transition between the ascending portion of the curve and the descending portion. As above, m and h are exponential and satisfy similar differential equations:

$$\frac{dm}{dt} = \alpha_m (1 - m) - \beta_m m, \tag{8.5.12}$$

$$\frac{dh}{dt} = \alpha_h (1 - h) - \beta_h h. \tag{8.5.13}$$

Further, the coefficients are functions of V interpolated as follows:

$$\begin{aligned}
\alpha_m &= \frac{0.1(25 - (V - V_r))}{e^{0.1(25 - (V - V_r))} - 1} \\
\beta_m &= 4e^{-(V - V_r)/18} \\
\alpha_h &= 0.07e^{-0.05(V - V_r)} \\
\beta_h &= \frac{1}{e^{0.1(30 - (V - V_r))} + 1}.
\end{aligned} \tag{8.5.14a–d}$$

Substituting equation (8.5.11) into equation (8.5.3) gives

$$i_{Na} = \bar{g}_{Na} m^3 h (V - E_{Na}). \tag{8.5.15}$$

The Hodgkin–Huxley space-clamped-axon equations produce action potentials.

In another series of experiments, Hodgkin and Huxley fixed electrodes along the entire length of the axon as before, but now the electrodes were used to measure the voltage as it varied during a depolarization event. These are called the *space-clamp* experiments. In addition to its role as a variable conductor of electrically charged particles, the axon membrane also functions as the dielectric of an electrical capacitor in that charged particles accumulate on either side of it. The

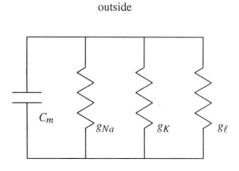

Figure 8.5.6 Axon membrane circuit equivalent

capacitance of the membrane was determined by Hodgkin and Huxley to be about 1 microfarad per square centimeter:

$$C_m = 1 \times 10^6 \quad \text{farad/cm}^2.$$

Electrically, the membrane may be depicted as in Figure 8.5.6. When space clamped, the sum of the membrane ionic currents serves to deposit charge on or remove charge from this membrane capacitor. Said differently, the effective current "through" the capacitor balances the membrane current made up of the sodium, potassium, and leakage components; the sum of these currents must be 0 by Kirchoff's Law,

$$i_C + i_{Na} + i_K + i_l = 0. \tag{8.5.16}$$

The effective current through a capacitor is given by

$$i_C = C_m \frac{dV}{dt}$$

where membrane capacitance C_m is measured in farads per square centimeter, $\frac{dV}{dt}$ is in volts/second, and i is in amperes per square centimeter. Substituting from equation (8.5.16) gives

$$C_m \frac{dV}{dt} = -(i_{Na} + i_K + i_l). \tag{8.5.17}$$

We now collect the various equations to give the Hodgkin–Huxley space-clamped equations. Substituting equations (8.5.5), (8.5.10), and (8.5.15) into equation (8.5.17) and recalling equations (8.5.7), (8.5.12), and (8.5.13) gives

$$\frac{dV}{dt} = \frac{-1}{C_m}(\bar{g}_{Na}m^3h(V - E_{Na}) + \bar{g}_Kn^4(V - E_K) + g_l(V - E_l)),$$

$$\frac{dn}{dt} = \alpha_n(1 - n) - \beta_nn$$

$$\frac{dm}{dt} = \alpha_m(1 - m) - \beta_mm$$

$$\frac{dh}{dt} = \alpha_h(1 - h) - \beta_hh \qquad\qquad (8.5.18a–d)$$

where the alpha's and beta's are given functions of V according to equations (8.5.8), (8.5.9), and (8.5.14(a)–(d)).

The following code solves this system of differential equations and produces an action potential such as Figure 8.5.7. The action potential is initiated by a pulse of current lasting 0.001 of a second.

The Hodgkin–Huxley propagation equations predict impulse speed.

In vivo, an axon is not clamped. Consequently, instead of the entire axon undergoing an action potential at the same time, an action potential is localized and propagates along the axon in time. Thus voltage is a function of position, x, along the axon as well as a function of time. Consider a small section of axon lying between x and $x + \Delta x$. The basic equation is the current in at x minus the current out at $x + \Delta x$, and minus the membrane current must equal the charge build-up on that section of membrane, that is, must equal the capacitance current. Hence

$$i(x) - i(x + \Delta x) - (i_{Na} + i_K + i_l)2\pi a\Delta x = C_m2\pi a\Delta x\frac{dV}{dt}.$$

In this we have taken the radius of the axon to be a and multiplied the per-square-centimeter quantities by the surface area of the section of axon in question. Divide by Δx, let $\Delta x \to 0$, and divide by $2\pi a$ to get,

$$\frac{-1}{2\pi a}\frac{\partial i}{\partial x} - (i_{Na} + i_K + i_l) = C_m\frac{dV}{dt}. \qquad (8.5.19)$$

Next suppose there were no membrane current. Then the voltage drop over the length of the axon is related to the current i along the axon by Ohm's Law $V = iR$. In this, R is the total resistance of the axoplasm (not membrane resistance). But in fact each section of axon of length Δx contributes to the overall resistance in proportion to its length, namely $R\Delta x/L$, where L is the total length of the axon. Thus if position along the axon is denoted by x, then resistance as a function of x increases linearly from 0 to R. In the meantime, voltage as a function of x falls from V to 0. In particular, Ohm's Law as applied to the section dx becomes

$$\frac{dV}{dx} = -i\frac{dR}{dx}, \qquad (8.5.20)$$

```
> Ena:=55: Ek:=-82: El:=-59: gkbar:=24.34: gnabar:=70.7:
> gl:=0.3: vrest:= -70: Cm:=0.001:
> alphan:=v->0.01*(10-(v-vrest))/(exp(0.1*(10-(v-vrest)))-1);
> betan:=v->0.125*exp(-(v-vrest)/80);
> alpham:=v->0.1*(25-(v-vrest))/(exp(0.1*(25-(v-vrest))) - 1);
> betam:=v->4*exp(-(v-vrest)/18);
> alphah:=v->0.07*exp(-0.05*(v-vrest));
> betah:=v->1/(exp(0.1*(30-(v-vrest)))+1);
> pulse:=t->-20*(Heaviside(t-0.001)-Heaviside(t-0.002));
> rhsv:=(t,v,n,m,h)->-(gnabar*m^3*h*(v - Ena)
  +gkbar*n^4*(v - Ek)+gl*(v - El)
  + pulse(t))/Cm;
> rhsn:=(t,v,n,m,h)->1000*(alphan(v)*(1-n) - betan(v)*n);
> rhsm:=(t,v,n,m,h)->1000*(alpham(v)*(1-m) - betam(v)*m);
> rhsh:=(t,v,n,m,h)->1000*(alphah(v)*(1-h) - betah(v)*h);
> with(share): readshare(ODE,plots):
> inits:=[0,vrest,0.315,0.042,0.608];
> output:=rungekutta([rhsv,rhsn,rhsm,rhsh],inits,0.0001,100);
> plot({makelist(output,1,2)});
```

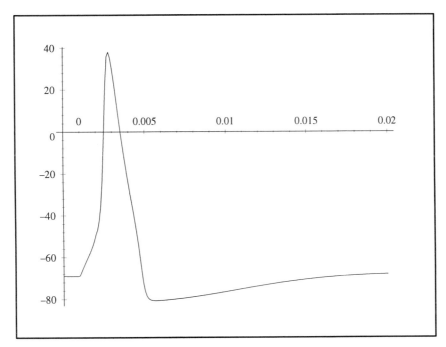

Figure 8.5.7   A simulated action potential

where the negative sign indicates that V is decreasing while R is increasing.

In this $\frac{dR}{dx}$ is the resistance per unit length, which we take to depend only on the cross-sectional area of the axon. And this dependence is in inverse proportion—that is, if the area is halved, then the resistance is doubled. Hence, letting $\bar{R}$ denote the resistance per unit length per unit area, then

$$\frac{dR}{dx} = \frac{\bar{R}}{\pi a^2}, \tag{8.5.21}$$

where a is the radius of the axon. Substituting equation (8.5.21) into equation (8.5.20) and solving for i gives

$$i = -\frac{\pi a^2}{\bar{R}} \frac{dV}{dx}. \tag{8.5.22}$$

In this, the negative sign can be interpreted to mean that current moves down the voltage gradient.

Differentiate equation (8.5.22) with respect to x and substitute equation (8.5.19) to get

$$\frac{a}{2\bar{R}} \frac{\partial^2 V}{\partial x^2} - (i_{Na} + i_K + i_1) = C_m \frac{dV}{dt}. \tag{8.5.23}$$

This equation, along with the equations for n, m, and h corresponding to the membrane ion currents, equations (8.5.7), (8.5.12), and (8.5.13), constitute the Hodgkin–Huxley propagation equations. Hodgkin and Huxley numerically obtain from these equations the value 18.8 meters/second for the propagation velocity c of an action potential. It is known to be 21 meters/second, so the agreement is quite good.

We will not solve this partial differential equation system here, instead we will show that the propagation velocity is proportional to the square root of the radius a and inversely proportional to the axon resistance $\bar{R}$. Hodgkin and Huxley note that if the action potential propagates along the axon unchanged in shape, then its shape as a function of x for fixed t is equivalent to its shape as a function of t for fixed x. This is formalized by the *wave equation* [11]

$$\frac{\partial^2 V}{\partial x^2} = \frac{1}{c^2} \frac{\partial^2 V}{\partial t^2}. \tag{8.5.24}$$

In this, c is the propagation velocity. Substituting the second derivative with respect to x from equation (8.5.24) into equation (8.5.23) gives

$$\frac{a}{2\bar{R}c^2} \frac{d^2 V}{dt^2} - (i_{Na} + i_K + i_1) = C \frac{dV}{dt}. \tag{8.5.25}$$

In this equation the only dependence on a occurs in the first term. Since the other terms do not depend on a, the first term must be independent of a as well. This can only happen if the coefficient is constant with respect to a,

$$\frac{a}{2\bar{R}c^2} = \text{constant}.$$

But then it follows that

$$c = (\text{constant})\sqrt{\frac{a}{\bar{R}}}.$$

Thus the propagation velocity is proportional to the square root of the axon radius and inversely proportional to the square root of axon resistance. The large size of the squid's axon equips it for fast responses, an important factor in eluding enemies.

Section 8.6

The Fitzhugh–Nagumo Two-Variable Action Potential System

The Fitzhugh–Nagumo two-variable model behaves qualitatively like the Hodgkin–Huxley space-clamped system. But being simpler by two variables, action potentials and other properties of the Hodgkin–Huxley system may be visualized as phase-plane plots.

The Fitzhugh–Nagumo phase–plane analysis demonstrates all-or-nothing response.

Fitzhugh [12] and Nagumo [13] proposed and analyzed the following system of two differential equations, which behaves qualitatively like the Hodgkin–Huxley space-clamped system:

$$\frac{dV}{dt} = V - \frac{1}{3}V^3 - w$$

$$\frac{dw}{dt} = c(V + a - bw). \tag{8.6.1}$$

In this, V plays the role of membrane potential, but w is a general "refactory" variable not representing any specific Hodgkin–Huxley variable. The parameters a, b, and c are the constants

$$a = 0.7 \qquad b = 0.8 \qquad c = 0.08.$$

The virtue of this sytem is in elucidating the regions of physiological behavior of membrane response.

The *phase plane* is the coordinate plane of the two dependent variables V and w. A curve $V = V(t)$ and $w = w(t)$ in the phase plane corresponding to a solution of the differential equation system for given initial values $V_0 = V(0)$, $w_0 = w(0)$ is called a *trajectory*.

Two special curves in the phase plane are the *isoclines*. These are the curves for which either dV/dt or dw/dt are zero. The w isocline, from (8.6.1),

$$V + a - bw = 0,$$

is a straight line with slope 1.25 and intercept 0.875. To the left of it, $dw/dt < 0$ and to the right $dw/dt > 0$. The V isocline

$$V - \frac{1}{3}V^3 - w = 0 \qquad \text{or} \qquad w = V(1 - \frac{1}{3}V^2)$$

is a cubic with roots 0, and $\pm\sqrt{3}$. Above it $dV/dt < 0$ and below it $dV/dt > 0$. The isoclines divide the phase plane into four regions or quadrants. In the first, above the cubic and right of the line, $dV/dt < 0$ and $dw/dt > 0$. Hence a trajectory in this quadrant will tend toward decreasing V and increasing w, upward and leftward. In Quadrant 2, above the cubic and left of the line, trajectories tend

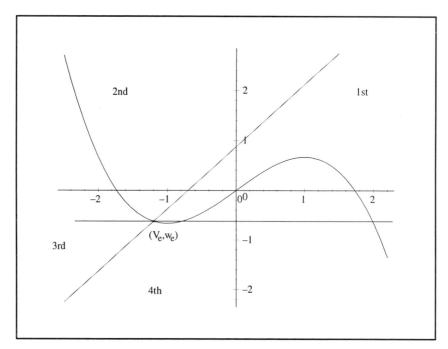

Figure 8.6.1   Direction field for equations (8.6.1).

downward and leftward. In Quadrant 3, below the cubic and left of the line, trajectories tend downward and rightward. In Quadrant 4, below the cubic and right of the line, the derivatives are $dV/dt > 0$ and $dw/dt > 0$ so trajectories tend upward and rightward. The Quadrants are shown in Figure 8.6.1.

The isoclines intersect at $V_e = -1.1994$ and $w_e = -0.6243$. At this point $dV/dt = 0$ and $dw/dt = 0$, so a trajectory at that point does not move at all, it is an *equilibrium*. For this particular system of differential equations it can be shown that trajectories starting anywhere eventually lead to this equilibrium. As a result it is known as a globally attracting stable equilibrium. It plays the role of the resting state in our description of an axon.

Consider the progress of a trajectory that is begun at the point (V_0, w_0) located to the right of the ascending portion of the cubic isocline. As this is in Quadrant 4, the trajectory will tend rightward until crossing the decending section of the same isocline. From there the trajectory will tend upward and leftward until crossing the w isocline. Proceeding from there leftward and downward, it next crosses the V isocline again. Finally it proceeds downward and rightward ending up at the equilibrium point. See Figure 8.6.2.

Concentrate on the behavior of V along this trajectory. It first increases in value until reaching the descending branch of the cubic isocline. This will be its maximum value. Crossing this isocline, V then decreases, eventually below equilibrium. Completing the trajectory, V increases slowly to equilibrium. But this describes the behavior of the membrane potential, V, during an action potential.

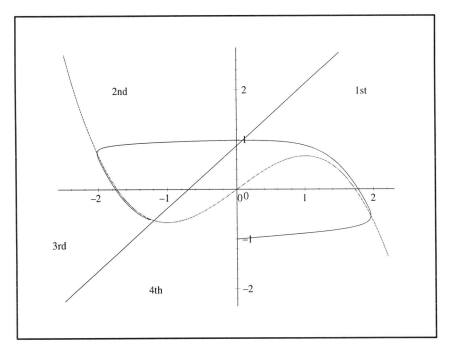

Figure 8.6.2 Solution curve for equations (8.6.1).

Next suppose a trajectory is begun immediately to the right of the equilibrium point (V_e, w_e), then a very different thing happens. Assume the starting point (V_0, w_0) is taken to so that $w_0 = w_e$ and $V_0 = V_e + \Delta V$. This starting point lies on the horizontal line through the equilibrium as shown in Figure 8.6.1 and which, in turn, intersects the cubic isocline at -0.786 and 1.98 besides -1.1994. Therefore, depending on the size of ΔV, the starting point falls in either Quadrant 1 or 4.

If it is in Quadrant 1, then the response trajectory returns more or less directly to equilibrium. This is analogous to a sub-threshold stimulation in an axon.

But if ΔV is so large that $V_0 = -0.64$, say, well inside Quadrant 4, then the response trajectory corresponds to an action potential.

Thus, this Fitzhugh–Nagumo model gives an all-or-none response to a stimulus, just as an axon does. The separating point along the line $w = -0.624$ lies between $V_0 = -0.65$ and $V_0 = -0.64$. More generally, there is an entire separating all-or-none curve, corresponding to different values of w, called a *separatrix*. It follows the ascending branch of the cubic isocline closely.

Refractory and enhanced regions

Other features of axon behavior are demonstrated with this model as well. During an action potential consider what happens if a stimulus occurs while the trajectory lies above the cubic isocline. Nothing! That is, such a stimulus causes the trajectory to jump horizontally to the right, but then it resumes its leftward horizontal movement. In particular, there is basically no change in the refractory variable w.

Now suppose a stimulus occurs while the action potential trajectory is descending in the third quadrant headed back to equilibrium. If the stimulus is large enough to cross the separatrix, then a new action potential can be initiated. Consequently, this region corresponds to the relative refractory region of an axon's behavior.

Finally, suppose a sub-threshold stimulation occurs from the equilibrium point. There is no action potential, but a second sub-threshold stimulation might be sufficient to cross the all-or-none separatrix and initiate an action potential. Therefore the region between the equilibrium and the separatrix corresponds to the enhanced state of an axon.

Section 8.7

Questions for Thought and Discussion

1. Discuss the roles of voltage-gated channels and diffusion processes in the transmission of information across neuronal synapses.
2. Starting with the number 2, number the following events in the order in which they occur. ("Site A" is an arbitrary mid-axonal location.)

neuronal membrane is depolarized at Site A
   by external stimulus .... ____1____
acetylcholine esterase breaks down neurotransmitter .... _____
K^+ channels open at Site A............................... _____
postsynaptic chemical-gated channels open................. _____
Na^+/K pump restores resting potential at Site A .... _____
interior of neuron at Site A is at positive potential
   with respect to exterior............................... _____

3. In what ways is the transmission of information by an action potential different from the transmission of electrical information by a copper wire?

References and Suggested Further Reading

1. **Neuronal biology and physiology:** Elaine N. Marieb, *Human Anatomy and Physiology*, The Benjamin/Cummings Publishing Company, Inc. Redwood City, California, 1992.

2. **Neuronal biology:** Peter H. Raven and George B. Johnson, *Biology*, Mosby-Year Book, St. Louis, MO, 3rd ed., 1992.

3. **Neuronal biophysics:** F. R. Hallett, P. A. Speight and R. H. Stinson, *Introductory Biophysics*, Halstead Press, John Wiley and Sons, Inc., New York, 1977.

4. **Molecular biology of neurons:** Bruce Alberts, Dennis Bray, Julian Lewis, Martin Raff, Keith Roberts and James D. Watson, *Molecular Biology of the Cell*, Garland Publishing, Inc. New York, 3rd ed., 1994.

5. **Hodgkin–Huxley experiments:** J. D. Murray, *Mathematical Biology*, Springer–Verlag, New York, 1989.

6. **Hodgkin–Huxley experiments:** A. L. Hodgkin, A. F. Huxley, Currents carried by sodium and potassium ions through the membrane of the giant axon of *Logio*, *J. Physio.* **116**, 449–472, 1952a.

7. **Hodgkin–Huxley experiments:** A. L. Hodgkin, A. F. Huxley, Components of membrane conductance in the giant axon of *Logio*, *J. Physio.* **116**, 473–496, 1952b.

8. **Hodgkin–Huxley experiments:** A. L. Hodgkin, A. F. Huxley, The dual effect of membrane potential on sodium conductance in the giant axon of *Logio*, *J. Physio.* **116**, 497–506, 1952c.

9. **Hodgkin–Huxley experiments:** A. L. Hodgkin, A. F. Huxley, A quantitative description of membrane current and its application to conduction and excitation in nerve, *J. Physio.* **117**, 500–544, 1952d.

10. **Hodgkin–Huxley experiments:** A. L. Hodgkin, A. F. Huxley, B. Katz, Measurement of current-voltage relations in the membrane of the giant axon of *Logio*, *J. Physio.* **116**, 424–448, 1952.

11. **Hodgkin–Huxley experiments:** R. V. Churchill, J. W. Brown, *Fourier Series and Boundary Value Problems*, 4th ed., McGraw-Hill, NY, 1987.

12. **Fitzhugh–Nagumo equations:** R. Fitzhugh, Impulses and physiological states in theoretical models of nerve membrane, *Biophy. J.* **1**, 445–466, 1961.

13. **Fitzhugh–Nagumo equations:** J. S. Nagumo, S. Arimoto, S. Yoshizawa, An active pulse transmission line simulating nerve axon, *IRE* **20**, 2061–2071, 1962.

Chapter 9

The Biochemistry of Cells

Introduction to this chapter

The purpose of this chapter is to present the structure of some of the molecules that make up a cell and to show how they are constructed under the supervision of hereditary elements of the cell. This will lead into a mathematical description of biological catalysis at the end of this chapter and is a necessary prelude to the discussion of the Human Immunodeficiency Virus in Chapter 10. As a result, this chapter contains a lot of biological information.

We will see that biological molecules can be created outside of a cellular environment, but only very inefficiently. Inside a cell, however, the information for biomolecules is encoded in the genetic material called nucleic acid. Thus, we will establish a direct relationship between the chemicals that constitute a cell and the cell's hereditary information.

The topical material of this chapter is organized along the lines of small to large. We begin by presenting a description of the atoms found in cells and then show how they are assembled into small organic molecules. Some of these small molecules can then be polymerized into large biochemical molecules, the biggest of which have molecular weights on the order of billions. These assembly processes are mediated by certain macromolecules which are themselves molecular polymers, and whose own assembly has been mediated by similar molecular polymers. Thus, we develop a key process in biology—self-replication.

Section 9.1

Atoms and Bonds in Biochemistry

Most of the atoms found in a cell are of common varieties: hydrogen, carbon, nitrogen and oxygen. They are, in fact, major components of air and dirt. What is it that makes them so fundamental to life? To answer this question we must examine the ways that these atoms form bonds with one another—because it is through molecular organization that we will characterize living systems.

A living system is a highly organized array of atoms, attached to one another by chemical bonds. The bonds may be strong, requiring considerable energy for their rearrangement. This leads to structures that are somewhat permanent and that can be changed only under special biochemical conditions. The bonds result from a process of "electron sharing." Carbon, nitrogen and oxygen atoms can form a practically unlimited array, held together by covalent bonds.

Alternatively, some chemical bonds are weak, the heat energy available at room temperature being sufficient to break them. Because of their weakness, the structures they form are highly variable, leading to material movement and regional uniformity (among other things). The most important weak bond is called a hydrogen bond: It is the electrical attraction between a hydrogen nucleus on one molecule and an asymmetrically oriented electron on a nitrogen or oxygen atom of the same molecule or a different one.

Organization is the key to living systems.

In Section 3.3 we pointed out that the individual processes found in living systems are also found in nonbiological situations. We emphasized that the "signature" of life was the organization, or integration, of those processes into a unified system. We now extend that concept to physical organization at the atomic and molecular levels.

Calcium, phosphorus, potassium, sulfur, sodium and chlorine account for about 3.9% of the atoms in our bodies.[1] Just four other elements make up the other 96%; they are hydrogen, carbon, nitrogen and oxygen. These four elements most abundant in our bodies are also found in the air and earth around us—as H_2O, CO_2, N_2, O_2 and H_2. Thus, if we want to explain why something has the special quality we call "life" it does not seem very fruitful to look for exotic ingredients; they aren't there. Where else might the explanation be?

An important clue can be found in experiments in which living systems are frozen to within a few degrees of $0°K$, so that molecular motion is virtually halted. Upon reheating, these living systems resume life processes. The only properties conserved in such an experiment are static structural ones. We can conclude that a critical property of life lies in the special ways that the constituent atoms of living systems are organized into larger structures. We should therefore suspect that the atoms most commonly found in our bodies have special bonding properties, such that they can combine with one another in many ways. This is indeed the case: Carbon, nitrogen, oxygen and hydrogen are capable of a virtually infinite number of different molecular arrangements. In fact, it has been estimated that the number of ways that the atoms C, H, O, N, P and S can be combined to make low molecular weight compounds ($MW < 500$) is in the billions! [1].

Of the large number of possible arrangements of C, N, O and H, the forces of evolution have selected a small subset, perhaps a thousand or so, on which to base life. Members of this basic group have then been combined into a vast array

---

[1]About 15 more elements are present in trace amounts.

of biomacromolecules. For example, the number of atoms in a typical biomacro-molecule might range from a few dozen into the millions, but those with more than a few hundred atoms are always polymers of simpler subunits.

Living systems are assemblages of common atoms, each part of the system having a very specific organization at all size levels. In other words, all living things can be thought of as regions of great orderliness, or organization. Death is marked by the disruption of this organization, either suddenly, as in the case of a bullet, or slowly, as in the case of degenerative disease. In any case, death is followed by decompositional processes that convert the body to gases, which are very disorganized.

Physicists use *entropy* as a measure of disorder; there is an important empir-ical rule, the Second Principle of Thermodynamics, which states that entropy in the universe increases in the course of every process. Living systems obey this rule, as they do all other natural chemical and physical principles. As an organism grows, it assembles atoms into an orderly, low-entropy arrangement; at the same time the entropy of the organism's surroundings increases by even more, to make the net entropic change in the universe positive. This net increase is to be found in such effects as the motion of air molecules induced by the person's body heat, in the gases he/she exhales and in the natural waste products he/she creates.

Nature is full of good examples of the critical role played by organization in living systems. Consider that a bullfighter's sword can kill a 600-pound bull and that 0.01 micrograms of the neurotoxin *tetrodotoxin* from a puffer fish can kill a mouse. The catastrophic effects of the sword and the toxin seem out of proportion to their masses. In light of the discussion above, however, we now understand that their effects are not based on mass at all, but instead on the disruption of critically organized structures, e.g., the nervous system [2].

Covalent bonds are strong interactions involving electron sharing.

A very strong attraction between two atoms results from a phenomenon called "electron-sharing;" it is responsible for binding atoms into biochemical molecules. One electron from each of two atoms becomes somewhat localized on a line be-tween the two nuclei. The two nuclei are electrostatically attracted to the electrons and therefore remain close to one another.

Figure 9.1.1 shows simple planetary models of two hydrogen atoms. (Later we will generalize our discussion to other atoms.) The radius of this orbit is about 0.05 nm, so the nuclei are about 0.1 nm apart. At some time each of the electrons will find itself at a point immediately between the two nuclei. When this happens, each of the two nuclei will exert the same electrical attraction on the electron, meaning that the electron can no longer be associated with a particular nucleus. There being no reason to "choose" either the right or the left nucleus, the electron will spend more time directly in between the two than in any other location.[2] The

---

[2]The idea that an electron is more likely to be found in one region of space than in another is built into the quantum mechanical formulation, which is outside the scope of this book. In the quantum mechanical formulation there are no orbits and the electron is represented as a probability cloud. The denser the cloud, the greater the probability of finding the electron there.

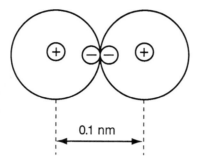

Figure 9.1.1 A model of the hydrogen molecule. Planetary orbits are shown, but the electrons are equally attracted to both nuclei and therefore spend most of their time in the region directly between the two nuclei. This interaction is called a *covalent bond*.

two electrons in the center then act like a kind of glue, attracting the nuclei to themselves and thus toward each other. A stable molecule is thereby formed; the attraction between its constituent atoms is called a *covalent bond.*

A covalent bond always contains two electrons because of an unusual electronic property: An electron spins on its own axis. For quantum mechanical reasons, an electron always pairs up with another electron having the opposite spin direction, leading to "spin pairing" in covalent bonds and in certain other situations. An atom or molecule with an odd number of electrons is called a *radical*; it is unstable, quickly pairing up with another radical via a covalent bond. For example, atomic hydrogen has a very transitory existence, quickly forming the diatomic hydrogen molecule H_2, in which the electrons' spins are paired. Thus, electrons in stable chemicals appear in pairs. For a further discussion of this topic, see Yeargers [3].

Covalent bonds are very stable. In order to break one, i.e., to dissociate a biomolecule, would require at least four electron volts of energy. For comparison, that much energy is contained by quanta in the ultraviolet region of the electromagnetic spectrum and exceeds that of the visible region of the solar spectrum. (In passing, this helps us to understand why sunlight is carcinogenic—its ultraviolet component alters the chemistry of chemical components of our skin.) If not for the fact that most of the sun's ultraviolet radiation is filtered out by the earth's atmosphere, life on earth would have to be chemically quite different from what it is.

Each kind of atom forms a fixed number of covalent bonds to its atomic neighbors; this number is called the *valence*. The following table gives the atomic numbers and valences of hydrogen, carbon, nitrogen and oxygen.

Atom	Symbol	Atomic No.	Valence
Hydrogen	H	1	1
Carbon	C	6	4
Nitrogen	N	7	3
Oxygen	O	8	2

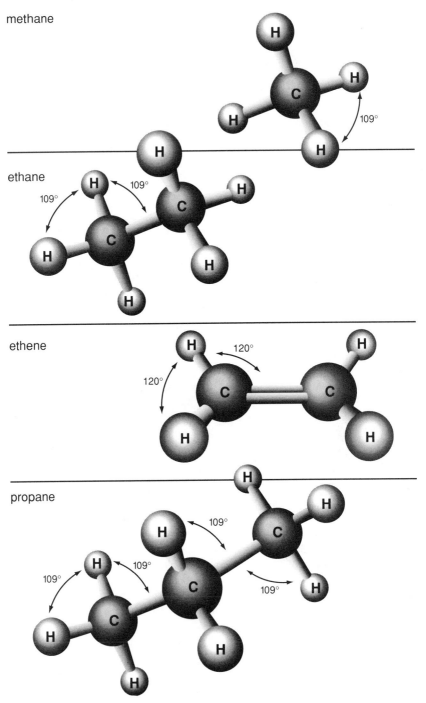

Figure 9.1.2   Three-dimensional models of several small hydrocarbons (containing only hydrogen and carbon atoms). Bond angles are shown.

Figure 9.1.2 shows the structures and names of several common organic molecules. The nominal bond angles are shown; they may fluctuate by several degrees depending on the exact composition of the molecule. In each case the length of the bond is about 0.1 nm, again depending on the constituent atoms. Note also that double bonds are possible, but only if they are consistent with the valences given above.

You can see from Figure 9.1.2 that there are only two basic bonding schemes: If the molecule has only single bonds, the bond angles are 109°, and if there is a double bond, the bond angles are 120°. Note that the former leads to a three-dimensional shape and the latter to a planar shape. This should become evident if you compare the structures of ethane and ethene.

Hydrogen bonds are weak interactions.

Figure 9.1.3 shows some more molecular models, containing oxygen and nitrogen. These molecules are electrically neutral: Unless ionized, they will not migrate toward either pole of a battery. Unlike hydrocarbons, however, their charges are not uniformly distributed. In fact, nitrogen and oxygen atoms in molecules have pairs of electrons (called lone pairs) that are arranged in a highly asymmetrical way about the nucleus. Figure 9.1.3 shows the asymmetrically oriented electrons of nitrogen and oxygen. There are three important points to be noted about these pictures: First, the reason that lone pair electrons are "paired" is that they have opposite spin directions from one another, as was described earlier. Second, the angles with which the lone pairs project outward are consistent with the 109° or 120° bond angles described earlier. Third, it must be emphasized that these molecules are electrically neutral—their charges are not uniformly distributed in space, but they total up to exactly zero for each complete molecule. The presence of lone pairs has important structural consequences to molecules that contain nitrogen or oxygen. Consider the water molecule shown in Figure 9.1.3. Two lone pairs extend toward the right and bottom of the picture, meaning that the right and lower ends of the molecule are negative. The entire molecule is neutral, so therefore the left and upper ends must be positive. We associate the negative charge with the lone pairs and the positive charge with the nuclei of the hydrogen atoms at the other end. Such a molecule is said to be *dipolar*.

Dipolar molecules can electrically attract one another, the negative end of one attracting the positive end of the other. In fact, a dipolar molecule might enter into several such interactions, called *hydrogen bonds* (H-bonds). Figure 9.1.4 shows the H-bonds in which a water molecule might participate. Note carefully that the ensemble of five water molecules is not planar.

Hydrogen bonds are not very strong, at least when compared to covalent bonds. They can be broken by energies of the order of 0.1 eV, an energy that is thermally available at room temperature. There are two mitigating factors, however, that make H-bonds very important in spite of the ease with which they can be broken. The first is their sheer number. Nitrogen and oxygen are very common atoms in living systems, as mentioned earlier, and they can enter into

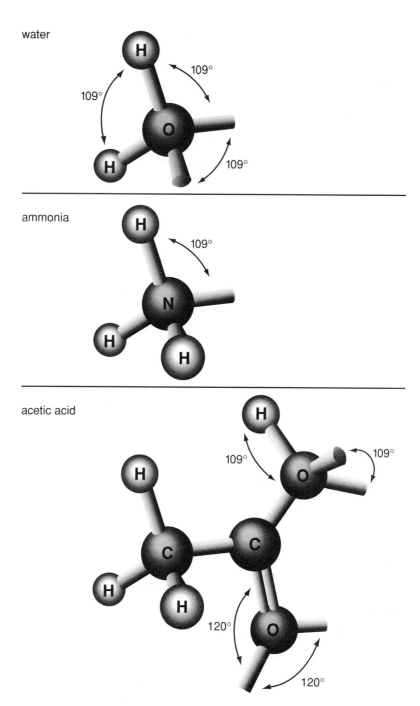

Figure 9.1.3 Three-dimensional models of molecules containing oxygen and nitrogen. The stubs originating on the oxygen and the nitrogen atoms, but not connected to any other atom, represent lone pairs, or asymmetrically oriented electrons.

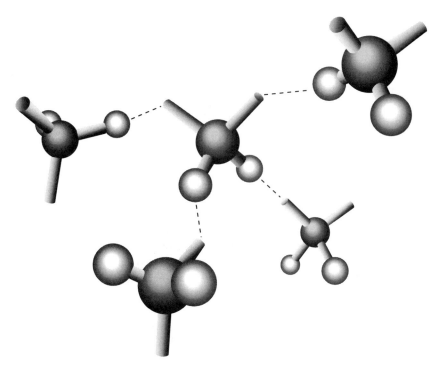

Figure 9.1.4 Three-dimensional model of one possible transient arrangement of water molecules in the liquid phase. The central molecule is hydrogen-bonded to four
other molecules, each of which is in turn hydrogen-bonded to four. The hydrogen
bonds are represented by dotted lines between the lone pairs and hydrogen protons.
This configuration will break up quickly at room temperature, and the molecules will
reform into other, similar configurations.

H-bonding with neighboring, complementary H-bonding groups. While each H-
bond is weak, there are so many of them that they can give considerable stability
to systems in which they occur.

The second factor complements the first: The weakness of H-bonds means
that the structures they stabilize can be altered easily. For example, every water
molecule can be held by H-bonds to four other water molecules (see Figure 9.1.4).
At 20–30 degrees Celsius there is just enough heat energy available to break these
bonds. Thus, H-bonds between water molecules are constantly being made and
broken, causing water to be a liquid at room temperature. This allows biological chemistry to be water-based at the temperatures prevailing on the earth. As a
second example, we shall see later that the genetic chemical DNA is partly held
together by H-bonds that have marginal stability at body temperature, a considerable chemical convenience for genetic replication, which requires partial disas-
sembly of the DNA.

Hydrogen bonding plays a critical role in a number of biological phenomena.
Solubility is an example: A molecule that is capable of forming hydrogen bonds
tends to be water soluble. We can understand this by substituting any other dipo-

lar molecule (containing one or more lone pairs) for the central water molecule of Figure 9.1.4. On the other hand, a molecule lacking lone pair electrons is not water soluble. Look at the propane molecule in Figure 9.1.2 and note that such *hydrocarbons* lack the ability to dissolve in water because they lack the necessary asymmetrical charges needed for H-bonding. We shall return to the topic of H-bonding when nucleic acids and heat storage are discussed later in this chapter. These topics, as well as other kinds of chemical bonding interactions, are discussed by Yeargers [3].

Section 9.2

Biopolymers

At the beginning of this chapter it was pointed out that the attribute we call "life" is due to the organization, not the rarity, of constituent atoms. Figures 9.1.2 and 9.1.3 showed that sequences of carbon, oxygen and nitrogen atoms, with their many bends and branches, can potentially combine to form elaborate three-dimensional macromolecules. What happens is that atoms combine to form molecular monomers having molecular weights of the order of a few hundred. In turn, these monomers are chained into linear or branched macromolecular polymers having molecular weights of up to a billion. The ability to create, organize and maintain these giant molecules is what distinguishes living things from non-living things.

Polysaccharides are polymers of sugars.

A typical *sugar* is *glucose*, shown in Figure 9.2.1(a). The chemical characteristics that make glucose a sugar are the straight chain of carbons, the multiple $-OH$ groups and the double-bonded oxygen. Most of the other sugars we eat are converted to glucose, and the energy is then extracted via the conversion of glucose to carbon dioxide. This process is called *respiration*; it will be described below. A more common configuration for a sugar is exemplified by the ring configuration of glucose, shown in Figure 9.2.1b.

The polymerization of two glucose molecules is a *condensation* reaction, shown in Figure 9.2.2. Its reverse is *hydrolysis*. We can extend the notion of sugar polymerization into long linear or branched chains, as shown by the arrows in Figure 9.2.2. The actual function of a polysaccharide, also called a *carbohydrate*, will depend on the sequence of component sugars, their orientations with respect to each other and whether the chains are branched.

Polysaccharides serve numerous biological roles. For example, plants store excess glucose as *starch*, a polysaccharide found in seeds like rice and wheat (flour is mostly starch). The structural matter of plants is mainly *cellulose*; it comprises most of what we call wood. When an animal accumulates too much glucose, it is polymerized into *glycogen* for storage in the muscles and liver. When we need glucose for energy, glycogen is hydrolyzed back to monomers. These and other functions of sugars will be discussed later in this and subsequent chapters.

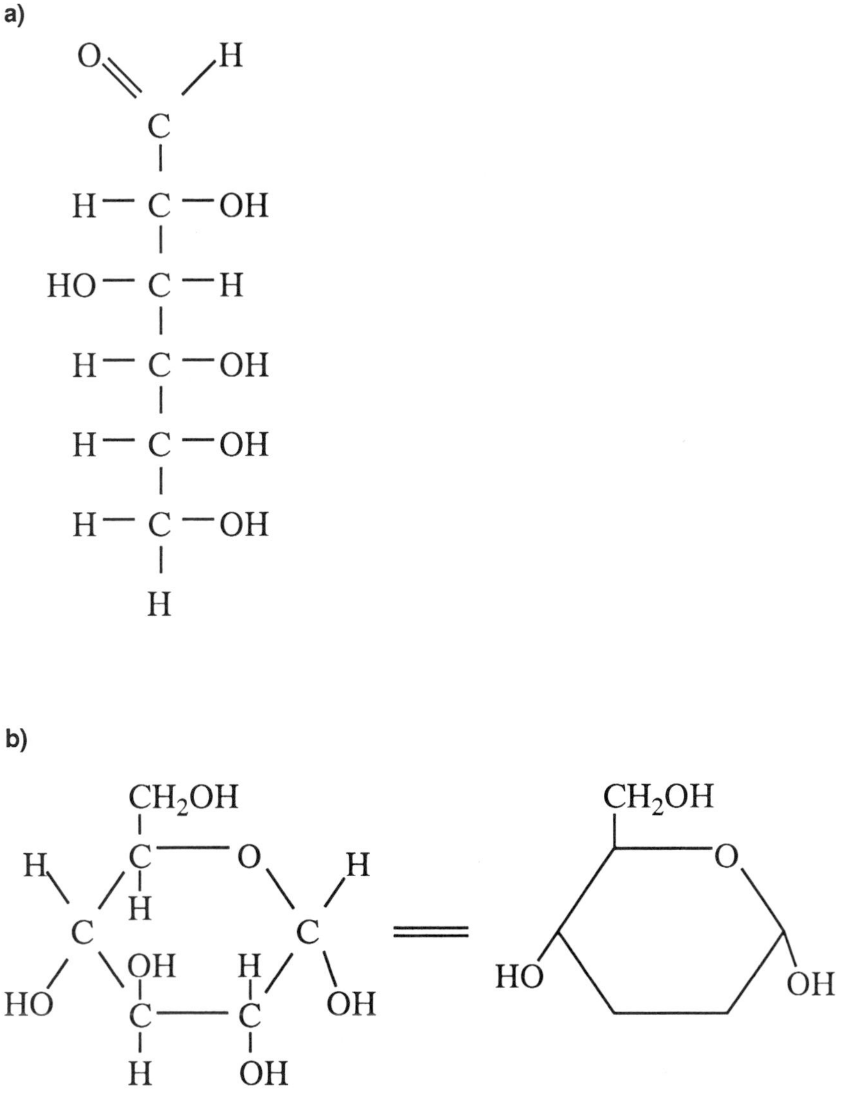

a)

b)

Figure 9.2.1 (a) A model of the linear form of the glucose molecule. (b) A model of the ring form of the glucose molecule. The right-hand version, which omits many of the identifying symbols, is the more common representation.

Lipids are polymers of fatty acids and glycerol.

Fatty acids, exemplified in Figure 9.2.3, are distinguished from each other by their lengths and the positions of their double bonds. Note the organic acid group (-COOH) at one end. Fatty acids with double bonds are said to be *unsaturated*;

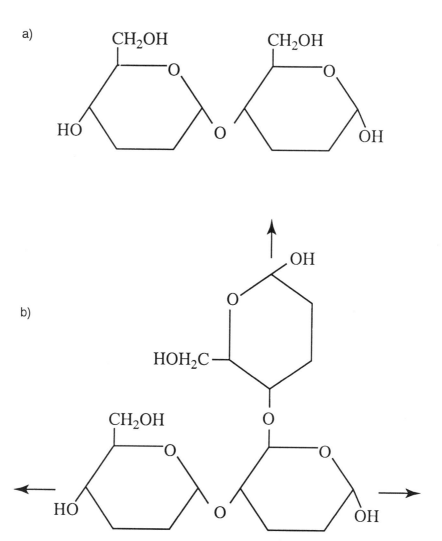

Figure 9.2.2   (a) A model of a disaccharide, consisting of two glucose molecules. (b) A model showing the three possible directions that the polysaccharide of (a) could be extended. A large polysaccharide, with many such branches, would be very complex.

polyunsaturated fatty acids are common in plants whereas saturated fatty acids, lacking double bonds, are common in animals. *Glycerol* and three fatty acids combine to form a *lipid*, or *fat*, or *triglyceride*, as pictured in Figure 9.2.4. The reverse reaction is again called hydrolysis.

Lipids are efficient at storing the energy of excess food that we eat; a gram of lipid yields about four times the calories of a gram of other foods, e.g., carbo-

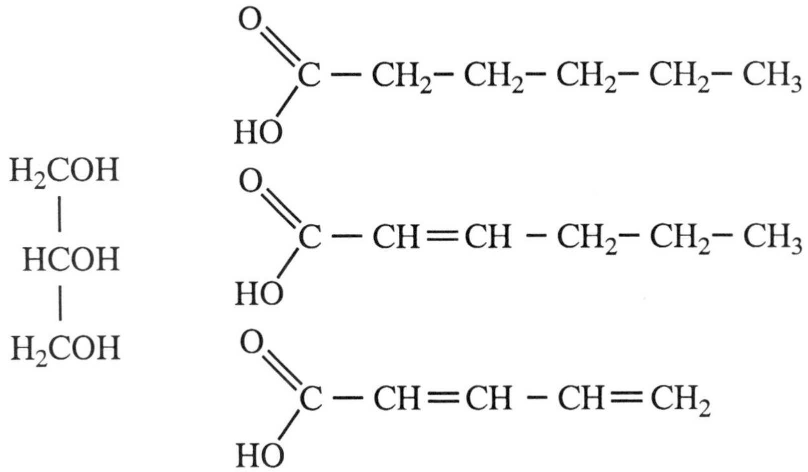

$$O \atop \diagdown C - CH_2 - CH_2 - CH_2 - CH_2 - CH_3 \atop HO \diagup$$

H_2COH

$|$

$HCOH$

$|$

H_2COH

$$O \atop \diagdown C - CH = CH - CH_2 - CH_2 - CH_3 \atop HO \diagup$$

$$O \atop \diagdown C - CH = CH - CH = CH_2 \atop HO \diagup$$

Figure 9.2.3 A model of a glycerol molecule (left) and three arbitrary fatty acids (right).

hydrates and proteins. Lipids are fundamental components of cell membranes: A common lipid of cell membranes is a phospholipid, pictured in Figure 9.2.5. You should now be able to put Figure 9.2.5 into the context of Figure 6.1.1. Note how the hydrocarbon regions of the phospholipid are in the interior of the membrane and how the hydrophilic oxygen groups (having lone pair electrons) are on the membrane's exterior, where they can hydrogen-bond to the surrounding water.

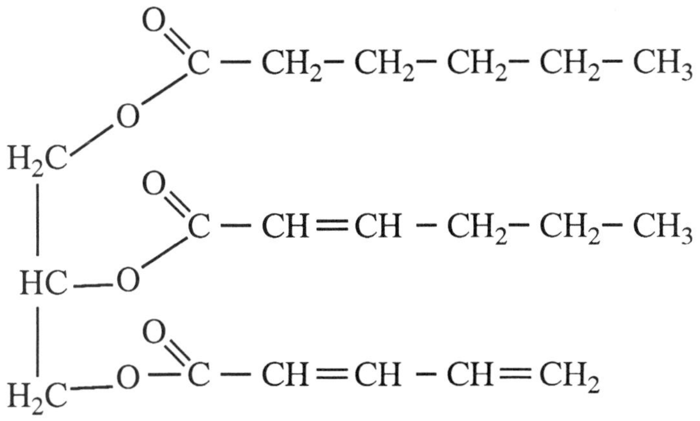

$$O \atop \diagdown C - CH_2 - CH_2 - CH_2 - CH_2 - CH_3 \atop O \diagup$$

$H_2C \diagup$

$|$

$HC - O$

$|$

H_2C

$$O \atop \diagdown C - CH = CH - CH_2 - CH_2 - CH_3 \atop O \diagup$$

$$O \atop \diagdown C - CH = CH - CH = CH_2$$

Figure 9.2.4 A model of a fat, or triglyceride. It consists of a glycerol and three fatty acids. (Compare Figure 9.2.3.)

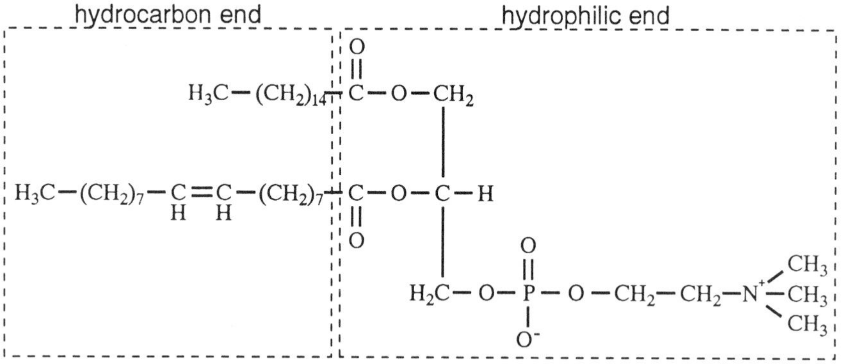

hydrocarbon end   hydrophilic end

$$H_3C-(CH_2)_{14}-\overset{\overset{\displaystyle O}{\displaystyle \|}}{C}-O-CH_2$$

$$H_3C-(CH_2)_7-\underset{H}{C}=\underset{H}{C}-(CH_2)_7-\underset{\underset{\displaystyle O}{\displaystyle \|}}{C}-O-C-H$$

$$H_2C-O-\overset{\overset{\displaystyle O}{\displaystyle \|}}{\underset{\underset{\displaystyle O^-}{\displaystyle |}}{P}}-O-CH_2-CH_2-N^+\begin{matrix} CH_3 \\ -CH_3 \\ CH_3 \end{matrix}$$

Figure 9.2.5 A phospholipid, or phosphoglyceride, found in cell membranes. Note that it has a hydrophilic end that is attracted to water and a hydrocarbon (hydrophobic) end that is repelled by water. The hydrophilic end faces the aqueous outside world or the aqueous interior of the cell. The hydrophobic end of all such molecules is in the interior of the membrane, where there is no water. This picture should be compared to the schematic lipids shown in Figure 6.1.1: The circles on the phopholipids of Figure 6.1.1 correspond to the right-hand box of this figure and the two straight lines of Figure 6.1.1 correspond to the two hydrocarbon chains in the left-hand box of this figure.

Nucleic acids are polymers of nucleotides.

Nucleic acids contain the information necessary for the control of a cell's chemistry. This information is encoded into the sequence of monomeric units of the nucleic acid, called *nucleotides*, and is expressed as chemical control through a series of processes called the Central Dogma of Genetics—to be described below. When a cell reproduces *asexually*, its nucleic acids are simply duplicated and the resultant molecules are partitioned equally among the subsequent daughter cells, thus assuring that the daughter cells will have the same chemical processes as the original cell. In *sexual reproduction*, nucleic acids from two parents are combined in fertilization, resulting in an offspring whose chemistry is related by sometimes complex rules to that of its parents.

There are two kinds of nucleic acids, *deoxyribonucleic acid* (DNA) and *ribonucleic acid* (RNA). The monomer of a nucleic acid is a *nucleotide*, which is composed of three parts: a sugar, one or more phosphate groups and a nitrogenous base. Figure 9.2.6 shows the components of a typical nucleotide.

DNA is a double helix. Figure 9.2.7 shows a model of the macromolecule, partially untwisted to reveal its underlying structure. Note that it is formed from two covalently linked, linear polymers, which are wrapped around each other. The two single strands are H-bonded to one another, as shown by dotted lines in the figure. Figure 9.2.8 shows the details of the H-bonding between DNA nucleotides.

The DNA molecule is very long compared to its width. The double helix is $2.0 \times 10^{-9} m$ wide, but about $10^{-3} m$ long in a bacterium and up to $1 m$ long in

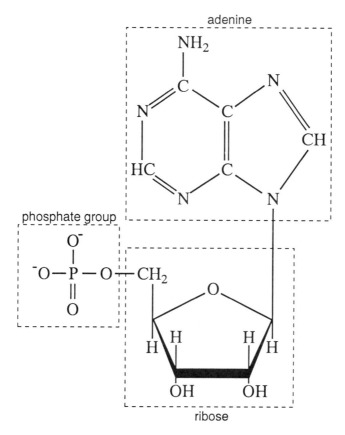

Figure 9.2.6 A typical nucleotide, consisting of a nitrogenous base (adenine), a sugar (ribose) and a phosphate group. Other nucleotides can have other nitrogenous bases, a different sugar and more phosphates.

a human. There are ten base pairs every $3.4 \times 10^{-7}m$ of length of double helix. Thus, a 1 meter-long DNA molecule has about 3×10^8 base pairs. If any of the four nucleotides can appear at any position, there could exist $4^{3 \times 10^8}$ possible DNA molecules of length $1m$. Obviously an incredible amount of information can be encoded into such a complex molecule. Note that DNA uses only a four-letter "alphabet," but can compensate for the small character set by writing very long "words."

There are some important structural details and functional consequences to be noted about Figures 9.2.7 and 9.2.8.

1. Each of the two single-stranded polymers of a DNA molecule is a chain of covalently linked nucleotides. All four possible nucleotides are shown, but there are no restrictions on their order in natural systems; any nucleotide may appear

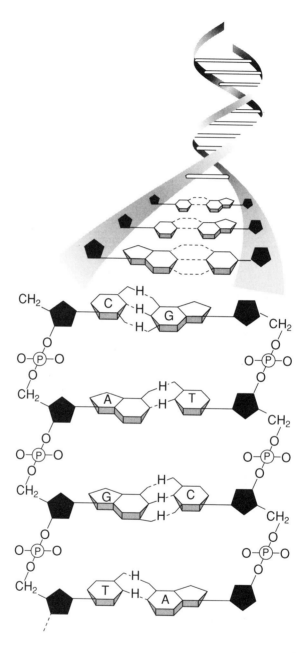

Figure 9.2.7 A DNA molecule, showing the arrangement of the nucleotide components into two covalent polymers, each of which is hydrogen-bonded to the other. Note that A (adenine) and T (thymine) are hydrogen-bonded to each other, and C (cytosine) and G (guanine) are hydrogen-bonded to each other. The hydrogen bonds are indicated by the dashes. (Redrawn from "Biology – The Unity and Diversity of Life," by Cecie Starr and Ralph Taggart, 6th ed., 1992; Wadsworth Publishing Company; Belmont, CA. Used with permission.)

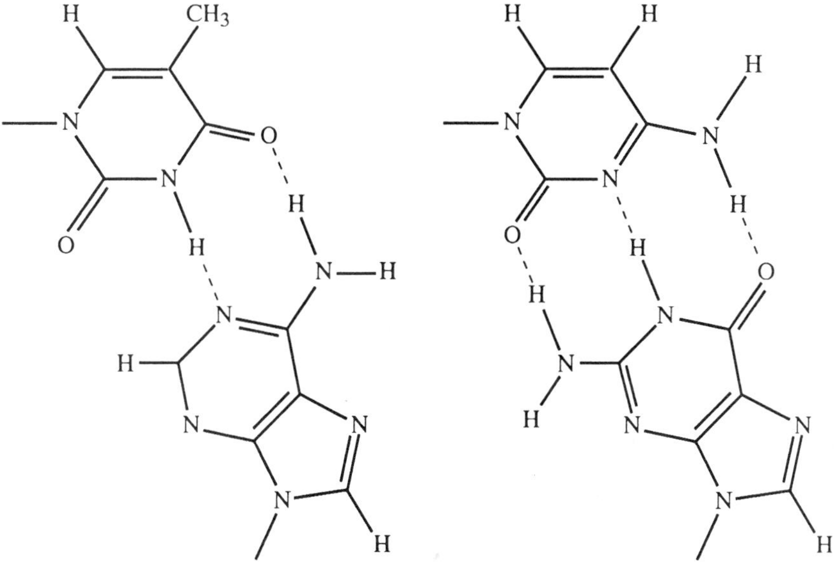

Figure 9.2.8 A detailed picture of the complementary hydrogen bonds between A and T (left pair), and between C and G (right pair). Compare this figure to the hydrogen-bonded groups in Figure 9.2.7. See text for details.

at any position on a given single strand. It is now possible experimentally to determine the sequences of long DNA chains.

2. Once a particular nucleotide is specified at a particular position on one strand, the nucleotide opposite it on the other strand is completely determined. Note that A and T are opposite one another, as are C and G; no other base pairs are allowed in DNA. (From now on we shall indicate the names of the nucleotides by their initials, e.g., A, T, C and G.) There are very important physical and biological reasons for this *complementary* property. The physical reason can be seen by a close examination of the H-bonds between an A and a T or between a C and a G in Figure 9.2.8. Recall that an H-bond is formed between a lone pair of electrons and a hydrogen nucleus and note that two such bonds form between A and T and that three form between C and G. There are no other ways to form two or more strong H-bonds between any of these nucleotides; thus, the ways shown in Figure 9.2.7 are the only possibilities. For example, A cannot effectively H-bond to C or G. You should note that the property of complementary H-bonding requires that the two single strands have different nucleotide sequences, but that the sequence of one strand is utterly determined by the other.

3. The helical configuration is a spontaneous consequence of H-bonding the two single strands together. Helicity disappears if the H-bonds are disrupted. Recall from the discussion of the structure of water in Section 9.1 that H-bonds have marginal stability at room temperature. We should therefore expect that the two strands of helical DNA can be separated, i.e., the helix can be *denatured*, without expending much energy. In fact, DNA becomes denatured at around 45–55°C, only about 8 to 18 degrees above body temperature. Once thermal denaturation has occured, however, the two strands can often spontaneously reassociate into their native double helical configuration if the temperature is then slowly reduced. This should be expected in light of complementary H-bonding between the two strands.

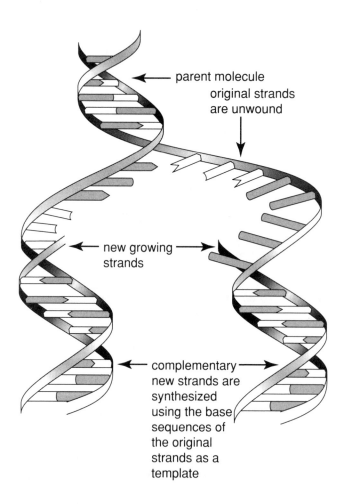

parent molecule original strands are unwound

new growing strands

complementary new strands are synthesized using the base sequences of the original strands as a template

Figure 9.2.9 A model of a replicating DNA molecule. The two strands of the parent double helix separate, and each one acts as a template for a new strand. Complementary hydrogen bonding assures that the two resulting double helices are exact copies of the original molecule. (Redrawn from "Biology," 1st ed., by Joseph Levine and Kenneth Miller; D. C. Heath Company; Lexington, MA, 1991.)

There is another important structural feature related to double helicity: Look at Figure 9.2.7 and note that each nucleotide is fitted into the polynucleotide in such a way that it points in the same direction along the polymer. It is therefore possible to associate directionality with any polynucleotide. In order for the two strands of any nucleic acid to form a double helix, they must have opposite directionalities, i.e., they must be antiparallel to each other.[3]

4. Complementry hydrogen bonding provides a natural way to replicate DNA accurately.   This is the biological reason for complementary H-bonding and is illustrated in Figure 9.2.9. The two strands of DNA are separated and each then acts as a *template* for a new, complementary strand. In other words, the sequence information in each old strand is used to determine which nucleotides should be inserted into the new, complementary strand. *This mechanism allows DNA to code for its own accurate replication*, which is a necessary requirement for a genetic chemical.

---

[3]For example, look at the location of the methyl group (-CH$_2$-) between the phosphate group and the ribose group. Note how it is in a different position on the two strands.

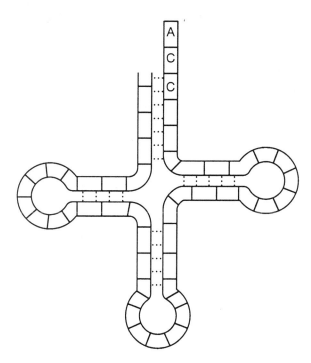

Figure 9.2.10   A model of a transfer RNA molecule. A single-stranded tRNA molecule folds back on itself and becomes double helical in the regions shown by the dotted hydrogen bonds. (The actual helicity is not shown on the figure.) Note that there are several non-helical (non-hydrogen-bonded) turns, at the bottom, right and left sides.

Occurring just prior to cell division, the process of DNA self-replication yields two double-stranded DNA molecules that are exact copies of the original. Then, during cell division, each of the daughter cells gets one of the copies. The two daughter cells thus each end up with the same genetic material that the original cell had, and should therefore also have the same life properties.

There are three classes of RNA molecules: The first is called *messenger RNA*, or mRNA. Each piece of mRNA averages about a thousand bases in length, but is quite variable. It is single-stranded and nonhelical. The second kind of RNA is *transfer RNA*, or tRNA. There are several dozen distinguishable members of this class; they contain in the range of 75 to 95 bases, some of which are not the familiar A, T, C and G. tRNA is single-stranded but is double helical. This unexpected shape is due to the folding over of the tRNA molecule, as shown in Figure 9.2.10. The third kind of RNA is *ribosomal RNA*, or rRNA. This molecule accounts for most of a cell's RNA. It appears in several forms in cellular organelles associated with protein synthesis, and it has molecular weights ranging from around a hundred up to several thousand. The functions of the various RNAs will be discussed shortly.

Proteins are polymers of amino acids.

The monomer of a protein is an amino acid, a synonym for which is *residue*. A protein polymer is often called a *polypeptide*. While many amino acids can exist, only twenty are found in proteins. They share the general structure shown in Figure 9.2.11. The group labelled *R* can take on twenty different forms, thus accounting for all members of the group.[4] The right end (-COOH) is the *carboxyl* end and the bottom (-NH$_2$) is the *amino* end.

Figure 9.2.12 shows how two amino acids are polymerized into a dipeptide (two residues). Note that the attachment takes place by combining the amino end of one residue with the carboxyl end of the other. The covalent bond created in this process is called a *peptide bond*, as shown in Figure 9.2.12.

---

[4]We will ignore the fact that one of the amino acids is a slight exception.

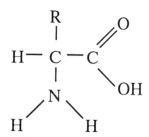

Figure 9.2.11   A model of an amino acid, which is the monomer of a protein. The label "*R*" stands for any one of twenty different groups. (The text mentions a slight exception.) Thus, twenty different amino acids may be found in proteins.

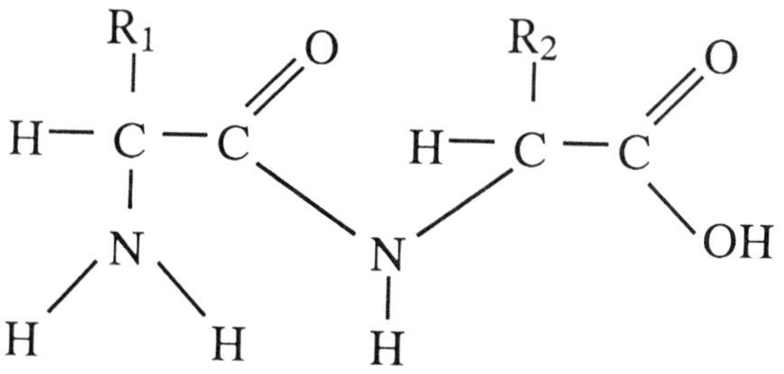

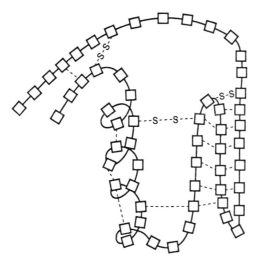

Figure 9.2.12 A pair of amino acids bonded covalently into a dipeptide. The labels R_1 and R_2 can be any of the twenty groups mentioned in the caption of Figure 9.2.11. Thus, there are 400 different dipeptides.

An interesting feature of a dipeptide is that, like an individual amino acid, it has both a carboxyl end and an amino end. As a result, it is possible to add other residues to the two ends of the dipeptide and thereby to extend the polymerization process as far as we like. It is quite common to find natural polypeptides of hundreds of residues and molecular weights over a hundred thousand. Figure 9.2.13 is an idealized picture of a polypeptide "backbone"; the individual amino acids are represented as boxes. Note that the polymer has a three-dimensional structure that includes helical regions, sheet-like regions and that the whole 3-D shape is maintained by H-bonds and disulfide (-S-S-) bonds. The disulfide bonds are covalent and the two amino acids that contribute the sulfur atoms are

Figure 9.2.13 A model of a single protein, or polypeptide, molecule. Each box corresponds to an amino acid. The resultant chain is held in a roughly ovate shape by sulfur-to-sulfur covalent bonds and by many hydrogen bonds, a few of which are indicated by dashed lines.

generally far from one another as measured along the polymer. They are brought into juxtaposition by the flexibility of the polymer and held there by the formation of the disulfide bond itself.

Our model of a protein is that of a long polymer of amino acids, connected by peptide bonds and folded into some kind of 3-D structure. At any location any of twenty different amino acids may appear. Thus, there are 20^{100} possible polypeptides of 100 amino acids in length. Nowhere near this number have actual biological functions, but the incomprehensibly large number of possible amino acid sequences allows living systems to use proteins in diverse ways. Some of these ways will be described next.

Some proteins are catalysts.

There exists a very important class of proteins, called *enzymes*, whose function it is to to speed up the rate of biochemical reactions in cells (see References [2] and [4]). In order to understand this function we must understand what is meant by "reaction rate": Suppose there is a chemical reaction described by $A \Longleftrightarrow B$, as shown in Figure 9.2.14. Let us suppose that initially there is lots of A and no B. As time passes some A is converted to B, and some B back to A. Eventually the relative amounts of A and B reach steady values, i.e., do not change with time. This final state is called an *equilibrium state*. The speed with which A is converted to B is the rate of the *reaction*. The observed rate evidently changes with time, starting out fast and reaching a net of zero at equilibrium, and therefore it is usually measured at the outset of the experiment, when there is lots of A and no B.

There are several very important biological consequences of enzymatic catalysis. First, the essential effect of a catalyst is to speed up the rate of a reaction. A biochemical catalyst, i.e., an enzyme, can speed up the rate of a biochemical reaction by as much as 10^{13} times. This enormous potential increase has some very important consequences to cellular chemistry: First, catalyzed biochemical reactions are fast enough to sustain life, but uncatalyzed reactions are not. Secondly, if a reaction will not proceed at all in the absence of a catalyst, then no catalyst can ever make it proceed. After all, speeding up a rate of zero by 10^{13} still gives a rate

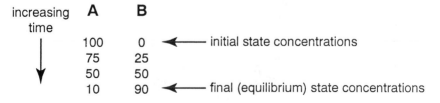

Figure 9.2.14   The progress of the reaction $A \leftrightarrow B$. The numbers give the amounts of the compounds A and B at various times. At the outset there is no B but, as time passes, the amount of B increases until A and B reach equilibrium at a ratio of B : $A = 9 : 1$.

of zero. Third, catalysts have no effect whatever on the relative concentrations of reactants and products at equilibrium, but they do affect the time the system takes to reach that equilibrium. Thus, enzymes do not affect the underlying chemistry or net energetic requirements of the system in which they participate. Fourth, enzymes are very specific as to the reactions that they catalyze, their activity usually being limited to a single kind of reaction. This observation can be combined with the first one above (enzymatic increase in reaction rate) to yield an important conclusion: Whether or not a particular biochemical reaction goes at a high enough rate to sustain life depends entirely on the presence of specific enzyme molecules that can catalyze that particular reaction. Thus, enzymes act like valves, facilitating only the reactions appropriate to a particular cell. No other reactions proceed fast enough to be significant and so they can be ignored.

The valve-like function of enzymes explains why a human and a dog can eat the same kind of food, drink the same kind of water and breathe the same air, yet not look alike. The dog has certain enzymes that are different from those of the human (and, of course, some that are the same). Thus, many biochemical reactions in a dog's cells proceed in a different direction from those in a human— in spite of there being the same initial reactants in both animals. Figure 9.2.15 shows how different metabolic paths can originate from the same starting point

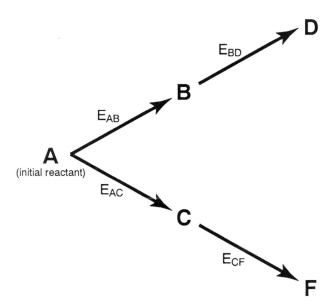

Figure 9.2.15 A diagram showing how enzymes can direct sequences of reactions. A is the initial reactant, and the pair of enzymes E_{AB} and E_{BD} would catalyze the conversion of A to D. Alternatively, the enzymes E_{AC} and E_{CF} would catalyze the conversion of A to F. It is the enzymes, not the initial reactant, that determine what the end product will be. Of course, this does not mean that there will always exist an enzyme that can catalyze a particular reaction; rather, there will almost always exist an enzyme that can catalyze the particular reactions needed by a given cell.

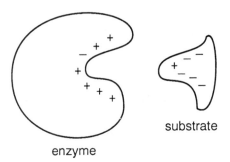

substrate

enzyme

Figure 9.2.16 A model of the "lock and key" mechanism for enzyme-substrate specificity. The enzyme and the substrate are matched to each other by having complementary shapes and electrical charge distributions.

because of different enzyme complements. The same reasoning explains why two people have different hair color, or numerous other differences.

The nature of the specificity of an enzyme for a single chemical reaction can be understood in terms of a "lock and key" mechanism: Suppose that we are again dealing with the reaction $A \Longleftrightarrow B$, catalyzed by the enzyme E_{AB}. The catalytic event takes place on the surface of the enzyme at a specific location, called the *active site*, as shown in Figure 9.2.16. The compound A, or *substrate*, has a shape and electrical charge distribution that are complementary to the active site. This assures that only the reaction $A \Longleftrightarrow B$ will be catalyzed. Note that this reaction is reversible, and that the enzyme catalyzes in both directions.

Now we are in a position to understand why the 3-D structure of an enzyme is so important. Refer back to Figure 9.2.13 and recall that H-bonds and disulfide bonds hold together amino acids that are far from one another in the primary amino acid sequence. Therefore, the active site may be composed of several amino acids that are separated along the polymeric chain by a hundred or more intervening amino acids, but that are held close together by virtue of the folded 3-D polypeptide structure. This means that anything that disturbs, or denatures, the folded structure may disrupt the active site and, therefore, destroy enzymatic activity. All that is necessary is to break the hydrogen and disulfide bonds that maintain the 3-D structure. We can now see why cells are sensitive to heat: Heating to about 50°C inactivates their enzymes, quickly reducing the rates of their reactions to almost zero. Later in this chapter we will return to the topic of enzymatic function.

Noncatalytic proteins.

The immense diversity of possible protein structures allows these macromolecules to be used for many biological purposes. Many of these have nothing to do with catalysis. We will divide these noncatalytic proteins into two somewhat arbitrary, but customary, categories and discuss them next.

Category 1 – Fibrous proteins. These are called "fibrous" because they consist of large numbers of polypeptides arranged in parallel to yield long, string-like arrays. Collagen, for example, is a fibrous protein found in skin and other organs. It consists of shorter protein molecules, each staggered one quarter-length from the next one and thus linked into very long strings). Collagen acts as a binder, the long fibers helping to hold our bodies together.

Other examples of fibrous proteins are found in muscle tissue. Each muscle cell contains large numbers of fibrous proteins that are capable of sliding past one another and exerting force in the process. Our muscles can then move our skeletons and, therefore, our bodies. What we call "meat" is just muscle cut from an animal and, of course, it contains a lot of protein.

Another example of a fibrous protein is keratin, which appears in several forms in hair and nails, among other places. Some keratins form ropes of multiple strands, held together by disulfide bonds. Other keratins form sheet-like structures. One important form of keratin is silk, a thread-like exudation used in the wrapping of the cocoon of the silkworm *Bombyx mori*.

Category 2 – Globular proteins. These proteins tend to be spherical or ovate and are often found dispersed, e.g., dissolved in solution. If aggregated, they do not form fibers. Enzymes are globular proteins, but we have already discussed them and we will therefore restrict our discussion here to noncatalytic globular proteins.

As an example, the polypeptide *hormones* are typical noncatalytic globular proteins. They were introduced in Chapter 8. Hormones are biochemical communicators: They are manufactured in *endocrine glands* in one part of the body and are moved by the bloodstream to another part of the body, where they exert their effects on *target tissues*. At their target tissues, hormones change the production and activity of enzymes and alter membrane permeability.

Insulin, a globular protein hormone, is produced by an organ called the pancreas and is released into the blood to be carried throughout the body. The function of insulin is to regulate the metabolism of glucose in the body's cells. Lack of insulin has powerful metabolic consequences: The disorder *diabetes mellitus* is associated with the loss of insulin-producing cells of the pancreas, increases in the glucose levels of blood and urine, malaise and even blindness.

Another class of noncatalytic globular proteins, introduced in Chapter 6, determines the selectivity of material transport by membranes. These proteins recognize and regulate the intercellular movements of specific compounds like amino acids and various sugars and ions like Na^+ and Cl^-. Called *transport proteins*, or *permeases*, they penetrate through membranes and have a sort of active site on one end to facilitate recognition of the material to be transported. They are not catalysts, however, in that the transported matter does not undergo a permanent chemical change as a result of its interaction with the transport protein.

Globular proteins are used to transport material in the body. One example, *hemoglobin*, which was introduced in Chapter 7, contains four polypeptide chains and four heme groups, the latter being organic groups with an iron atom. Hemoglobin is found in red blood cells, or *erythrocytes*. The principal use of hemoglobin is to carry oxygen from the lungs to the sites of oxygen-utilzing metabolism in the body.

Globular proteins are key molecules in our immune systems. A group of blood cells, called *lymphocytes*, are able to distinguish between "self" and "non-self", and therefore to recognize foreign material, like pathogens, in our bodies. These foreign substances are often proteins but may be polysaccharides and nucleic acids; in any case, if they stimulate immune responses they are called *antigens* (Ag). Antigens stimulate lymphoctes to produce a class of globular proteins, called *antibodies* (Ab) or *immunoglobulins*, that can preferentially bind to Ag, leading to the inactivation of the Ag. The immune response will be discussed in some detail in Chapter 10.

Of particular importance to us in that chapter will be the globular proteins found in a covering, or *capsid*, of viruses. Viruses have very elementary structures, the simplest being a protein coat surrounding a core of genetic material. Viruses are so small that the amount of genetic material they can contain is very limited. Thus, as an information-conserving mechanism, they use multiple copies of the same one or two polypeptides to build their protein coverings. Thus, a typical virus may have an outer coat consisting of hundreds of copies of the same globular protein.

Section 9.3

Molecular Information Transfer

This section is a discussion of molecular genetics. The ability of DNA to guide its own self-replication was described in an earlier section. In this section we will see how genetic information of DNA, coded into its polymeric base sequence, can be converted into base-sequence information of RNA. The base-sequence information of RNA can then be converted into amino acid-sequence information of proteins. The amino acid sequence of a protein determines its 3-D shape and therefore its function, i.e., participation in O_2-transport in erythrocytes, selection of material to cross a membrane or catalysis of a specific biochemical reaction. The net process is contained in the following statement: DNA is the hereditary chemical because it provides an informational bridge between generations via self-replication, and it ultimately determines cellular chemistry. These processes are schematically condensed into the Central Dogma of Genetics:

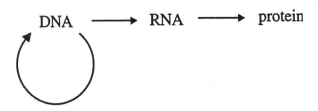

It is very important to recognize that the arrows of the Central Dogma show the direction of information flow, not the direction of chemical reactions. Thus,

DNA passes its information on to RNA—the DNA is not chemically changed into RNA.[5]

Information flow from DNA to RNA is called transcription.

Recall that enzymes determine which reactions in a cell effectively take place. For organisms other than certain viruses, DNA is the source of the information that determines which enzymes will be produced. In any case, there is an intermediary between DNA and proteins—it is RNA. This is expressed in the Central Dogma presented above (see References [5] and [6]).

RNA production is shown schematically in Figure 9.3.1. The sequence of the single covalent strand of RNA nucleotides is determined by complementary H-bonding with one strand of a DNA molecule; in other words, the single, or coding, strand of DNA acts as a template for RNA production. Note the similarity between the use of a single-stranded DNA template for DNA production and the use of a single-stranded DNA template for RNA production. The differences are that RNA uses a different sugar and substitutes uracil in place of thymine.

The process of RNA production from DNA, called *transcription*, requires that the DNA molecule become denatured over a short portion of its length, as shown in Figure 9.3.1. This is a simple matter energetically because all that is required is to break a small number of H-bonds. The O-shaped denatured region moves along the DNA molecule, the double helix opening up at the leading edge of the "O" and closing at its trailing edge. RNA molecules, as mentioned earlier, are usually less than a thousand or so nucleotides long. Thus, RNA replication normally begins at many sites in the interior of the DNA molecule, whose length may be on the order of millions of nucleotides.

Information flow from RNA to enzymes is called translation.

The process of protein production from RNA code brings together, one by one, all three kinds of RNA: ribosomal, messenger and transfer. The three varieties are transcribed from the DNA of the cell and exported to sites away from the DNA. Here subcellular structures called *ribosomes* are constructed, in part using the rRNA. Ribosomes are the sites of protein synthesis, but the actual role of the rRNA is not well understood. Several dozen different kinds of transfer RNA are transcribed from DNA. They all have a structure similar to that shown in Figure 9.2.10, which shows that tRNA is single-stranded, but is helical by virtue of the folding of the polymer onto itself. This requires that some regions on the strand have base sequences that are complementary to others, but in reverse linear order. (Recall from Figure 9.2.8 that a nucleic acid double helix requires that the two strands be antiparallel.) The various kinds of tRNA differ in their constituent bases and overall base sequences; the most important difference for us, however,

---

[5]We will modify the Central Dogma somewhat in Chapter 10.

is the base sequence in a region called the *anticodon*, indicated in the figure. The anticodon is actually a loop containing three bases which, because of the looping, are not H-bonded to any other bases in the tRNA molecule.

Let us consider the anticodon more closely. It contains three nucleotides that are not hydrogen-bonded to any other nucleotides. The number of such trinucleotides, generated at random, is $4^3 = 64$, so we might expect that there could

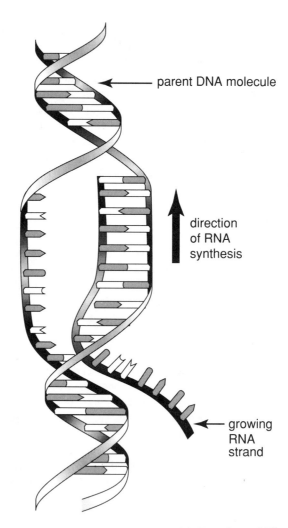

Figure 9.3.1 A model showing the polymerization of RNA, using a DNA template. The DNA opens up to become temporarily single-stranded over a short section of its length, and one of the two DNA strands then codes for the RNA. Complementary hydrogen bonding between the DNA nucleotides and the RNA nucleotides assures the correct RNA nucleotide sequence. (Redrawn from "Biology," 1st ed., by Joseph Levine and Kenneth Miller; D. C. Heath Company, Lexington, MA, 1991. Used with permission.)

be 64 different kinds of tRNA, if we considered only the anticodons. Actually, fewer than that seem to exist in nature, for reasons to be discussed shortly. The anticodon bases are not H-bonded to any other bases in the tRNA molecule, but are arranged in such a 3-D configuration that they could H-bond to three bases on *another* RNA molecule.

All tRNA molecules have a short "pigtail" at one end that extends beyond the opposite end of the polymer. This pigtail always ends with the sequence CCA. An amino acid can be covalently attached to the terminal adenine, giving a tRNA-amino acid molecule, as shown in Figure 9.3.2. A given type of tRNA, identified by its anticodon, can be attached to one, and only one, specific type of amino acid. No other pairings are possible. When we see such specificity in biochemistry we should always suspect that enzymes are involved. In fact, there are enzymes whose catalytic function is to link up an amino acid with its correct tRNA. A tRNA molecule that is attached to its correct amino acid is said to be "charged."

Messenger RNA is transcribed in strings of about 1000 or so nucleotides, but that is only an average figure—much mRNA is considerably longer or shorter. The reason for this variability is that each piece of mRNA is the transcription product of one or a few genes on DNA. Thus, the actual length of a particular

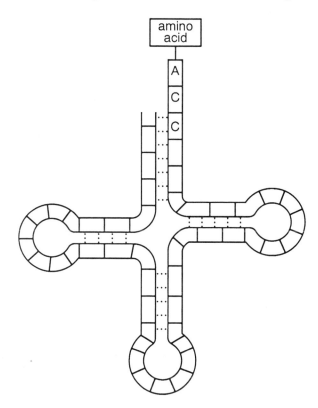

Figure 9.3.2 A model of a tRNA molecule, with an amino acid attached to one end. An enzyme assures that the tRNA molecule becomes covalently attached to its correct amino acid. Compare this figure with Figure 9.2.10.

piece of mRNA corresponds to an integral number of DNA genes and, of course, that leads to a great deal of variability in length. After being exported from the DNA, the mRNA travels to a ribosome where it becomes reversibly attached to the ribosome.

The next part of this discussion is keyed to Figure 9.3.3. (a) One end of a piece of mRNA is attached to a ribosome, the area of association covering at least six mRNA nucleotides. (b) A tRNA molecule, with its correct amino acid attached, forms complementary H-bonds between its anticodon and the first three nucleotides of the mRNA. The latter trinucleotide is called a *codon*. Note that codon–anticodon recognition mates up not only the correct anticodon with its correct codon but, in the process, also matches up the correct amino acid with its codon. (c) Next, a second charged tRNA hydrogen bonds to the second mRNA codon. (d) A peptide linkage forms between the two amino acids, detaching the first amino acid from its tRNA in the process.

Let us review what has happened so far: A sequence of DNA nucleotides comprising a small integral number of genes has been transcribed into a polymer of mRNA nucleotides. The sequence of the first six of these nucleotides has subsequently been translated into the sequence of two amino acids. There is a direct informational connection mapping the sequence of the original six DNA nucleotides into the sequence of the two amino acids. The correctness of this mapping is controlled by two physical factors: First, complementarity between DNA and mRNA and between mRNA and tRNA and, second, by specific enzymatic attachment of tRNA to amino acids.

Now returning to Figure 9.3.3, the ribosome moves three nucleotides down the mRNA and a third charged tRNA attaches to the mRNA at the third codon. (e) A third amino acid is then added to the growing polypeptide chain. The translation process continues and eventually a complete polypeptide chain is formed. The nucleotide sequence of the DNA has been converted into the primary structure of the polypeptide. Note how the conversion of nucleotide sequence to amino acid sequence was a transfer of information, not a chemical change of DNA to protein.

Figure 9.3.3 is really a pictorial representation of the Central Dogma. The overall process yields proteins, including enzymes of course. These enzymes determine what chemical reactions in the cell will proceed at a rate consistent with life. Two very important observations come out of this discussion: First, the chemistry of a cell is ultimately determined by the sequence of DNA nucleotides and, second, because of this, the replication and partitioning of DNA during cell division assures that daughter cells will have the same chemistry as the parent cell. We can extend the latter conclusion: The union of a sperm and an egg in sexual reproduction combines genetic material from two parents into a novel combination of DNAs in a new organism, thus assuring that the offspring has both chemical similarities to, and chemical differences from, each of the parents.

A gene is enough nucleic acid to code for a polypeptide.

The word "gene" is often loosely used to mean "a site of genetic information." A more exact definition from molecular biology is that a gene is a sequence of nu-

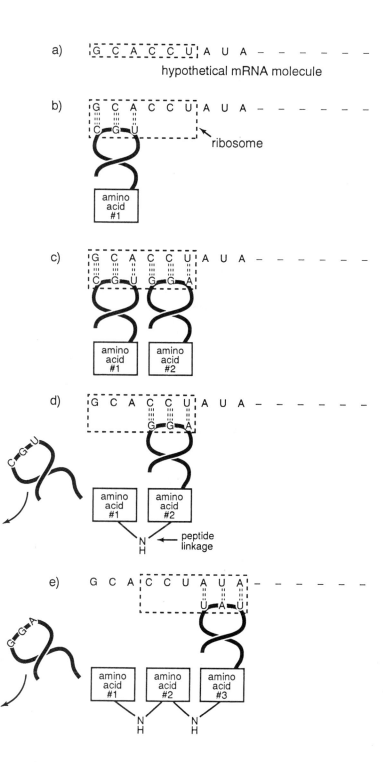

Figure 9.3.3 A simple model of the polymerization of a polypeptide, using DNA information and RNA intermediaries. (a) A ribosome attaches to the end of an mRNA molecule. (b) A molecule of tRNA, with its correct amino acid attached, hydrogen-bonds to the mRNA. The hydrogen-bonding is between the first three nucleotides of the mRNA (a codon) and the three tRNA nucleotides in a turn of tRNA (an anticodon). Each tRNA has several such turns, as depicted in Figures 9.2.10 and 9.3.2, but only one is the anticodon. (c) A second tRNA then hydrogen-bonds to the mRNA, thus lining up two amino acids. (d) The two amino acids are joined by a covalent bond, a process that releases the first tRNA. (e) The ribosome moves down the mRNA molecule by three nucleotides and a third tRNA then becomes attached to the mRNA. The process continues until an intact protein is formed. Note how the amino acid sequence is ultimately dictated by the nucleotide sequence of the DNA.

cleotides that codes for a complete polypeptide. This definition, however, requires the elaboration of several points:

1. If a functioning protein's structure contains two separately created polypeptides then, by definition, two genes are involved.
2. As will be discussed below, some viruses eliminate DNA from their replicative cycle altogether. Their RNA is self-replicating. In those cases, their genes are made of RNA.
3. The nucleotide sequence for any one complete gene normally lies entirely on one strand of DNA, called the *coding strand*. However, not all genes need lie on the same one strand; transcription may jump from one strand to the other between gene locations. Further, there may even be overlapping genes on the same strand.

The concept of coding

The so-called *genetic code* can be presented in a chart showing the correspondence between RNA trinucleotides (codons) and the amino acids they specify. Such charts, for all 64 possible codons, are available in virtually all biochemistry, genetics and introductory biology texts.

Several interesting features emerge from considering such a table. There are 64 codons potentially available to specify 20 amino acids. It turns out, however, that there are only about half that many distinctive tRNA molecules, indicating that some tRNAs can bind to more than one codon. This redundancy is explained by the *wobble hypothesis*: Examination of tRNA structure shows that the nucleotide at one end of the anticodon has only a loose fit to the corresponding codon nucleotide—it wobbles. Thus, H bonding specificity is relaxed at this position and some tRNAs can bind to more than one codon.[6]

Not all possible codons specify an amino acid. Three of them are *termination*, or *stop, codons*. They do not specify any amino acid; rather, they signal the ribosome to cease translation and to release the completed polypeptide. This is especially useful if one piece of mRNA codes for two adjacent genes: Termination codons signal the translation machinery to release the first polypeptide before starting on the translation of the second one. Without the termination codons the ribosome would continue to add the amino acids of the second polypeptide to the end of the first one, negating the biological functions of both.

The nature of mutations

Mutations are changes in the nucleotide sequence of DNA. A base change in a codon would probably result in a new amino acid being coded at that point. For example, sickle cell anemia results from a single incorrect amino acid being

---

[6]Recall that polynucleotides have directionality; thus, the two ends of a codon or anticodon are distinct. Only the one drawn at the right-hand end wobbles.

inserted into the protein fraction of hemoglobin. Suppose a nucleotide pair were deleted: Virtually every amino acid encoded thereafter (downstream) would be incorrect. Evidently the severity of a deletion, or an addition for that matter, depends on how close to the start of transcription it occurs.

Section 9.4

Enzymes and Their Function

Two important concepts that have been presented in this chapter are the Central Dogma of Genetics and the role of enzymes in facilitating specific chemical reactions in a cell. DNA, via RNA, codes for a specific set of cellular enzymes (among other proteins). Those enzymes can catalyze a specific set of chemical reactions and thereby determine the biological nature of the cell.

In this section we will take a closer look at the way that enzymes work. Our approach will be a thermodynamic one, following the path of solar energy into biological systems, where it is used to create orderly arrangements of atoms and molecules in a cell. We will show how enzymes select from among the many possible configurations of these atoms and molecules to arrive at those which are peculiar to that type of cell.

The sun is the ultimate source of energy used by biological systems.

The sun is the ultimate source of energy available to drive biological processes. (We ignore the tiny amounts of energy available from geothermal sources.) Its contributions are two-fold: First, solar energy can be captured by green plants and incorporated into chemical bonds, from which it can be then obtained by animals that eat the plants and each other. Second, solar energy heats the biosphere and thus drives biochemical reactions, vitually all of whose rates are temperature-dependent. Both of these considerations will be important in the discussion to follow.

Entropy is a measure of disorder.

A highly disordered configuration is said to have high *entropy*. The most disordered of two configurations is the one that can be formed in the most ways. To show how this definition conforms to our everyday experience, consider the possible outcomes of tossing three coins: HHH, HHT, HTH, THH, HTT, THT, TTH, TTT. There is only one way to get all heads, but there are six ways to get a mixture of heads and tails. Thus, a mixture of heads and tails is the more disordered configuration. The condition of mixed heads and tails has high entropy (is a disorderly outcome), and the condition of all heads has low entropy (is an orderly outcome). Note that all eight specific configurations have the same probability (1/8), but that six of them contain at least one head and one tail.

Given that there generally are more disordered outcomes than there are ordered outcomes, we would expect that disorder would be more likely than order. This, of course, is exactly what we see in the case of the coins: Throw three coins and a mixture of heads and tails is the most common result, whereas all heads is a relatively uncommon result.

The universe is proceeding spontaneously from lower to higher entropy.

An empirical rule, the Second Principle of Thermodynamics, states that the entropy of the universe increases in every process. For instance, if a drop of ink is placed in a beaker of water, it will spontaneously spread throughout the water. There are few ways to put all the ink into the one spot in the water and many ways to distribute it throughout the water, so we see that the entropy of the water/ink mixture increases. As other examples, consider what happens when the valve on a tank of compressed gas is opened or when a neatly arranged deck of cards is thrown up into the air. In each case, entropy increases.

The Second Principle does not preclude a decrease in entropy in some local region. What it does require is that if entropy decreases in one place it must increase somewhere else by a greater absolute amount. There is no reason why the ink, once dispersed, cannot be reconcentrated. The point is that reconcentration will require some filtration or adsorption procedure that uses energy and generates heat. That heat will cause air molecules to move, and rapidly moving air molecules have more entropy (are more disordered) than slowly moving molecules. Likewise, the air can be pumped back into the tank and the cards can be picked up and resorted, both of which processes require work, which generates heat and, therefore, entropy.

Living systems are local regions of low entropy; their structures are highly organized, and even small perturbations in that organization can mean the difference between being alive and not being alive. From the earlier discussion we can see that nothing in the Second Principle forbids the low entropy of living systems, as long as the entropy of the universe increases appropriately during their formation.

Entropy increases in a process until equilibrium is reached.

Recall the examples of the previous section: The ink disperses in the water until it is uniformly distributed; the gas escapes the tank until the pressure is the same inside and outside of the tank; the cards flutter helter-skelter until they come to rest on a surface. In each case the process of entropy-increase continues to some end-point and then stops. That end-point is called an *equilibrium state.*

Any equilibrium can be disrupted; more water can be added to the ink, the room containing the gas can be expanded and the table bearing the cards can drop away. In each case the system will then find a new equilibrium. Thus, we can regard equilibria as temporary stopping places along the way to the maximal universal entropy predicted by the Second Principle.

Free energy is energy available to do useful work.

Every organism needs energy for growing, moving, reproducing and all the other activities we associate with being alive. Each of these activities requires organized structures. To maintain this organization, or low entropy, requires that the living system expend energy, much as energy was required to reconcentrate the ink or to resort the cards in the earlier examples.

Free energy is energy that can do useful work. In living systems, "useless" work is that which causes a volume change or which increases entropy. Whatever energy is left over is "free" energy. Living systems do not change their volume much, so entropy is the only significant thief of free energy in a cell. Therefore, free energy in a cell decreases when entropy increases. To a good approximation, we can assume that a living system begins with a certain amount of potential energy obtained from sunlight or food; some energy will then be lost to entropy production, and the remainder is free energy.[7]

To a physical chemist, the convenient thing about free energy is that it is a property of the system alone, thus excluding the surroundings. In contrast, the Second Principle requires that one keep track of the entropy of the entire universe. As a result, it is usually easier to work with free energy than with entropy. We can summarize the relationship between the two quantities as they pertain to living systems by saying that entropy of the universe always increases during processes and that a system in equilibrium has maximized its entropy, whereas the free energy of a system decreases during processes and, at equilibrium, the system's free energy is minimized.

Free energy flows, with losses, through biological systems.

Thermonuclear reactions in the sun liberate energy, which is transmitted to the earth as radiation, which is absorbed by green plants. Some of the sun's radiation then heats the plant and its surroundings, and the rest is incorporated into glucose by *photosynthesis*. In photosynthesis some of the free energy of the sun is used to create covalent bonds among parts of six carbon dioxide molecules, forming glucose, the six-carbon sugar, as shown in the following (unbalanced) reaction:[8]

$$CO_2 + H_2O \xrightarrow{\text{light energy}} \text{glucose} + H_2O + O_2.$$

The plant, or an animal that eats the plant, then uses some of the free energy of the glucose to add a phosphate group to adenosine diphosphate (ADP) in the process called *respiration*.

---

[7]If you have studied physical chemistry, you will recognize this quantity specifically as Gibbs' free energy [4].

[8]The reason that water appears on both sides of the reaction equation is that the two water molecules are not the same: One is destroyed and the other is created in the reaction. The reaction shown is a summary of the many reactions that comprise photosynthesis.

$$\text{Glucose} + H_2O + O_2 \qquad\qquad CO_2 + H_2O$$

$$\text{ATP} \qquad\qquad \text{ADP + phosphate}$$

The resultant adenosine triphosphate now has some of the energy that originated in the sun. The ATP can then move around the cell by diffusion or convection and drive various life processes (moving, growing, repair, driving Na/K pumps, etc.).

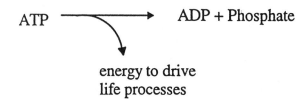

$$\text{ATP} \longrightarrow \text{ADP + Phosphate}$$

$$\downarrow$$

energy to drive
life processes

To recapitulate: Sunlight drives photosynthesis, in which carbon dioxide is combined to make glucose. The latter thus contains some of the energy that originated in the sun. In respiration, the plant or an animal that eats the plant then converts some of the free energy in the glucose into free energy of ATP. Finally, at a site where it is needed, the ATP gives up its free energy to drive a biological process, e.g., contraction of a muscle.

At every step along the way from sun to, e.g., muscle movement, entropy is created and free energy is therefore lost. By the time an animal moves its muscle, only a small fraction of the original free energy the green plant got from the sun remains. If a subsequent carnivore should eat the herbivore, still more free energy would be lost. After the carnivore dies, decomposing organisms get the last of whatever free energy is available to living systems.

The heat generated in biochemical reactions can help drive other reactions.

The earlier discussion pointed out that free energy, ultimately derived from the sun, is used to drive the processes we associate with being alive. As these processes occur entropy is generated. Although the resultant heat energy will eventually be lost to the surroundings, it can be stored for a short while in the water of the cell and thus be used to maintain or increase the rates of cellular chemical reactions.

In order to understand how heat energy can promote chemical reactions we need to digress a bit. If a process were able to occur spontaneously (increasing entropy; decreasing free energy), why would it not have already occurred? Water should spontaneously flow from a lake to the valley below, as shown in Figure 9.4.1(a). This has not happened because there is a dam in the way, but a siphon would take care of that without any net outlay of energy. (Figure 9.4.1(b)) The latter point is critical: The water going up the siphon requires the same amount of energy that it gets back in going down the siphon.[9] From that point on, the water can fall to the valley, developing exactly as much kinetic energy as it would have if the dam had not existed in the first place.

---

[9]We are ignoring friction here.

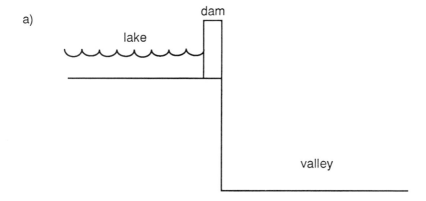

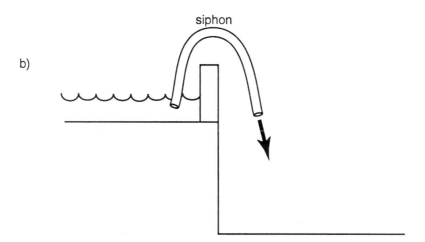

Figure 9.4.1   (a) A lake holds back water above a valley; thus, the water has a certain amount of potential energy with respect to the valley. (b) The water can get past the dam via a siphon, but the energy of the water with respect to the valley is not changed by the trip through the siphon. In other words, the energy yielded by the water in falling to the valley is independent of the path it takes. (We are assuming that friction is negligible.)

The example of the dam is a macroscopic analog to biochemical processes. For example, in respiration a cell takes up glucose, a high free energy compound, and converts it to CO_2, a low free energy compound. This process, on thermodynamic grounds, should therefore be spontaneous. In fact, we can demonstrate a spontaneous change of glucose to CO_2 by putting some glucose into an open dish, from which it will disappear over a period of days to weeks, via conversion to CO_2 and H_2O. The reason the process in the dish takes so long is that there is an intermediate state (in reality, several) between glucose and CO_2 and H_2O, as shown in the free energy diagram in Figure 9.4.2. The intermediate state is called a *transition state*, and it is the analog of the dam in Figure 9.4.1. Before the sugar can change to the gas, releasing its free energy, the transition state must be overcome, i.e., a certain amount of *activation energy* is needed to move the system into the transition state.[10] This energy is returned on the other side of the transition state, after which the chemical system behaves as if the transition state were not there. The examples of the dam and the glucose suggest a general conclusion: The net change in free energy between two states is independent of any intermediate states.

---

[10]Figure 9.4.2 is, of course, only a model. The actual conversion of glucose to carbon dioxide in an open dish would involve numerous intermediate compounds, some of which would be real transition states and some of which would be more-or-less stable compounds. For instructive purposes we represent the system as having a single transition state.

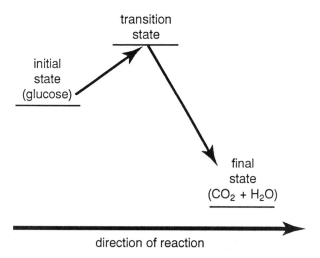

Figure 9.4.2   A free energy diagram of the conversion glucose $\rightleftarrows$ CO_2 + H_2O. There is a transition state between the initial and final states. Even though the conversion of glucose to CO_2 and H_2O is energetically downhill, it will not be a spontaneous conversion because of the transition state.

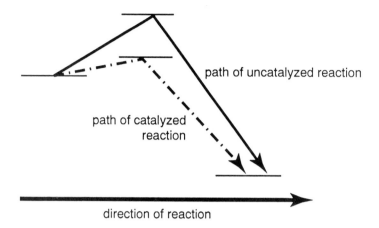

direction of reaction

Figure 9.4.3 The effect of enzymic catalysis on the height of a transition state. The enzyme lowers the energy of the transition state but, as in Figure 9.4.1, the overall change in energy is independent of the path. Lowering the transition state does, however, permit the reaction to proceed spontaneously in the presence of a little thermal energy.

Transition states are the rule, not the exception, and the biochemical reactions of living systems are typical in that those that release free energy must first be activated. There are two sources of activation energy available to cells, however: First, most cells exist at 0–40°C and, second, heat energy is generated by the normal inefficiency of cellular processes.[11] This heat energy is stored in H-bond vibrations in the water of the cell, at least until it is finally lost to the external environment. While this heat energy is in the cell it is available to push systems into their transition states, thus promoting chemical reactions. After serving its activation function, the heat energy is returned unchanged.

The preceding discussion explains how heat serves a vital cellular function in providing activation energy to drive cellular biochemical reactions. This, however, does not close the subject, because activation energy is tied in with another observation: The glucose in a dish changes to CO_2 and H_2O over a period of months, and the same change can occur in a cell in seconds or less. Yet, the temperatures in the dish and in the cell are the same, say 37°C. The difference is that the reactions in the cell are catalyzed by enzymes.

In brief, the catalytic function of an enzyme is to reduce the energy of the transition state and thereby to lessen the amount of heat energy needed by the system to meet the activation energy requirement. In this manner the enzyme speeds up the rate at which the reaction proceeds from the initial state (100% reactant) toward the final, equilibrium state (perhaps 100% product). Figure 9.4.3 is a free energy diagram for a biochemical system in its catalyzed and uncatalyzed conditions. The enzyme catalyst lowers the activation energy and makes it much easier

---

[11]Direct sunlight is also used by many "cold-blooded" animals to heat up their bodies.

for the initial state to be converted into the transition state, and thus into the final state. The dependence of reaction rate on activation energy is exponential; thus, a small change in activation energy can make a very big difference in reaction rate. For comparison, enzymatic catalysis potentially can speed up the rates of reactions by as much as 10^{13} times.

How much energy is actually available? At 30°C the average amount of heat energy available is about 0.025 eV per molecule, but the energy is unevenly distributed, and some substrate molecules will have more and some will have less. Those that have more will often have enough to get to the transition states made accessible by enzymatic catalysis.

Section 9.5

Rates of Chemical Reactions

Stoichometric rules are not sufficient to determine the equilibrium position of a reversible chemical reaction; but adding reaction rate principles makes the calculation possible. Primarily, rate equations were designed to foretell the speed of specific reactions, and, in this capacity, they predict an exponentially decaying speed, as reactants are consumed, characterized by the reaction's rate constant. But in fact, the equilibrium position of a reversible reaction is reached when the rate of formation equals the rate of dissociation. Therefore equilibrium positions, as well as reaction rates, are determined by a combination of the forward and reverse rate constants.

Irreversible (uni-directional) reactions are limited by the first reactant to be exhausted.

Consider the irreversible bimolecular reaction

$$A + B \quad \longrightarrow \quad X + Y, \tag{9.5.1}$$

in which one molecule each of reactants A and B chemically combine to make one molecule each of products X and Y. It follows that the rate of disappearance of reactants equals the rate of appearance of products. The *Conservation of Mass* principle takes the form

$$\frac{dX}{dt} = \frac{dY}{dt} = -\frac{dA}{dt} = -\frac{dB}{dt}.$$

If M_0 denotes the initial number of molecules of species M, by integrating each member of this chain of equalities from time 0 to time t, we get

$$X(t) - X_0 = Y(t) - Y_0 = -A(t) + A_0 = -B(t) + B_0. \tag{9.5.2}$$

Equation (9.5.2) gives the amount of each species in terms of the the others, so if any one of them is known, then they all are. But in order to know the amount of any one of them, we must know how fast the reaction occurs. This is answered by the *Law of Mass Action* (due to Lotka): The rate of at which two or more chemical species simultaneously combine is proportional to the product of their concentrations. Letting $[M]$ denote the concentration of species M, the mass action principle states that the rate at which product is formed is equal to

$$k[A][B] \qquad\qquad (9.5.3)$$

where the constant of proportionality k is characteristic of the reaction.

So far our considerations have been completely general, but now we must make some assumptions about where the reaction is occuring. We suppose this to occur in a closed reaction vessel, such as a beaker with a fixed amount of water. In this case, concentration is the number of molecules divided by, for all species, the same fixed volume of medium.[12] We allow the possibility that one or more of the products, X or Y, be insoluble and precipitate out of solution. This is one of the main reasons that a bimolecular reaction may be irreversible. For such an insoluble species, "concentration" means the ratio of its number of molecules over the volume of the medium, even though it is not dissolved. While a product may precipitate out without disturbing the reaction, the reactants must remain dissolved. We now use the notation $m(t)$, to mean this extended notion of concentration of species M.

Combining the mass action principle with equation (9.5.2) we get

$$\frac{dx}{dt} = kab$$
$$= k(a_0 + x_0 - x)(b_0 + x_0 - x) \qquad\qquad (9.5.4)$$

with initial value $x(0) = x_0$. The stationary points of equation (9.5.4) are given by setting the right-hand side to zero and solving to get (see Section 2.4)

$$x = a_0 + x_0, \qquad \text{or} \qquad x = b_0 + x_0. \qquad\qquad (9.5.5)$$

The first of these says that the amount of X will be its original amount plus an amount equal to the original amount of A. In other words, A will be exhausted. The second equation says the reaction stops when B is exhausted.

Suppose, just for argument, that $a_0 < b_0$. Then also $a_0 + x_0 < b_0 + x_0$. While $x(t) < a_0 + x_0$, the right-hand side of equation (9.5.4) is positive, therefore the derivative is positive, so x increases. This continues until x asymptotically

---

[12]By contrast, for an open reaction vessel, such as the heart or a chemostat, the concentrations are determined by that of the inflowing reactants.

reaches a_0+x_0, whereupon the reaction stops. The progression of the reaction as a function of time is found by solving equation (9.5.4) which is variables separable:

$$\frac{dx}{(a_0 + x_0 - x)(b_0 + x_0 - x)} = k\,dt.$$

Note the similarity of this equation to the Lotka–Voterra system of Section 4.4. The left-hand side can be written as the sum of simpler fractions

$$\frac{1}{(a_0 + x_0 - x)(b_0 + x_0 - x)} = \frac{1}{b_0 - a_0}\frac{1}{a_0 + x_0 - x}$$
$$- \frac{1}{b_0 - a_0}\frac{1}{b_0 + x_0 - x}.$$

Thus equation (9.5.4) may be rewritten as

$$\left[\frac{1}{a_0 + x_0 - x} - \frac{1}{b_0 + x_0 - x}\right] dx = (b_0 - a_0)k\,dt.$$

Integrating gives the solution

$$- \ln(a_0 + x_0 - x) + \ln(b_0 + x_0 - x) = (b_0 - a_0)kt + q$$
$$\ln\left(\frac{b_0 + x_0 - x}{a_0 + x_0 - x}\right) = (b_0 - a_0)kt + q,$$

where q is the constant of integration. Now this may be solved for in terms of x,

$$x = \frac{(a_0 + x_0)Qe^{(b_0-a_0)kt} - (b_0 + x_0)}{Qe^{(b_0-a_0)kt} - 1}, \tag{9.5.6}$$

where $Q = e^q$ is a constant. This equation is graphed in Figure 9.5.1. For the purpose of drawing the figure, we choose the constants, solve the resulting differential equation and draw its graph as follows:

```
> k:=1; a0:=2; b0:=3; x0:=1/2;
> dsolve({diff(x(t),t)=k*(a0+x0-x(t))*(b0+x0-x(t)),x(0)=x0}, {x(t)});
> simplify(");
> x:=unapply(rhs("),t);
> plot([t,x(t),t=0..4],t=-1..3,tickmarks=[3,3],labels=['t','x(t)']);
```

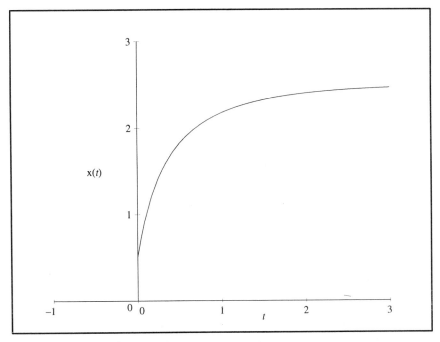

Figure 9.5.1 A typical solution to equation (9.5.4).

The result is

$$x(t) = \frac{1}{2}\frac{-15 + 14e^{-t}}{-3 + 2e^{-t}}.$$

Example: Suppose 2 moles of silver nitrate ($AgNO_3$) are mixed with 3 moles of hydrochloric acid (HCl). A white precipitate, silver chloride, is formed, and the reaction tends to completion,

$$AgNO_3 + HCl \longrightarrow AgCl \downarrow + HNO_3.$$

From above, asymptotically, the reaction stops when the 2 moles of silver nitrate have reacted, leaving 2 moles of silver chloride precipitate and 1 mole of hydrochloric acid unreacted.

Kinetics for reversible reactions work the same way.

Now assume reaction (9.5.1) is reversible

$$A + B \; \rightleftarrows \; X + Y \tag{9.5.7}$$

with the reverse reaction also being bimolecular. This time there is a backward rate constant, k_{-1} as well as a forward one, k_1. From the mass action principle applied to the reverse reaction, we have

$$\text{rate of conversion of X} + \text{Y} = k_{-1}[X][Y].$$

Under normal circumstances, the forward and backward reactions take place independently of each other and consequently the net rate of change of any species, say X, is just the sum of the effects of each reaction separately. It follows that the net rate of change in X is given by

$$\frac{dx}{dt} = (\text{conversion rate of A+B}) - (\text{conversion rate of X+Y})$$
$$= k_1[A][B] - k_{-1}[X][Y], \tag{9.5.8}$$
$$= k_1(a_0 + x_0 - x)(b_0 + x_0 - x) - k_{-1}x(y_0 - x_0 + x),$$

where equation (9.5.2) has been used in the last line. Circumstances under which the forward and backward reactions are not independent include precipitation of one of the species, as we have seen above. Another occurs when one of the reactions is highly exothermic. In that case, conditions of the reaction radically change, such as the temperature.

The analysis of equation (9.5.8) goes very much like that of equation (9.5.4). The stationary points are given as the solutions of the $\frac{dx}{dt} = 0$ equation

$$0 = k_1(a_0 + x_0 - x)(b_0 + x_0 - x) - k_{-1}x(y_0 - x_0 + x)$$
$$= (k_1 - k_{-1})x^2 - (k_1(a_0 + b_0 + 2x_0) + k_{-1}(y_0 - x_0))x \tag{9.5.9}$$
$$+ k_1(a_0 + x_0)(b_0 + x_0)$$

As one can see, if $k_1 \neq k_{-1}$, this is a quadratic equation and therefore has two roots, say $x = \alpha$, and $x = \beta$, which may be found using the quadratic formula, $\frac{1}{2a}(-b \pm \sqrt{b^2 - 4ac})$. The right-hand side of equation (9.5.8) thus factors into the linear factors

$$\frac{dx}{dt} = (k_1 - k_{-1})(x - \alpha)(x - \beta). \tag{9.5.10}$$

Again, just as above, this variable separable differential equation is easily solved but the nature of the solution depends on whether the roots are real or complex, equal or distinct. To decide about that, we must examine the discriminant of the quadratic formula, $b^2 - 4ac$. By direct substitution of the coefficients from equation (9.5.9) into the discriminant and then simplifying, we get

$$b^2 - 4ac = k_1^2(a_0 - b_0)^2 + 2k_1 k_{-1}(a_0 + b_0 + 2x_0)(y_0 - x_0) + k_{-1}^2(y_0 - x_0)^2. \tag{9.5.11}$$

The first and last terms are squares and so are positive (or zero). We see that, if $y_0 \geq x_0$, then the discriminant is always positive or zero and the two roots are real. Since X was an arbitrary choice, we can always arrange that $y_0 \geq x_0$, so we assume this is so.

Unless the initial concentrations are equal, $a_0 = b_0$ and $y_0 = x_0$, the roots will be distinct. We assume without loss of generality that

$$\alpha < \beta. \tag{9.5.12}$$

Then, in a similar way to the derivation of equation (9.5.6), the solution of equation (9.5.10) is

$$\ln\left(\frac{x-\beta}{x-\alpha}\right) = (\beta - \alpha)(k_1 - k_{-1})t + q,$$

where q is the constant of integration. This may be solved in terms of x,

$$x = \frac{\beta - Qe^{rt}}{1 - Qe^{rt}} \tag{9.5.13}$$

where Q is a constant and

$$r = (\beta - \alpha)(k_1 - k_{-1}).$$

If the discriminant is zero, then $\beta = \alpha$ and in that case the solution is

$$\frac{-1}{x-\alpha} = (k_1 - k_{-1})t + q,$$

or

$$x = \alpha - \frac{1}{(k_1 - k_{-1})t + q}$$

where q is again the constant of integration.

Exercises

1. Suppose that $A + B \to C$, that the initial concentrations of A, B, and C are 1/2, 1/3, and 0, respectively, and that the rate constant is k.
   a. Show that this leads to the differential equation in $z(t) = [C(t)]$ given by

$$z' = k\left(\frac{1}{2} - z\right)\left(\frac{1}{3} - z\right), \qquad z(0) = 0.$$

   b. Solve this equation.

c.   Show that the corresponding equation for $x(t) = [A(t)]$ is

$$x' = kx\left(\frac{1}{6} - x\right), \qquad x(0) = \frac{1}{2}.$$

d.   Solve this equation. Show by adding the solutions x and z that the sum is constant.
e.   At what time is 90% of the steady state concentration of C achieved?
f.   Suppose that k is increased 10%. Now rework part e.

2.   Suppose that $A + B \leftrightarrow C + D$ is a reversible reaction, the initial concentrations of A and B are 4/10 and 5/10, respectively, and that the initial concentrations of C and D are 0. Take $k_1 = 10$ and $k_{-1} = 5/2$.
a.   Show that this leads to the differential equation

$$y' = 10(0.4 - y)(0.5 - y) - \frac{5y^2}{2}, \qquad y(0) = 0.$$

b.   What is the equilibrium level of [C]? Draw two graphs, one where $k_{-1} = 5/2$ and one where $k_{-1} = 5/4$.

Section 9.6

Enzyme Kinetics

Enzymes serve to catalyze reactions in living systems, enabling complex chemical transformations to occur at moderate temperatures, many times faster than their uncatalyzed counterparts. Proteins, serving as the catalysts, are first used and then regenerated in a multi-step process. Overall, the simplest enzyme-catalyzed reactions transform the enzyme's specific substrate into product, possibly with the release of a by-product. Referred to as enzyme saturation, these reactions are typically rate limited by the amount of enzyme itself. The degree to which saturation occurs relative to substrate concentration is quantified by the Michaelis–Menten constant of the enzyme-substrate pair.

Enzyme catalyzed reactions are normally rate-limited by enzyme saturation.

The importance of enzyme catalyzed reactions along with a general description of the biochemical principles of enzyme catalysis was given in Section 9.4. Here we will consider an enzyme, E, which acts on a single substrate, S, and converts it to an alternate form which is regarded as the product P. The enzyme performs this function by temporarily forming an enzyme-substrate complex, C, which then decomposes into product plus enzyme:

$$S + E \rightleftharpoons C$$
$$C \longrightarrow P + E. \tag{9.6.1}$$

The regenerated enzyme is then available to repeat the process.[13] Here we will work through the mathematics of enzyme kinetics. The general principles of chemical kinetics discussed in the previous section apply to enzyme kinetics as well. However, due to the typically small amount of enzyme compared to substrate, the conversion rate of substrate to product is limited when the enzyme becomes *saturated* with substrate as enzyme-substrate complex.

As in the previous section, we let m denote the concentration of species M. The forward and reverse rate constants for the first reaction will be denoted k_1 and k_{-1} respectively while the rate constant for the second will be taken as k_2. The rate equations corresponding to the reactions (9.6.1) are[14]

$$\frac{dc}{dt} = k_1 es - k_{-1}c - k_2 c$$

$$\frac{ds}{dt} = -k_1 es + k_{-1}c$$

$$\frac{de}{dt} = -k_1 es + k_{-1}c + k_2 c \qquad (9.6.2)$$

$$\frac{dp}{dt} = k_2 c.$$

Note that complex C is both formed and decomposed by the first reaction and decomposed by the second. Similarly, enzyme E is decomposed and formed by the first reaction and formed by the second. The first three equations are independent of the formation of product P, and so, for the present, we can ignore the last equation. As before, we denote by subscript 0 the initial concentrations of the various reactants. In particular, e_0 is the initial, and therefore total, amount of enzyme since it is neither created nor destroyed in the process.

By adding the first and third equation of system (9.6.2) we get

$$\frac{dc}{dt} + \frac{de}{dt} = 0.$$

Integrating this and using the initial condition that $c_0 = 0$, we get

$$e = e_0 - c. \qquad (9.6.3)$$

We may use this to eliminate e from system (9.6.2) and get the following reduced system:

$$\frac{dc}{dt} = k_1 s(e_0 - c) - (k_{-1} + k_2)c$$

$$\frac{ds}{dt} = -k_1 s(e_0 - c) + k_{-1}c. \qquad (9.6.4)$$

---

[13]Compare this scheme to Figure 9.4.3. S+E constitutes the initial state, C is the transition state and P+E is the final state.
[14]The units of k_1 are different than those of k_{-1} and k_2 since the former is a bimolecular constant while the latter are uni-molecular.

In Figure 9.6.1 we show some solutions of this system of differential equation. For the purpose of drawing the figure, we take the constants to be

```
> k1:=1/10; km1:=1/10; k2:=1/10; e0:=4/10; (km1+k2)/k1;
```

The equations are non-linear and cannot be solved in closed form. Consequently we use numerical methods to draw these graphs. It should be observed that the level S, graphed as $s(t)$, drops continuously toward zero. Also, the intermediate substrate C, graphed as $c(t)$, starts at zero, rises to a positive level, and gradually settles back to zero. In the exercises we establish that this behavior is to be expected.

```
> with(plots): with(DEtools):
> enz:=[diff(c(t),t)=k1*s(t)*(e0-c(t))-(km1+k2)*c(t),
    diff(s(t),t)=-k1*s(t)*(e0-c(t)) + km1*c(t)];
> J:=DEplot2(enz,[c,s],0..100,[0,0,8/10],stepsize=1,
    scen  [t,s],labels=['t', ' ']):
> K:=DEplot2(enz,[c,s],0..100,[0,0,8/10],stepsize=1,
    scene=[t,c],labels=['t', ' ']):
> L:=textplot({[75,0.3,'s(t)'],[60,0.1,'c(t)']}):
> display({J,K,L});
```

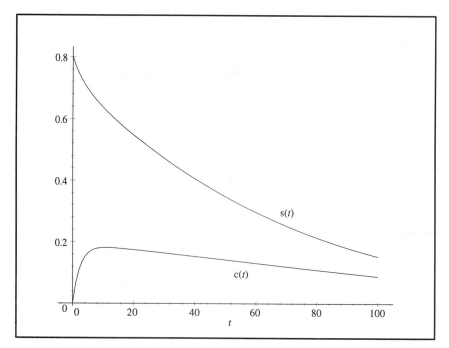

Figure 9.6.1   Solutions for Equation (9.6.2)

In the exercises we provide techniques to draw what may be a more interesting graph: Figure 9.6.2. In particular, we draw graphs of $s(t)$, $p(t)$, and $e(t)$. The first two of these are, in fact, the most interesting as they demonstrate how much of S is left and how much of P has been formed. The addition of a graph for $e(t)$ illustrates that during the intermediate phase, some of the enzyme is tied up in the enzyme-substrate complex, but as the reaction approaches equilibrium, the value of $e(t)$ returns to its original value.

From Figure 9.6.1, notice that the concentration of complex rises to a relatively invariant ("effective") level which we denote as c_{Eff}. This is found by setting $\frac{dc}{dt} = 0$ in system (9.6.4) and solving for c,

$$0 = k_1 s(e_0 - c) - (k_{-1} + k_2)c$$

or

$$s(e_0 - c) = \frac{k_{-1} + k_2}{k_1} c.$$

The combination k_M of rate constants

$$k_M = \frac{k_{-1} + k_2}{k_1} \tag{9.6.5}$$

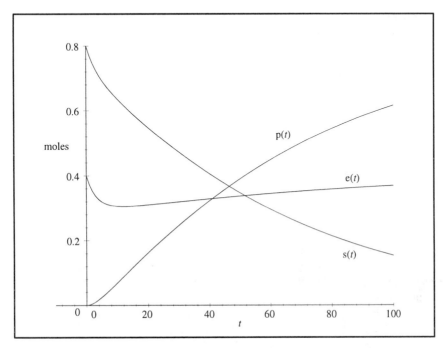

Figure 9.6.2 Solutions for Equation (9.6.1)

is known as the *Michaelis–Menten constant*; it has units moles per liter. Solving for c above we get

$$c = \frac{s e_0}{k_M + s},$$ (9.6.6)

which is seen to depend on the amounts of substrate S. But if s is much larger than k_M, then the denominator of equation (9.6.6) is approximately just s and we find the invariant level of complex to be

$$c_{Eff} \approx e_0.$$ (9.6.7)

Thus, most of the enzyme is tied up in enzyme-substrate complex.

By the *velocity* v of the reaction we mean the rate, $\frac{dp}{dt}$, at which product is formed. From equation (9.6.2), this is equal to $k_2 c$. When the concentration of substrate is large, we may use c_{Eff} as the concentration of complex and derive the maximum reaction velocity

$$v_{max} = k_2 e_0.$$ (9.6.8)

Likewise, from equations (9.6.6) and (9.6.8), the initial reaction velocity, v_0, is given by

$$v_0 = \left.\frac{dp}{dt}\right|_{t=0} = k_2 \frac{s e_0}{k_M + s}$$

$$= \frac{v_{max} s}{k_M + s}.$$ (9.6.9)

In this, s is the initial substrate concentration, $s = [S]_0$. Equation (9.6.9) is the *Michaelis–Menten equation*, the rate equation for a one-substrate, enzyme-catalyzed reaction. Its graph is shown in Figure 9.6.3.

The value of k_M for an enzyme can be experimentally found from Figure 9.6.3. At low substrate concentrations $k_M + s \approx k_M$ and so the graph approximates the line $v_0 = (v_{max}/k_M)s$ near $s = 0$. On the other hand, at high substrate concentrations the reaction rate approaches v_{max} asymptotically because, at these concentrations, the reaction is essentially independent of substrate concentration. By experimentally measuring the initial reaction rate for various substrate concentrations, a sketch of the graph can be made. Working from the graph, the substrate level which gives $\frac{1}{2}v_{max}$ initial velocity is the value of k_M, seen as follows.

```
> fcn:=s->vmax*s/(kM+s); vmax:=10: kM:=15:
> crv:=plot([x,fcn(x),x=0..150],x=-20..160,y=-1..12,tickmarks=[0,0]):
> asy:=plot(10,0..150,tickmarks=[0,0]):
> midline:=plot(5,0..15.3,tickmarks=[0,0]):
> vertline:=plot([15.3,y,y=0..5],tickmarks=[0,0]):
> a:=.0: A:=0.0: b:=13: B:=13*vmax/kM:
> slope:=x->A*(x-b)/(a-b)+B*(x-a)/(b-a):
> slopeline:=plot(slope,a..b):
> txt1:=textplot({[130.5,10.3,'-vmax-'],[-10.5,5,'1/2 vmax']},align=LEFT):
> txt4:=textplot({[b+14,B+0.5,'slope = vmax/kM']}):
> txt5:=textplot({[15.3,-0.3,'-kM-']},align=BELOW):
> with(plots):
> display({crv,asy,midline,vertline,slopeline,txt1,txt4,txt5});
```

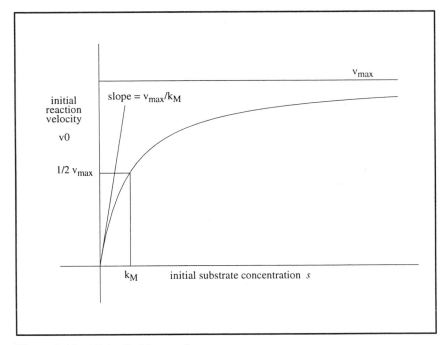

Figure 9.6.3   Michaelis-Menten plot

From equation (9.6.9) with $v_0 = v_{\max}/2$,

$$\frac{1}{2}v_{\max} = \frac{v_{\max}s}{k_M + s},$$

and, solving for k_M gives

$$k_M = s.$$

Thus, we interpret k_M as the substrate concentration at which the reaction rate is half maximal. By inverting the Michaelis–Menten equation (9.6.9) we get

$$\frac{1}{v_0} = \frac{k_M + s}{v_{max}s}$$

$$= \frac{k_M}{v_{max}}\frac{1}{s} + \frac{1}{v_{max}}.$$

(9.6.10)

This is the *Lineweaver–Burk equation* and shows that a least squares fit may be made to the *double-reciprocal plot* of $\frac{1}{v_0}$ vs $\frac{1}{s}$. This has the advantage of allowing an accurate determination of v_{max}.

Another transform of the Michaelis–Menten equation that allows the use of least squares is obtained from equation (9.6.10) by multiplying both sides by $v_0 v_{max}$; this yields

$$v_0 = -k_M\frac{v_0}{s} + v_{max}.$$

(9.6.11)

A plot of v_0 against $v_0/[S]$ is called the *Eadie–Hofstee plot*; it allows the determination of k_M as its slope and v_{max} as its intercept.

Exercises

1. Our intuition for the long-range forecast for (9.6.1) is that some of the reactants that move from S to C move on to P. But the assumption is that the second reaction is only one-way so that the products will never move back toward S.[15] This suggests S will be depleted. We conjecture that $s_\infty = 0$ and $c_\infty = 0$. We confirm this with the notions of stability that we studied in Section 2.5.

   a. Find all the stationary solutions by observing that setting $\frac{ds}{dt} = 0$ and $\frac{dc}{dt} = 0$ leads to the equations

   $$k_1 s(e_0 - c) - (k_{-1} + k_2)c = 0$$

   $$-k_1 s(e_0 - c) + k_{-1}c = 0.$$

   While it is clear that $s = 0$ and $c = 0$ is a solution, establish that this is the only solution for the equations as follows:

   > solve(k1*s*(e0-c)-(km1+k2)*c=0,c);

---

[15]In the context of a free energy diagram (Figures 9.4.2 and 9.4.3), the one-way nature of the process C → P is due to a lack of sufficient free energy in the environment to cause the reverse reaction P → C.

Substitute this into the second equation and set the resulting equation equal to zero. Argue that s must be zero and c must be zero.

```
> subs(c=",-k1*s*(e0-c) + km1*c);
> numer(")/denom(" ") = 0;
```

b. Establish that $s = c = 0$ is an attracting stationary point by finding the linearization about this only stationary point. (Recall Section 4.4.)

```
> jacobian([k1*s*(e0-c)-(km1+k2)*c, -k1*s*(e0-c) + km1*c],[c,s]);
> subs({c=0,s=0},");
> eigenvals(");
```

c. Verify that the eigenvalues of the linearization are

$$-\frac{1}{2}\left((k_{-1} + k_2 + k_1 e_0) \pm \sqrt{(k_{-1} + k_2 + k_1 e_0)^2 - 4k_2 k_1 e_0}\right)$$

and that both these are negative. Argue that this implies $\{0,0\}$ is an attracting stationary point for $\{c(t), s(t)\}$.

2. Draw the graph of Figure 9.3.3. With *Maple V.* release 3, we use a program called ODE from the share package to get numerical solutions for this systems. Here is the syntax for using this program and for drawing Figure 9.3.3. The routine uses a Runge–Kutta package by calling the program with `rungekuttahf`. Remove the final f from the call command if the machine on which the program is running does not have a math co-processor.

```
> with(share): readshare(ODE,plots); with(plots):
> k1:=1/10: k2:=1/10: km1:=1/10: s0:=8/10: e0:=4/10:
> rss:=(t,s,c,p,e)->-k1*e*s + km1*c;
  rsc:=(t,s,c,p,e)->k1*e*s - (km1+k2)*c;
  rsp:=(t,s,c,p,e)->k2*c;
  rse:=(t,s,c,p,e)->-k1*e*s + (km1 + k2)*c;
> init:=[0,s0,0,0,e0];
> output:=rungekuttahf([rss,rsc,rsp,rse],init,1,100):
> J:=plot({makelist(output,1,2),makelist(output,1,4),
  makelist(output,1,5)},view=[-10..100,0..0.8]):
> K:=textplot({[90,0.14,'s(t)'],[90,0.42,'e(t)'],[68,0.55,'p(t)'],
  [-10,0.5,'moles ']},view=[-10..100,0..0.8]):
> display({J,K});
```

3. Draw the graph of the solution $c(t)$ for in equation (9.3.2) with constants chosen so that $k_M \approx 1$ and $S = 10$. The point to observe is that $c(t) \approx e_0$ for large values of t.

```
> with(share): readshare(ODE,plots); with(plots):
> k1:=1: k2:=1/10: km1:=1/40: s0:=10: e0:=4/10:
  (km1+k2)/k1: s0/(" +s0);
> rss:=(t,s,c,p,e)->-k1*e*s + km1*c;
  rsc:=(t,s,c,p,e)->k1*e*s - (km1+k2)*c;
  rsp:=(t,s,c,p,e)->k2*c;
  rse:=(t,s,c,p,e)->-k1*e*s + (km1 + k2)*c;
> init:=[0,s0,0,0,e0];
> output:=rungekuttahf([rss,rsc,rsp,rse],init,1/10,100):
> plot(makelist(output,1,3));
```

4. Suppose that $A + B \to C$, that the initial concentrations of A, B, and C are 2,
    3, and 0, respectively, and that the rate constant is k.
    a. The concentration of C is sampled at $t = 3/2$ and is found to be 3/5.
        What is an approximation for k?
    b. Instead of determining the concentration of C at just $t = 3/2$, the con-
        centration of C is found at five times:

Time	Concentration
0.5	.2
1.0	.4
1.5	.6
2.0	.8
2.5	1.

    Estimate K. Plot your data and the model your K predicts on the same
graph.
5. We have stated in Section 9.2 that the addition of an enzyme to a reaction
    could potentially speed the reaction by a factor of 10^{13}. This problem gives a
    glimpse of the significance of even a relatively small increase in the reaction
    rate.
    Suppose that we have a reaction

$$A \leftrightarrow B \to C.$$

    Suppose also that $k_{-1} = k_2 = 1$, that the initial concentration of A is $a_0 = 1$,
    and the initial concentrations of B and C are zero.
    a. Show that the differential equations model for this system is

$$\frac{da}{dt} = -k_1 a(t) + k_1 b(t)$$

$$\frac{db}{dt} = k_1 a(t) - (k_{-1} + k_2) b(t)$$

$$\frac{dc}{dt} = k_2 b(t).$$

b. Find $a(t)$, $b(t)$, and $c(t)$ for $k_1 = 1$ and for $k_1 = 10$. Plot the graphs for the three concentrations in both situations.

```
> with(linalg):
> k1:=1; km1:=1; k2:=1;
> A:=matrix([[-k1,km1,0],[k1,-km1-k2,0],[0,k2,0]]);
> u:=evalm(exponential(A,t) &* [1,0,0]):
> a:=unapply(u[1],t);
  b:=unapply(u[2],t);
  c:=unapply(u[3],t):
> plot({a(t),b(t),c(t)},t=0..7);
> solve(c(t)=.8,t);
```

c. Take $k_1 = 1, 10, 20, 30, 40$, and 50. Find T_k such that $c(T_k) = .8$ for each of these k's. Plot the graph of the pairs $\{k, T_k\}$. Find an analytic fit for these points.

Section 9.7

Questions for Thought and Discussion

1. Draw the structural formulas ("stick models") for (a) butane, the four-carbon hydrocarbon, having all carbons in a row, and no double bonds; (b) iso-propanol, having three carbons, an -OH group on the middle carbon, and no double bonds; (c) propene, with one double bond.
2. Relate this set of reactions to a free energy level diagram: $A + E \leftrightarrow B \to C + E$, where E is an enzyme. What effect does E have on the energy levels?
3. Assume this reaction: $A \leftrightarrow C \leftrightarrow B$, where the intermediate state C has a *lower* free energy than A or B. Knowing what you do about the behavior of free energy, what would happen to the free energy difference between A and B if the free energy of C were changed?
4. A mechanical analog of the situation in Question 3 is a wagon starting at A, rolling downhill to C and then uphill to B. There is a frictional force on the wagon wheels. What do you think will be the effect of varying the depth of C?
5. Describe the chemical differences between RNA and DNA. What are their biological (functional) differences?
6. Outline the process of information flow from DNA to the control of cellular chemistry.
7. Name six kinds of proteins and describe their functions.

References and Suggested Further Reading

1. **Thermodynamics:** Harold J. Morowitz, *Energy Flow in Biology*, Academic Press, New York, 1968.

2. **Biochemical structure:** Lubert Stryer, *Biochemistry*, W.H. Freeman and Company, 2nd ed., 1981.

3. **Biochemical structure and thermodynamics:** Edward K. Yeargers, *Basic Biophysics for Biology*, CRC Press Inc., Boca Raton, FL., 1992.

4. **Thermodynamics and enzyme function:** P. W. Atkins, *Physical Chemistry*, W.H. Freeman and Co. New York, 3rd ed., 1986.

5. **Chemical genetics:** David T. Suzuki, Anthony J. F. Griffiths, Jeffrey H. Miller and Richard C. Lewontin, *An Introduction to Genetic Analysis*, W. H. Freeman and Co., New York, 3rd ed., 1986.

6. **Chemical genetics:** James D. Watson, Nancy W. Hopkins, Jeffrey W. Roberts, Joan A. Steitz and Alan M. Weiner, *Molecular Biology of the Gene*, The Benjamin/Cummings Publishing Co., Inc., Menlo Park, CA., 4th ed., 1987.

Chapter 10

A Biomathematical Approach
to HIV and AIDS

Introduction to this chapter

Acquired Immunodeficiency Syndrome (AIDS) is medically devastating to its victims and wreaks financial and emotional havoc on everyone, infected or not. The purpose of this chapter is to model and understand the behavior of the causative agent of AIDS—the Human Immunodeficiency Virus (HIV). This will necessitate discussions of viral replication and immunology. By the end of this chapter the student should have a firm understanding of the way that HIV functions and be able to apply that understanding to a mathematical treatment of HIV infection and epidemiology.

Viruses are very small biological structures whose reproduction requires a host cell. In the course of viral infection the host cell is changed or even killed. The host cells of HIV are specific and very unique: They are cells of our immune system. This is of monumental importance to the biological and medical aspects of HIV infection and its aftermath. HIV infects several kinds of cells, but perhaps its most devastating cellular effect is that it kills helper T lymphocytes. Helper T lymphocytes play a key role in the process of gaining immunity to specific pathogens; in fact, if one's helper T lymphocytes are destroyed, the entire specific immune response fails. Note the irony: HIV kills the very cells that are required by our bodies to defend us from pathogens, including HIV itself! The infected person then contracts a variety of (often rare) diseases to which uninfected persons are resistant, and that person is said to have AIDS.

Section 10.1

Viruses

Viruses are small reproductive forms with powerful effects. A virus may have only four to six genes, but those genes enable it to take over the synthetic machinery of a normally functioning cell, turning it into a small biological factory producing thousands of new viruses. Some viruses add another ability: They can insert their nucleic acid into that of the host cell, thus remaining hidden for many host cell generations prior to viral reproduction.

HIV is an especially versatile virus. It not only inserts its genetic information into its host's chromosomes, but it then causes the host to produce new HIV. Thus, the host cells, which are immune system components, produce a steady stream of HIV particles. Eventually this process kills the host cells and the patient becomes incapable of generating critical immune responses.

A virus is a kind of parasite.

Each kind of virus has its own special anabolic ("building up") needs which, because of its genetic simplicity, the virus may be unable to satisfy. The host cell then must provide whatever the virus itself cannot. This requires a kind of biological matching between virus and host cell analogous to that between, say, an animal parasite and its host. Host specificity is well-developed in viruses: As examples, the rabies virus infects cells of our central nervous system, cold viruses affect cells of our respiratory tract and the feline leukemia virus affects certain blood cells of cats (see Reference [1]).

The basic structure of a virus is a protein coat around a nucleic acid core.

Simple viruses may have only four to six genes, but most viruses have many more than that. In the most general case the viral nucleic acid, either DNA or RNA, is surrounded by a protein coat, called a *capsid* (see Figure 10.1.1). In addition,

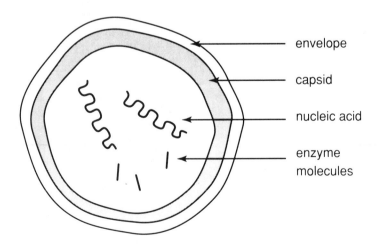

Figure 10.1.1 A generalized drawing of a virus. In a given real case the envelope and/or enzyme molecules may be absent and the nucleic acid may be DNA or RNA.

many viruses have outer layers, or *envelopes*, which may contain carbohydrates, lipids and proteins. Finally, inside the virus there may be several kinds of enzymes along with the nucleic acid.

A virus cannot reproduce outside a host cell, which must provide viral building materials and energy. All the virus provides is instructions via its nucleic acids and, occasionally, some enzymes. As a result, viruses are not regarded as living things.

Viral nucleic acid enters the host cell and redirects the host cell's metabolic apparatus to make new viruses.

A virus attaches to its specific host's outer covering, host–virus specificity being assured by host-to-viral molecular recognition. The molecules involved are proteins or *glycoproteins*, a sugar–protein combination. At this point the viral nucleic acid enters the host cell, the precise means of entry depending on the nature of the virus (see Figure 10.1.2). For instance, viruses called *bacteriophages* infect bacteria. Bacteriophages have no envelope and seem to inject their nucleic acid into the bacterium, leaving the viral protein capsid outside. Alternatively, nucleic acids from viruses that infect animals can enter the host cell by *fusion*, in which a virus joins its envelope to the cell membrane of the host cell and the entire viral capsid is drawn into the host cell. Fusion is facilitated by the fact that the viral envelope is chemically similar to the cell membrane. The capsid is then enzymatically removed, thus exposing its contents—the viral nucleic acid and possibly certain viral-specific enzymes.

What happens next depends on the identity of the virus, but it will ultimately lead to viral multiplication. Viral replication requires the production of viral-specific enzymes, capsid proteins and, of course, viral nucleic acid. The synthesis of these components is carried out by using the host cell's anabolic machinery and biochemical molecules. To do this, the host cell's nucleic acid must be shut down at an early stage in the infection, after which the viral nucleic acid takes control of the cellular machinery. It is said that the host cell's metabolic apparatus is changed from "host-directed" to "viral-directed." An analog can be found in imagining a computer-controlled sofa-manufacturing plant. We disconnect the original (host) computer and install a new (viral) computer that redirects the existing construction equipment to use existing materials to manufacture chairs instead of sofas.

Typically a virus uses the enzymes of the host cell whenever possible, but there are important situations where the host cell may lack a critical enzyme needed by the virus. For example, some viruses carry single-stranded nucleic acids, which must become double-stranded shortly after being inserted into the host. The process of forming the second strand is catalyzed by a particular polymerase enzyme, one that the host lacks. The viral nucleic acid can code for the enzyme, but the relevant gene is on the nucleic acid strand that is complementary to the one strand the virus carries. Thus the gene is unavailable until the viral nucleic acid becomes double stranded—but of course the nucleic acid cannot become double-stranded until the enzyme is available! The virus gets around

a) bacterial host

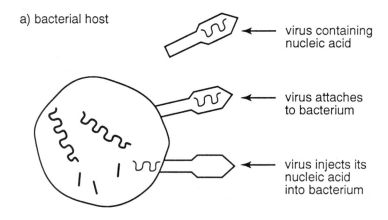

virus containing
nucleic acid

virus attaches
to bacterium

virus injects its
nucleic acid
into bacterium

b) animal cell host

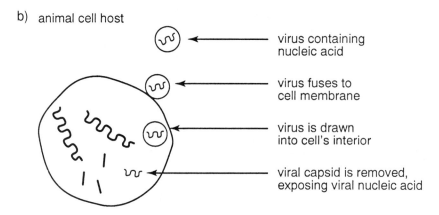

virus containing
nucleic acid

virus fuses to
cell membrane

virus is drawn
into cell's interior

viral capsid is removed,
exposing viral nucleic acid

Figure 10.1.2   Some models of viral infection. (a) A virus whose host is a bacterium recognizes some molecular feature of the correct host, attaches to it and injects
its nucleic acid into it. (A virus whose host is a bacterium is called a *bacteriophage*.)
(b) A virus whose host is an animal cell recognizes some molecular feature of the
correct host cell and is then drawn into the host cell, where the capsid is removed.

this problem by carrying one or more copies of the actual enzyme molecule in its
capsid and injecting them into the host at the time it injects the nucleic acid.[1]

_____

[1]Recall from Chapter 9 that, in a given segment of DNA, only one of the two DNA strands
actually codes for RNA. That strand is called the *coding strand*. In the example given above, the
coding strand would be the strand formed after infection. Thus its genes would not be available until
after the nucleic acid became double-stranded.

As the virus' various component parts are constructed, they are assembled into new, intact viruses. The nucleic acid is encapsulated inside the protein capsid, perhaps accompanied by some critical viral enzymes. The assembly of the capsid is spontaneous, like the growth of a crystal. The newly assembled viruses then escape from the host cell and can start the infection process anew.

Many RNA viruses do not use DNA in any part of their life cycle.

The Central Dogma was presented in Chapter 9 to show the path of genetic information flow.

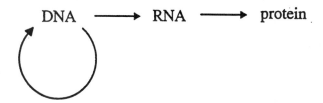

Note that, because RNA is complementary to DNA, it should be possible to skip the DNA part of the scheme. All that is necessary to justify this assertion is to demonstrate that RNA can code for its own self-replication. While this does not seem to happen in cellular systems, it is well-known in viruses: Viral RNA replicates just like DNA does, using complementary base-pairing. After replication, the RNA is packaged into new viruses.[2]

Our revised statement of the Central Dogma, accounting for RNA self-replication, now looks like this.

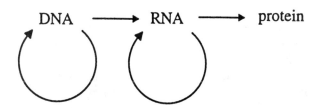

There are several variations in the host–cell–escape mechanism for viruses.

Some viruses merely replicate their nucleic acid, translate out the necessary proteins, encapsulate and then burst out of the host cell an hour or two after infection. This bursting process kills the host cell and is called *lysis*; the virus is said to be *lytic*.

---

[2]There are single-stranded and double-stranded RNA viruses, just like there are single- and double-stranded DNA viruses. HIV is a single-stranded RNA virus—its conversion to double-stranded form will be described in Section 10.3.

Other viruses, said to be *lysogenic*, delay the lytic part of their reproductive process. For example, the DNA of some DNA viruses is inserted into the host cell body and then into the host's DNA. Thus, when the host's DNA is replicated at cell division, so is the viral DNA. The inserted viral DNA is called a *provirus*, and it can remain inserted in the host DNA for many cell generations. Sooner or later, depending on the lysogenic virus, host and culture conditions, the provirus begins to replicate its nucleic acid and produces RNA, which then produces viral proteins. New viruses are then assembled and lyse the host to get out.

There is an alternative to lysis in the escape process: When the viruses exit the host cell, they may instead *bud off* from the host cell, in a process that is the reverse of fusion. In the process they take a piece of the cell membrane for an envelope, but do not kill the host cell. Cells which release viruses by budding can therefore act as virtually unending sources of new viruses. This, in fact, is the behavior of certain blood cells infected with HIV.

Section 10.2

The Immune System

Our bodies fight off pathogens by two means. One is a general defense system that removes pathogens without much regard to their biological nature; stomach acid is such a system.

Of more concern to us in our considerations of HIV is a second, specific response to pathogens (and other foreign substances); this response is tailored to each infective agent. Specialized blood cells called lymphocytes have the ability to recognize specific molecular parts of pathogens and to mount a chemical response to those fragments. Initially, we have at most only a few lymphocytes that can recognize each such fragment but, upon contact with the fragment, the lymphocyte will start to divide extensively to provide a clone of cells. Thus there results a large clone of identical lymphocytes, all of which are chemically "tuned" to destroy the pathogen.

In this section we describe the means by which lymphocytes respond to foreign substances to keep us from getting diseases and from being poisoned by toxins. This subject is of great importance to our understanding of HIV because certain lymphocytes are hosts for HIV. Thus, HIV infection destroys an infected person's ability to resist pathogens.

Some responses to pathogens are innate, or general.

We possess several general mechanisms by which we can combat pathogens. These mechanisms have a common property: They are essentially nondiscriminatory. Each one works against a whole class of pathogens and does not need to be adapted for specific members of that class. For example, tears and egg white contain an enzyme that lyses the cell walls of certain kinds of bacteria. Stomach acid kills many pathogens that we eat. Damaged tissue attracts blood clotting agents and dilates capillaries to allow more blood to approach the wound. Finally, there are blood cells that can simply engulf pathogens; these cells are *granulocytes* and *macrophages*.

The problem with the innate response is that it cannot adapt to new circumstances, whereas many pathogens are capable of rapid genetic change. Thus, many pathogens have evolved ways to circumvent the innate response. For such pathogens, we need an immune response that can change in parallel with the pathogen (see References [1] and [2]).

Blood cells originate in bone marrow and are later modified for different functions.

Humans have bony skeletons, as do dogs, robins, snakes and trout, but sharks and eels have cartilaginous skeletons. In the core, or *marrow*, of our bones is the blood-forming tissue, where all of our blood cells start out as *stem cells*. Repeated division of stem cells results in several paths of cellular specialization, or *cell lines*, as shown in Figure 10.2.1. Each cell line leads to one of the various kinds of mature blood cells described in Section 7.6. One cell line becomes red blood cells. Another line generates cells involved in blood clotting. Still other lines have the ability to engulf and digest pathogens. Finally, there is a cell line that generates cells capable of specifically adapted defenses to pathogenic agents. These cells are called *lymphocytes*.

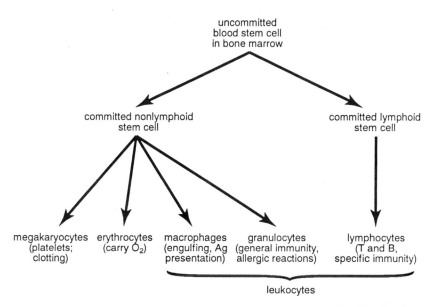

Figure 10.2.1 A flow chart showing the development of mammalian blood cells from their generalized state to their final, differentiated state.

Some immune responses are adaptive, or specific, to the pathogen.

Our immune system is capable of reactions specifically tailored to each foreign substance, or *antigen*; in other words, each and every antigen elicits a unique response. At first glance we might think that the finite amount of information that a cell can contain would place a ceiling on the number of specific responses possible. We will see that the restriction is not important because the specific immune system works against as many as 10^{12} distinct antigens![3]

Certain cell lines, derived from bone marrow stem cells, mature in our lymphatic system to become *lymphocytes*. For example, T-lymphocytes, or *T cells*, mature in the thymus gland, which is found prominently under the breastbone of fetuses, infants and children. B-lymphocytes, or *B cells*, mature in bone marrow. These two kinds of lymphocytes are responsible for the adaptive immune responses, but play different and complementary roles.

T cells are responsible for the cell-mediated immune response.

We will be especially interested in two groups of T cells: *helper T and cytotoxic T cells* (see Figure 10.2.2). After they mature in the thymus of neonatal and prenatal animals, these T cells are *inactive*. On their outer surfaces, inactive T cells have recognition proteins that can bind to antigens (via hydrogen bonds and other interactions). This binding cannot take place, however, unless some preliminary steps have already occurred. First, one or more *antigen-presenting cells, or macrophages*, must ingest the pathogen. Second, the antigen-presenting macrophages must then break off various molecular pieces of the pathogen and move them to their own surface, i.e., *present* the various antigenic fragments (called *epitopes*) to inactive T cells. This presentation activates the T cells and causes them to divide repeatedly into clones, each of which consists of identical, active helper T or cytotoxic T cells. In fact, there should result a clone of active helper and cytotoxic T cells for each of the various epitopes that the antigen-presenting cells display, one clone originating from each activated T cell.[4] An important point: The active helper T cells are required in the activation of the cytotoxic T cells. The active cytotoxic T cells then approach and kill cells infected with the pathogen, thus killing the pathogen at the same time. The cytotoxic T cell recognizes the infected cells because the infected cells, like macrophages, present epitopes on their surfaces. The T cell response is often called *cell-mediated* immunity because the effect requires the direct and continued involvement of intact T cells.

---

[3]The size of this number, even its order of magnitude, is subject to some debate. In any case, it is *very* big.

[4]When antigen-presenting cells cut up a pathogen, many different antigenically active epitopes may result. Potentially, each epitope can activate a different T cell upon presentation. Thus, a single infecting bacterium could activate many different T cell clones.

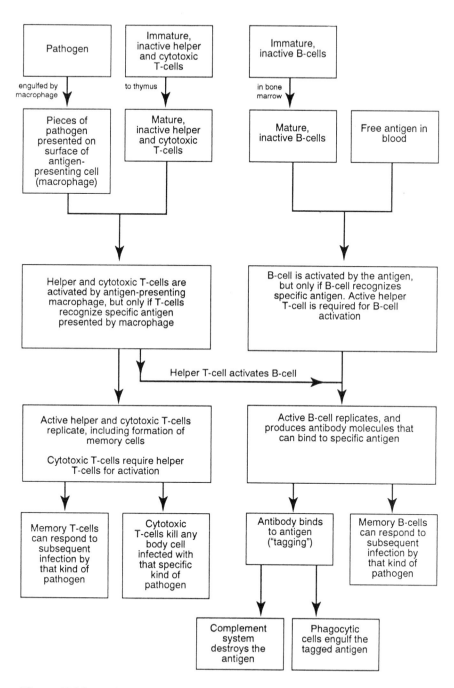

Figure 10.2.2   A flow chart showing the events and interactions surrounding the specific immune response. The cell-mediated response begins at the top left and the humoral response begins at the top center. The two responses interact at the center of the page. The details are described in the text.

The concept of an adaptive response, or *immunological specificity*, is associated with the recognition of an infected antigen-presenting cell by a helper T or cytotoxic T cell. An inactive T cell will be activated only if its specific receptors recognize the specific antigenic fragment being presented to it. Evidence suggests that the surface receptors of each individual inactive T cell are unique, numerous and of a single kind. Because there are upwards of a trillion or so different inactive T cells in our bodies, the presented parts of virtually every pathogen should be recognized by at least a few of the T cells.

B cells are responsible for the humoral immune response.

Like T cells, B cells are inactive at the time they mature and have recognition proteins on their surfaces. As with helper T cells, these surface receptors vary from cell to cell and can recognize antigens. However, while helper T cells require that the antigen appear on an antigen-presenting cell, B cells can recognize an antigen that is free in the liquid fraction of the blood. When an inactive B cell recognizes and binds to the antigen to which its surface proteins are complementary, the B cell is then activated, and it subsequently divides many times to form a clone of identical active B cells, sometimes called *plasma cells*. Active B cells then secrete large quantities of a single kind of protein molecule, called an *antibody*, into the blood. These antibodies are able to bind to the antigen, an act that "labels" the antigen for destruction by either of two mechanisms: A set of chemical reactions collectively called *complement*, can kill certain antibody-tagged bacteria, or tagged antigens can attract macrophages, which devour the antigen. The B cell response is often called the *humoral* immune response, meaning "liquid-based."

The concept of specificity for B cell activation arises in a way similar to that for T cells, namely in the recognition of the antigen by B cell surface receptors. Evidently all, or most, of our approximately one trillion inactive B cells have different surface receptors. The recognition by a B cell of the exact antigen for which that particular B cell's surface is "primed" is an absolute requirement for the activation of that B cell. Fortunately, most pathogens, bacteria and viruses for example, have many separate and distinct antigenic regions; thus, they can trigger the activation of many different B cells.

Intercellular interactions play key roles in adaptive immune responses.

The specificity of both T and B cell interactions with pathogens cannot be overemphasized; no adaptive immune response can be generated until receptors on these lymphocytes recognize the one specific antigen to which they can bind.

Note how T and B cells provide interlocking coverage: The cytotoxic T cells detect the presence of intracellular pathogens (by the epitopes that infected cells present) and B cells can detect extracellular pathogens. We would therefore expect T cells to be effective against already-infected cells and the B cells to be effective against toxins, such as snake venom, and free pathogens, such as bacteria, in the blood.

Our discussion so far has emphasized the individual roles of T and B cells. In fact, correct functioning of the adaptive immune system requires that these two kinds of cells interact with each other. It was pointed out earlier that the activation of cytotoxic T cells requires that they interact with active helper T cells. In fact, helper T cells are also needed to activate B cells, as shown in Figure 10.2.2. Note the pivotal role of active helper T cells: They exercise control over cell-mediated immunity *and* humoral immunity as well, which covers the entire adaptive immune system.

Lymphocytes diversify their receptor proteins as the cell matures.

At first glance, the Central Dogma of Genetics would seem to suggest that the information for the unique surface protein receptor of each inactive lymphocyte should originate in a different gene. In other words, every inactive lymphocyte would merely express a different surface receptor gene. In each person there seem to be about 10^{12} unique inactive lymphocytes and therefore there would have to be the same number of unique genes! Actually, independent estimates of the *total* number of genes in a human cell indicate that there are only about 10^5.

The many variant forms of lymphocyte surface receptor proteins originate as the cell matures and are the result of the random scrambling of genetic material— which leads to a wide variety of amino acid sequences without requiring the participation of a lot of genetic material. As an example, Figure 10.2.3 shows a length of hypothetical DNA that we will use to demonstrate the protein diversification process. We imagine the DNA to consist of two contiguous polynucleotide strings, or classes, labeled A and B. Each class has sections 1 through 4. The protein to be coded by the DNA will contain two contiguous polypeptide strings, one coded by a single section of A and one coded by a single section of B. Thus, there are 16 different proteins that could result. To generate a particular protein the unneeded sections of genetic material will be enzymatically snipped out, either at the DNA stage or the mRNA stage. The protein that ultimately results in the figure is derived from DNA sections A2 and B4. The selection of A2 and B4 was random; any of the other 15 combinations were equally likely.

In a real situation, namely the DNA coding for one of the proteins in B cell antibodies, there are about 240 sections distributed among four classes. Of these, perhaps seven sections are actually expressed in a given cell, meaning that there are thousands of combinations of seven sections that were not expressed in that cell. These other combinations will be expressed in other B cells, thereby generating a large number of lymphocytes with different surface proteins.

There are still other ways that lymphocytes generate diverse recognition proteins. For example, B cells form antibodies by combining two completely separate polypeptides, each of which uses the random choice method described in the previous two paragraphs. Further, when maturing, the nucleic acid segments that

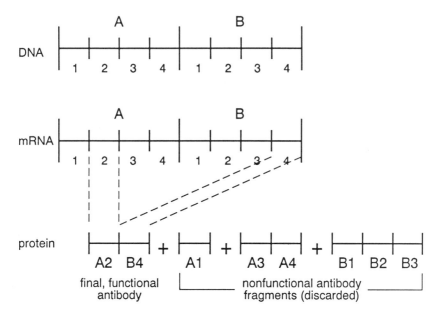

Figure 10.2.3 A simplified picture of the creation of a specific antibody by a single lymphocyte. The final antibody molecule is coded from one section each of DNA regions A and B. Because the two sections are picked at random there are 16 possible outcomes. This figure shows how many possible antibodies could be coded from a limited amount of DNA. In a real situation there would be many sections in many DNA regions, and the final, functional antibody would contain contributions coded by several regions.

code for lymphocyte surface recognition proteins mutate very rapidly, more so than do other genes. All of this leads to the great variability in recognition proteins that is so crucial to the functioning of the adaptive immune system, and it does so while requiring a minimum amount of DNA for its coding.

The adaptive immune system recognizes and tolerates "self" (clonal deletion).

The whole idea behind the immune system is to recognize foreign material and rid the body of it. On the other hand, it would be intolerable for a person's adaptive immune system to treat the body's own tissues as foreign. In order to prevent such rejection of self-products, or *autoimmune reactions*, the adaptive system must have some way to distinguish "self" from "non-self." This distinction is created during fetal development and continues throughout postnatal development.

The organ systems of a human fetus, including the blood-forming organs, are formed during the *organogenetic* period of fetal development, as discussed in Chapter 7. Most organogenesis is completed at least a month or two before birth, the remaining fetal period being devoted to enlargement and maturation of the fetus. Embryonic (immature) lymphocytes, which are precursors to inactive T and B cells, are present during the time of organogenesis. Each one will have a unique kind of recognition protein across its surface, inasmuch as such proteins are essentially generated at random from cell to cell. We could thus expect that among these embryonic lymphocytes there would not only be those that can bind to foreign substances, but also many that can bind to the embryo's own cells. The *clonal deletion* model explains how these self-reactive lymphocytes can be neutralized: Because there are no pathogens in a fetus, the only cells capable of binding to lymphocytes would be the fetuses' own cells. Therefore, embryonic B or T cells that bind to *any* cell in the fetus are killed or inactivated. Only self-reacting embryonic lymphocytes should be deleted by this mechanism. This reduces the possibility of maturation of a lymphocyte that could subsequently generate an autoimmune response.

There is good evidence for clonal deletion: Mouse embryos can be injected early *in utero* with a virus or other material that would normally be antigenic in a postnatal mouse. After birth, the treated mouse is unable to respond immunologically to subsequent injections of the virus. The mouse has developed an *acquired immunological tolerance* to the antigen. What has happened at the cellular level is that any embryonic mouse lymphocytes that reacted with the prenatally injected virus were killed or inactivated by clonal deletion—the virus was treated as "self." Thus, there can be no mature progeny of these lymphocytes after birth to react to the second exposure to the virus.

There is another mechanism for killing self-reacting lymphocytes.

Clonal deletion reduces the possibility of an autoimmune response, but does not eliminate it. Recall that clonal deletion requires that self-products meet up with embryonic lymphocytes; mature lymphocytes will not do. The fact is that some embryonic lymphocytes slip through the clonal deletion net by not meeting the self-products that would have inactivated them. In addition, lymphocytes seem to mutate frequently, a process that postnatally may give them receptors that can react with self-products. Finally, the thymus gland, while much reduced in adults, continues to produce a limited number of new T cells throughout life. These new cells, with receptors generated at random, may be capable of reacting with self-products.

There is a mechanism for getting rid of mature T cells that can react with their own body's cells: Recall that a T cell is activated when an infected antigen-presenting cell presents it with a piece of antigen. In fact, this activation has another requirement: The antigen-presenting cell must also display a *second* receptor, one that is found only on *infected* antigen presenters. If a mature T cell should bind to an uninfected antigen presenter, one lacking the second receptor, the T cell itself is inactivated (because that binding is a sign that the T cell receptors are complementary to uninfected self-products). On the other hand, if a mature T cell binds to an infected antigen presenter, the infection being signaled by the second receptor, that binding is acceptable, and the normal activation of the T cell ensues.

Inactive lymphocytes are selected for activation by contact with an antigen (clonal selection).

The clonal deletion system described above results in the inactivation or killing of immature T and B cells if they react with any antigen. This process provides the individual with a set of lymphocytes that can react only with non-self-products. These surviving T and B cells then remain in our blood and lymphatic circulatory systems in an inactive state until they come into contact with the antigens to which their surface receptors are complementary. This will be either as free, extracellular antigens for B cells, or on an antigen-presenting cell in the case of T cells.

Once the proper contact is made, the lymphocyte is activated and begins to divide rapidly to form a clone of identical cells. But what if the correct antigen never appears? The answer is an odd one—namely, the lymphocyte is never activated and remains in the blood and lymphatic systems all of our life. What this means is that only a tiny fraction of our lymphocytes ever become activated in our lifetimes; the rest just go around and around our circulation or remain fixed in lymph nodes. This process of activating only those few lymphocytes whose activity is needed, via contact with their appropriate antigens, is called *clonal selection*.

The notion of clonal selection suggests an immense amount of wasted effort on the part of the immune system. For example, each of us has one or more lymphocytes capable of initiating the rejection of a skin transplant from the seventieth president of the United States (in about a century), and others that would react against a cold virus that people contracted in Borneo in 1370 AD. None of us will ever need those capabilities, but we have them nevertheless. It might seem that a simpler mechanism would have been the generation of a single generic kind of lymphocyte and then its adaptation to each individual kind of antigen. This process is called the *instructive mechanism*, but it is not what happens.

The immune system has a memory.

Most people get mumps or measles only one time. If there are no secondary complications these diseases last about a week, which is the time it takes for the activation of T and B cells by a pathogen and the subsequent destruction of the pathogen. Surely these people are exposed to mumps and measles many times in their lives, but they seem to be unaffected by the subsequent exposures. The reason for this is well-understood: First, they may have antibodies from the initial exposure and, second, among the results of T and B cell activation are "memory" cells, whose surface recognition proteins are complementary to the antigenic parts of the activating pathogen (refer to Figure 10.2.2). These memory cells remain in our blood and lymphatic systems for the rest of our lives, and if we are infected by the same pathogen again, they mount a response just like the original one, but much more intensely and in a much shorter time. The combination of pre-existing antibodies from the initial exposure and the intense, rapid secondary response by memory cells usually results in our being unaware of the second exposure.

Why then do we get so many colds if the first cold we get as babies generates memory cells? The answer lies in two facts: The adaptive immune response is very specific, and the cold virus mutates rapidly. The memory cells are as specific for antigen as were their original inactive lymphocyte precursors. They will recognize only the proteins of the virus that caused the original cold; once having gotten a cold from that particular strain of cold virus, we won't be successfully infected by it again. The problem is that one effect of cold virus mutation is that viral coat proteins (the antigens) change their amino acid sequences. Once that happens, the memory cells and antibodies from a previous infection don't recognize the new, mutated strain of the virus and therefore can't respond to it. The immune response must start all over, and we get a cold that requires a week of recovery (again). If it is possible to say anything nice about mumps, chicken pox and such diseases, it is that their causative agents do not mutate rapidly and we therefore get the diseases only once, if at all. We shall see in the next section that rapid mutation characterizes HIV, allowing the virus to stay one step ahead of the specific immune system's defenses.

Vaccinations protect us by fooling the adaptive immune system.

The idea behind immunization is to generate the immune response without generating the disease. Thus, the trick is to inactivate or kill the pathogen without damaging its antigenic properties. Exposure to this inactive pathogen then triggers the immune responses described earlier, including the generation of memory cells. During a subsequent exposure, the live, active pathogen binds to any pre-existing antibody *and* activates memory cells; thus, the pathogen is inactivated before disease symptoms can develop. As an example, vaccination against polio consists of swallowing live-but-inactivated polio virus. We then generate memory cells that will recognize active polio viruses if we should be exposed to them at a later date.

Exposure to some pathogenic substances and organisms is so rare that vaccination of the general population against them would be a waste of time and money. Poisonous snake venom is a case in point: The active agent in snake venom is a destructive enzyme distinctive to each kind of snake genus or species, but fortunately almost no one ever gets bitten by a snake. Snake venom is strongly antigenic, as we would expect a protein to be, but the symptoms of snake bite appear so rapidly that the victim could die long before the appropriate lymphocytes could be activated. Unless the snake-bite victim already has pre-existing antibodies or memory T cells against the venom, say from a previous survivable exposure to the venom, he or she could be in a lot of trouble. The way around this problem is to get another animal, like a horse, to generate the antibodies by giving it a mild dose of the venom. The anti-venom antibodies are then extracted from the horse's blood and stored in a hospital refrigerator until a snake-bite victim arrives. The antibodies are then injected directly into the bitten area, to tag the antigenic venom, leading to its removal.

A snake-bite victim probably won't take the time to identify the species of the offending reptile, and each snake genus or species can have an immunologically distinctive venom. To allow for this, hospitals routinely store mixtures of antibodies against the venoms of all the area's poisonous snakes. The mixture is injected at the bite site, where only the correct antibody will react with the venom—the other antibodies do nothing and eventually disappear without effect.[5] This kind of immunization is said to be passive and it has a very important function in prenatal and neonatal babies, who get passive immunity via interplacental transfer of antibodies and from the antibodies in their mother's milk. This protects the babies until their own immune systems can take over.

Section 10.3

HIV and AIDS

The Human Immunodeficiency Virus defeats the immune system by infecting, and eventually killing, helper T cells. As a result neither the humoral nor the cell-mediated specific immune responses can function, leaving the patient open to opportunistic diseases.

As is true of all viruses, HIV is very fussy about the host cell it chooses. The problem is that its chosen hosts are immune system cells, the very same cells that are required to fend it off in the first place. Initially the victim's immune system responds to HIV infection by producing the expected antibodies, but the virus stays ahead of the immune system by mutating rapidly. By a variety of mechanisms, some poorly understood, the virus eventually wears down the immune system by killing helper T cells, which are required for the activation of killer T cells and B cells. As symptoms of a low T cell count become manifested, the patient is said to have AIDS.

In this section we will describe the reproduction of HIV as a prelude to a mathematical treatment of the behavior of HIV and the epidemiology of AIDS.

The Human Immunodeficiency Virus (HIV) infects T cells and macrophages, among others.

The outer coat of HIV is a two-layer lipid membrane, very similar to the outer membrane of a cell (see Figure 10.3.1). Projecting from the membrane are sugar-protein projections, called gp120. These gp120 projections recognize and attach to a protein called CD4, which is found on the surfaces of helper T cells, macrophages and monocytes (the latter are macrophage precursors). The binding

---

[5]Note that the unneeded antibodies do not provide a "memory" because there is no activation of lymphocytes, and hence no memory cells.

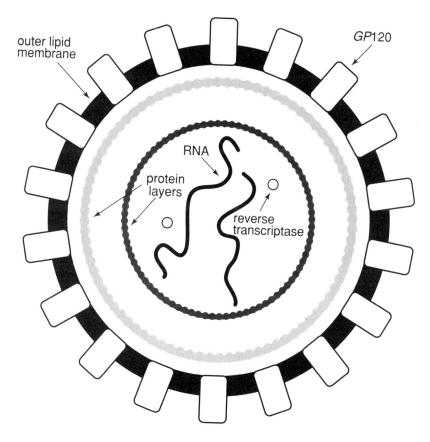

Figure 10.3.1 A model of the Human Immunodeficiency Virus (HIV). The outer membrane of the HIV is derived from the outer membrane of the host cell. Thus, an antibody against that part of the HIV would also act against the host cell. Note that the HIV carries copies of the reverse transcriptase enzyme.

of gp120 and CD4 leads to the fusion of the viral membrane and the cell membrane. Then, the viral capsid is brought into the blood cell (see References [3] and [4]).

HIV is a retrovirus.

The HIV capsid contains two identical single strands of RNA (no DNA). The capsid is brought into the host cell by fusion between the viral envelope and the cell membrane, as described in Section 10.1. The capsid is then enzymatically removed. The HIV RNA information is then converted into DNA information, a step that is indicated by the straight left-pointing arrow in the following Central Dogma flow diagram:[6]

---

[6]This is our final alteration to "dogma."

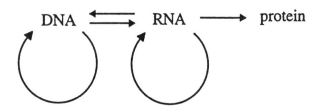

The conversion of RNA information into DNA information involves several steps and is called *reverse transcription*. First, the single-stranded HIV RNA acts as a template for the creation of a strand of DNA. This process entails complementary H-bonding between RNA nucleotides and DNA nucleotides, and it yields a hybrid RNA-DNA molecule. The RNA is then removed and the single-strand of DNA acts as a template for the creation of a second, complementary, strand of DNA. Thus, a double helix of DNA is manufactured, and it carries the HIV genetic information.

The chemical process of covalently polymerizing DNA nucleotides and de-polymerizing RNA nucleotides, like most cellular reactions involving covalent bonds, requires enzymatic catalysis to be effective. The enzyme that catalyzes reverse transcription is called *reverse transcriptase*. Reverse transcriptase is found inside HIV, in close association with the viral RNA, and it enters the host cell right along with the RNA, ready for use. Once HIV DNA is formed it is then spliced into the host cell's own DNA; in other words, it is a provirus.

In a general sense, a provirus becomes an integral part of the host cell's genetic material; for instance, proviruses are replicated right along with the host cell's genome at cell division. It should therefore not be surprising that the physiology and morphology of the host cell changes as a result of the incorporated provirus. For example, one important consequence of HIV infection is that gp120 projections appear on the lymphocyte's surface.

Once in the form of a provirus, HIV starts to direct the host cell's anabolic machinery to form new HIV. As the assembled viruses exit the host cell by budding, they pick up a part of the cell's outer lipid bilayer membrane, along with some of the gp120 placed there by the provirus. The newly-formed virus is now ready to infect a new cell.

The budding process does not necessarily kill the host cell. In fact, infected macrophages seem to generate unending quantities of HIV. T cells do eventually die in an infected person but, as explained below, it is not clear that they die from direct infection by the virus.

The flow of information from RNA to DNA was omitted when the Central Dogma was first proposed because, at the time, no one believed that information flow in that direction was possible. As a consequence, subsequent evidence that it existed was ignored for some years—until it became overwhelming. The process of RNA-to-DNA informational flow is still called "reverse transcription," the key enzyme is called "reverse transcriptase" and viruses in which reverse transcription is important are still called "retroviruses," as though something were running backward. Of course, there is nothing actually "backward" about such processes; they are perfectly normal in their natural context.

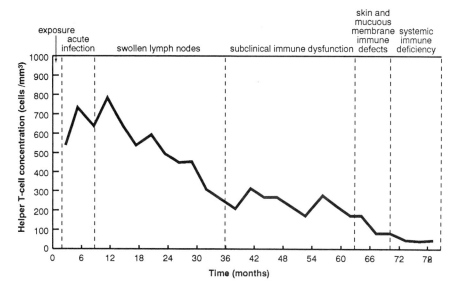

Figure 10.3.2 A graph of the helper T lymphocyte count of an HIV-infected person. Clinical symptoms are indicated along the top of the figure. Note the correlation between the decrease in T cell count and the appearance of the clinical symptoms. (Redrawn from "HIV Infection: The Classical Picture," by Robert Redfield and Donald Burke, *Scientific American*, October 1988, Vol. 259, no. 4; copyright © 1988 by Scientific American, Inc. All rights reserved.)

HIV destroys the immune system instead of the other way around.

As Figure 10.3.2 shows, the number of helper T cells in the blood drops from a normal concentration of about 800 per ml. to zero over a period of several years following HIV infection. The reason for the death of these cells is not well-understood, because budding usually spares the host cell and, besides, only a small fraction of the T cells in the body ever actually become infected by the HIV in the first place. Nevertheless, all the body's helper T cells eventually die. Several mechanisms have been suggested for this apparent contradiction: Among them, the initial contact between HIV and a lymphocyte is through the gp120 of the HIV and CD4 of the T cell. After a T cell is infected, gp120 projections appear on its own surface, and they could cause that infected cell to attach to the CD4 receptors of other, *uninfected* T cells. In this way, one infected lymphocyte could attach to many uninfected ones and disable them all. In fact, it has been observed that, if cells are artificially given CD4 and gp120 groups, they clump together into large multinuclear cells (called *syncitia*).

A second possible way that helper T cells might be killed is suggested by the observation that the infected person's lymph nodes atrophy. The loss of those parts of the lymphatic system may lead to the death of the T cells.

Third, a normal function of helper T cells is to stimulate killer T cells to kill viral-infected cells. It may be that healthy helper T cells instruct killer T cells to kill infected helper T cells. Eventually, this normal process could destroy many of the body's T cells as they become infected although, as noted earlier, only a small fraction of helper T cells ever actually become infected.

Fourth, it has been demonstrated that if an inactive, HIV-infected lymphocyte is activated by antigen, it yields greatly reduced numbers of memory cells. In fact, it seems that the activation process itself facilitates the reproduction of HIV by providing some needed stimulus for the proper functioning of reverse transcriptase.

HIV infection generates a strong initial immune response.

It is shown in Figure 10.3.3 that the immune system initially reacts vigorously to HIV infection, producing antibodies as it should.[7] Nonetheless, the circulating helper T cell count soon begins an irreversble decrease toward zero, as discussed above. As helper T cells die off, the ability of the adaptive immune system to combat any pathogen, HIV or other, also vanishes.

In Section 10.4 we will describe a mathematical model for the interaction between helper T cells and HIV.

The high mutability of HIV demands continued response from the adaptive immune system.

Mutations occur commonly in HIV RNA and the reason is reasonably well-understood: Reverse transcriptase lacks a *"proof-reading"* capacity. This proof-reading ability is found in enzymes that catalyze the polymerization of DNA from a DNA template in the "forward" direction of the Central Dogma. Thus, the occasional mismatching of H bonds between nucleotides, say the pairing of A opposite G, gets corrected. On the other hand, reverse transcriptase, which catalyzes DNA formation from an RNA template, seems not to be able to correct base-pairing errors, and this leads to high error rates in base placement—as much as one mismatched base out of every 2000 polymerized. The two RNA polynucleotides of HIV have between 9000 and 10,000 bases distributed among about nine genes, so this error rate might yield up to five changed bases, and perhaps three or four altered genes, per infection.

We are concerned here especially with the effects of mutated viral surface antigens, e.g., proteins and glycoproteins, on immune system recognition. Every new antigenic version of these particular viral products will require a new round of helper T cell activation to defeat the virus. The problem there is that, as pointed out earlier, activation of an HIV-infected helper T cell seems to help the HIV inside it to replicate and, further, leads to the formation of stunted memory T cell clones. Thus, each new antigenic form of HIV causes the immune system to

---

[7]The presence of antibodies against HIV is the basis for the diagnosis of HIV infection. Note that it takes several months to get a measurable response.

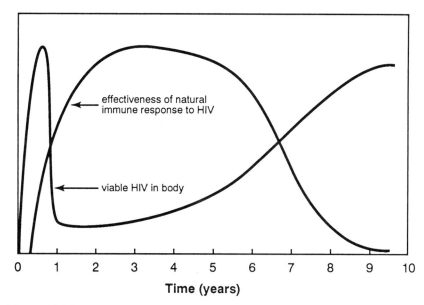

Figure 10.3.3 A graph of immune response and viral appearance versus time for an HIV-infected person. The initial infection generates a powerful immune response. That response, however, is later overwhelmed by the virus, which kills the helper T lymphocytes that are required by the humoral and cell-mediated immune responses. (Redrawn from "HIV Infection: The Clinical Picture," by Robert Redfield and Donald Burke, *Scientific American*, October 1988, Vol. 259, no. 4; copyright © 1988 by Scientific American, Inc. All rights reserved.)

stimulate HIV replication, while simultaneously hindering the immune system's ability to combat the virus. The HIV stays just ahead of the immune system, like a carrot on a stick, affecting helper T cells before the T cells can respond properly, and then moving on to a new round of infections. One could say, "The worse it gets, the worse it gets!"

The mutability of HIV has another unfortunate effect for its victims. Current therapy emphasizes drugs that interfere with the correct functioning of reverse transcriptase; AZT is an example. The high mutation rate of HIV can generate new versions of reverse transcriptase and, sooner or later, a version will appear that the drug cannot affect.

In Section 10.5 we will model the mutability of HIV and its eventual overwhelming of the immune system.

HIV infection leads to Acquired Immunodeficiency Syndrome (AIDS).

A person infected with HIV is said to be "HIV-positive." Such people may be asymptomatic for a considerable time following infection, or the symptoms may be mild and transient; the patient is, however, infectious. Eventually, the loss of helper T cells will leave the person open to infections, often of a rare kind (see Figure 10.3.2). As examples, pneumonia caused by a protozoan called *Pneumocystis carinii* and a cancer of blood vessels, called Kaposi's sarcoma, are extremely rare in the general population, yet they frequently are found in HIV-positive people. Everyone is exposed to the pathogens that cause these diseases, but do not get the disease if their immune system is working properly. When HIV-positive persons exhibit unusual diseases as a result of low helper T cell counts, they are said to have AIDS.

Section 10.4

An HIV Infection Model

A model for HIV infection involves four components: Normal T cells, latently infected T cells, infected T cells actively replicating the virus, and the virus itself. Any proposed model should incorporate the salient behavior of these components and respect biological constraints. In this section we present such a model and show that it has a stationary solution. This model was developed and explored by Perelson, Kirschner and co-workers.

T cell production attempts to maintain a constant T cell serum concentration.

In this section we will be presenting a model for T cell infection by HIV, as described in Section 10.2 (see References [5–8]). This model tracks four components; three types of T cells and the virus itself, and therefore requires a four-equation system for its description. As a preliminary step toward understanding the full system of equations, we present first a simplified version; namely the equation for T cells in the absence of infection. In forming a mathematical model of T cell population dynamics based on the discussion of Section 10.2, we must incorporate the following assumptions.

- Some immunocompetent T cells are produced by the lymphatic system; over relatively short periods of time, their production rate is constant and independent of the number of T cells present. Over longer periods of time their production rate adjusts to help maintain a constant T cell concentration, even in adulthood. Denote this *supply rate* by s.
- T cells are produced through clonal selection if an appropriate antigen is present, but the total number of T cells cannot increase unboundedly. Model this using a logistic term, $rT(1 - T/T_{max})$, with per capita growth rate r (cf., Section 3.4).
- T cells have a finite natural lifetime after which they are removed from circulation. Model this using a death rate term, μT, with a fixed per capita death rate μ.

Altogether, the differential equation model is

$$\frac{dT}{dt} = s + rT\left(1 - \frac{T}{T_{max}}\right) - \mu T. \tag{10.4.1}$$

In this, T is the T cell population in cells per cubic millimeter.

We want the model to have the property that solutions, $T(t)$, which start in the interval $[0, T_{max}]$ stay there. This will happen if the derivative dT/dt is positive when $T = 0$ and negative when $T = T_{max}$. From equation (10.4.1),

$$\left.\frac{dT}{dt}\right|_{T=0} = s,$$

and since s is positive, the first requirement is fulfilled. Next, substituting $T = T_{max}$ into equation (10.4.1), we get the condition that must be satisfied for the second requirement,

$$\left.\frac{dT}{dt}\right|_{T=T_{max}} = s - \mu T_{max} < 0,$$

or, rearranged,

$$\mu T_{max} > s. \tag{10.4.2}$$

The biological implication of this statement is that when the number of T cells have reached the maximum value T_{max}, then there are more cells dying than are being produced by the lymphatic system.

Turning to the stationary solutions of system (10.4.1), we find them in the usual way, by setting the right hand side to zero and solving for T:

$$-\frac{r}{T_{max}}T^2 + (r - \mu)T + s = 0.$$

The roots of this quadratic equation are

$$T = \frac{T_{max}}{2r}\left((r - \mu) \pm \sqrt{(r - \mu)^2 + 4s\frac{r}{T_{max}}}\right). \tag{10.4.3}$$

Since the product $4sr/T_{max}$ is positive, the square root term exceeds $|r - \mu|$,

$$\sqrt{(r - \mu)^2 + 4sr/T_{max}} > |r - \mu|,$$

and therefore one of the roots of the quadratic equation is positive while the other is negative. Only the positive root is biologically important, and we denote it by T_0, as the "zero virus" stationary point (see below). We now show that T_0 must lie between 0 and T_{max}. As already noted, the right hand side of equation (10.4.1) is positive when $T = 0$ and negative when $T = T_{max}$. Therefore it must have a root between 0 and T_{max}; this is our positive root T_0 calculated from equation (10.4.3) by choosing the $+$ sign. We will refer to the difference $p = r - \mu$ as the T cell *proliferation rate*; in terms of it, the globally attracting stationary solution is given by

$$T_0 = \frac{T_{max}}{2r} \left(p + \sqrt{p^2 + 4s\frac{r}{T_{max}}} \right). \tag{10.4.4}$$

This root T_0 is the only (biologically consistent) stationary solution of equation (10.4.1).

Now consider two biological situations.

Situation 1: Supply Rate Solution. In the absence of an infection, or at least an environmental antigen, the clonal production rate r can be small, smaller than the natural deathrate μ, resulting in a negative proliferation rate p. In this case the supply rate s must be high in order to maintain a fixed T cell concentration of about 1000 per cubic millimeter. Data in Reference [6] confirm this.

Table 10.4.1   Parameters for Situation 1

Parameter	Description	Value
s	T cell from precursor supply rate	$10/\text{mm}^3/\text{day}$
r	normal T cell growth rate	$.03/\text{day}$
T_{max}	maximum T cell population	$1500/\text{mm}^3$
μ	T cell death rate	$.02/\text{day}$

With these data, calculate the stationary value of T_0 using equation (10.4.3) as follows.

```
> f:=T->s+r*T*(1-T/Tmax)- mu *T;
> s:= 10; r:=.03; mu:=.02; Tmax:=1500;
> fzero:=solve(f(T) = 0,T);
> T0:=max(fzero[1],fzero[2]);
> T0num:=evalf(T0);
```

Next calculate and display trajectories from various starting points.

```
> deq:=diff(T(t),t)=f(T(t));
> with(DEtools):
> inits:={[0,0],[0,T0/4],[0,T0/2],[0,(T0+Tmax)/2],[0,Tmax]};
> phaseportrait(deq,T(t),0..25,inits,stepsize=1);
```

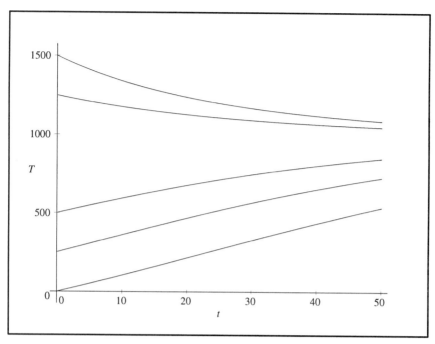

Figure 10.4.1 Time vs. number of T cells per cubic millimeter

Situation 2: Clonal Production Solution. An alternate scenario is that adult thymic atrophy has occurred, or a thymectomy has been performed. As a hypothetical and limiting situation, take s to equal zero and ask how r must change to maintain a comparable T_0. Use these parameters:

Table 10.4.2 Parameters for Situation 2

Parameter	Description	Value
s	T cell from precursor supply rate	$0/mm^3$/day
r	normal T cell growth rate	.06/day
T_{max}	maximum T cell population	$1500/mm^3$
μ	T cell death rate	.02/day

```
> s:= 0; r:=.06; mu:=.02; Tmax:=1500;
> fzero:=solve(f(T)=0,T);
> T0:=evalf(fzero[1],fzero[2]);
```

As above, T_0 is again about 1000 T cells per cubic millimeter. Trajectories in this second situation are plotted in Figure 10.4.2; contrast the convergence rate

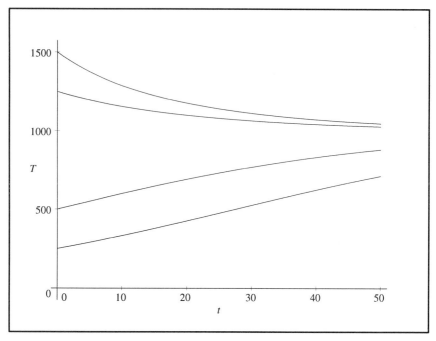

Figure 10.4.2 Time vs. T cell count with a reduced thymus function

to the stationary solution under this clonal T cell production situation with the supply rate convergence of Situation 1.

Remark: Contrasting these Situations shows that upon adult thymic atrophy or thymectomy, the response of the T cell population is much slower. This suggests that one would find differences in the dynamics of T cell depletion due to an HIV infection in people of different ages. Clearly, there is a need for r, the T cell growth rate, to be large in compensation when the supply rate, s, is small. How can one influence one's value of r? The answer should be an inspiration for continuing biological and medical research.

A four-equation system is used to model T cell–HIV interaction.

To incorporate an HIV infection into the above model, we follow the approach taken by Perelson, Kirschner, and De Boer [6] and differentiate three kinds of T cells: Besides the normal variety, whose number is denoted by T as before, there are T cells infected with provirus, but not producing free virus. Designate the number of these *latently* infected T cells by T_L. In addition, there are T cells that are infected with virus and are *actively* producing new virus. Designate the number of these by T_A. The interaction between virus, denoted by V, and T cells is reminiscent of a predator–prey relationship; a mass action term is used to quantify the interaction (see Section 4.4). However, only the active type T cells produce virus, while only the normal T cells can be infected.

We now present the model and follow with a discussion of its four equations separately:

$$\frac{dT}{dt} = s + rT\left(1 - \frac{T + T_L + T_A}{T_{max}}\right) - \mu T - k_1 V T, \qquad (a)$$

$$\frac{dT_L}{dt} = k_1 V T - \mu T_L - k_2 T_L, \qquad (b)$$

$$\frac{dT_A}{dt} = k_2 T_L - \beta T_A, \qquad (c)$$

$$\frac{dV}{dt} = N\beta T_A - k_1 V T - \alpha V. \qquad (d)$$

(10.4.5)

The first equation is a modification of equation (10.4.1) with the inclusion of an infection term having mass action parameter k_1. When normal T cells become infected, they immediately become reclassified as the latent type. In addition, note that the sum of all three types of T cells count toward the T cell limit, T_{max}.

The first term in the second equation corresponds to the reclassification of newly infected normal T cells. These cells disappear from equation (a) but then reappear in equation (b). In addition, equation (b) includes a per capita death rate term and a term to account for the transition of these latent-type cells to active-type with rate parameter k_2.

The first term of equation (c) balances the disappearance of latent T cells upon becoming active, with their appearance as active-type T cells. It also includes a per capita death rate term with parameter β corresponding to the lysis of these cells after releasing vast numbers of replicated virus. It is clear that T cells active in this sense perish much sooner than do normal T cells, therefore β is much larger than μ,

$$\beta \gg \mu. \qquad (10.4.6)$$

Finally the fourth equation accounts for the population dynamics of the virus. The first term, $N\beta T_A$, comes from the manufacture of virus by the active type T cells, but the number produced will be huge for each T cell. The parameter N, a large value, adjusts for this many-from-one difference. The second term reflects the fact that as a virus invades a T cell, it drops out of the pool of free virus particles. The last term, with per capita rate parameter α, corresponds to loss of virus through the body's defense mechanisms.

Remark: Note that in the absence of virus, i.e., $V = 0$, then both $T_L = T_A = 0$ as well and, setting these values into system (10.4.5), we see that this new model agrees with the old one, equation (10.4.1).

The T cell–HIV model respects biological constraints.

We want to see that the model is constructed well enough that no population goes negative or goes unbounded. To do this, we first establish that the derivatives, $\frac{dT}{dt}$, $\frac{dT_L}{dt}$, $\frac{dT_A}{dt}$, and $\frac{dV}{dt}$ are positive whenever T, T_L, T_A, or $V = 0$, respectively. This would mean that each population will increase, not decrease, at low population sizes.

But from equation (10.4.5a), if $T = 0$, then

$$\frac{dT}{dt} = s > 0;$$

if $T_L = 0$, then equation (10.4.5b) gives

$$\frac{dT_L}{dt} = k_1 V T > 0;$$

likewise if $T_A = 0$, then from equation (10.4.5c)

$$\frac{dT_A}{dt} = k_2 T_L > 0;$$

and, finally, equation (10.4.5d) becomes, when $V = 0$,

$$\frac{dV}{dt} = N\beta T_A > 0.$$

We have assumed all the parameters are positive, and so these derivatives are also positive as shown.

Following Perelson, Kirschner, and De Boer [6], we next show that the total T cell population as described by this model remains bounded. This total, T_Σ is defined to be the sum $T_\Sigma = T + T_L + T_A$ and satisfies the differential equation obtained by summing the right-hand side of the first three equations in system (10.4.5)

$$\frac{dT_\Sigma}{dt} = s + rT\left(1 - \frac{T_\Sigma}{T_{\max}}\right) - \mu T - \mu T_L - \beta T_A. \qquad (10.4.7)$$

Now suppose $T_\Sigma = T_{\max}$. Then from equation (10.4.7),

$$\frac{dT_\Sigma}{dt} = s - \mu T - \mu T_L - \beta T_A + \mu T_A - \mu T_A$$

and combining the second, third and last terms as $-\mu T_{\max}$, this gives

$$\frac{dT_\Sigma}{dt} = s - \mu T_{\max} - (\beta - \mu) T_A < s - \mu T_{\max},$$

where equation (10.4.6) has been used to obtain the inequality. Recalling condition (10.4.2), we find that

$$\frac{dT_\Sigma}{dt} < 0 \qquad \text{if } T_\Sigma = T_{\max}$$

proving that T_Σ cannot increase beyond $T_{\max}$.

In summary, the system (10.4.5) has been shown to be consistent with the biological constraints that solutions remain positive and bounded.

The T cell infected stationary solution is stable.

To find the stationary points of the T cell–HIV model, that is, equation (10.4.5), we must set the derivatives to zero and solve the resulting (non-linear) algebraic system in four equations and four unknowns. Solving the third equation, namely $0 = k_2 T_L - \beta T_A$, for T_A gives $T_A = (k_2/\beta)T_L$, which may in turn be substituted for all its other occurences. This reduces the problem to three equations and three unknowns. Continuing in this way we arrive at a polynomial in, say, T, whose roots contain the stationary points. We will not carry out this approach here. Instead we will solve this system numerically, below, using derived parameter values. However in Reference [6] it is shown symbolically that the uninfected stationary point T_0 (10.4.4) is stable (see Section 2.4) if and only if the parameter N satisfies

$$N < \frac{(k_2 + \mu)(\alpha + k_1 T_0)}{k_2 k_1 T_0}.$$

By defining the combination of parameters on the right-hand side as N_{crit}, we may write this as

$$N < N_{\text{crit}} \qquad \text{where} \quad N_{\text{crit}} = \frac{(k_2 + \mu)(\alpha + k_1 T_0)}{k_2 k_1 T_0}. \qquad (10.4.8)$$

In Table 10.4.1 we give values of the parameters of the system (10.4.5) as determined by Reference [6].

Table 10.4.3 Parameters of the HIV Infection Model

Parameter	Description	Value
s	T cell from precursor supply rate	$10/\text{mm}^3/\text{day}$
r	normal T cell growth rate	$.03/\text{day}$
$T_{\max}$	maximum T cell population	$1500/\text{mm}^3$
μ	normal/latently infected T cell death rate	$.02/\text{day}$
β	actively infected T cell death rate	$.24/\text{day}$
α	free virus death rate	$2.4/\text{day}$
k_1	T cells infection rate by free virus	$2.4 \times 10^{-5}\ \text{mm}^3/\text{day}$
k_2	latent to active T cell conversion rate	$3 \times 10^{-3}/\text{day}$
N	virus produced by an active T cell	taken as 1400 here

This model reflects the clinical picture as presented in Greene [9].

Exercises

1. In the uninfected situations, for both $s = 0$ and $s = 10$ derive the numerical solution T for $f(T) = 0$. Which of the roots for this equation is in the interval $[0, T_{max}]$?

2. In the virus-free situation, give a biological interpretation for r. Suppose that r is increased to r_n so that

$$\frac{r_n - r}{r} = .10.$$

   That is, r is increased by 10%. What is the percentage of increase of the steady state of T cells corresponding to a 10% increase in r?

3. With the parameters as stated for the infected situation, what is the numerical value for each of these: T_{max}, the uninfected steady-state of T cells, the infected steady-state of T cells, and N_{crit}. Is N_{crit} more or less that the N used in these parameters? What are the implications of this last answer?

4. Sketch a graph of how T, T_L, T_A, and V evolve during the first year and move toward equilibrium. Continue the graph for two more years. Here is *Maple* syntax that will accomplish this integration of the equations.

```
> RT:=(t,T,TL,TA,V)->s - mu*T+r*T*(1-(T+TL+TA)/Tmax)-k1*V*T;
> RTL:=(t,T,TL,TA,V)->k1*V*T-mu *TL-k2*TL;
> RTA:=(t,T,TL,TA,V)->k2*TL-b*TA;
> RV:=(t,T,TL,TA,V)->N*b*TA-k1*V*T-a*V;
> s:=10;r:=.03; Tmax:=1700;mu:=.02;
   b:=.24; a:=2.4; k1:=.000024; k2:=.003; N:=1400;
> with(share): readshare(ODE,plots):
> initl:=[0,1000,0,0,.001];
> sol:=rungekuttahf([RT,RTL,RTA, RV],
   initl,0.5,365):
> plot(makelist(sol,1,3),makelist(sol,1,4), makelist(sol,1,5));
> initl:=sol[365];
> plot(makelist(sol,1,2));
> sol:=rungekuttahf([RT,RTL,RTA, RV],
   initl,.99,700):
> plot(makelist(sol,1,2));
```

Section 10.5

A Model for a Mutating Virus

The model of the previous section illustrated the interaction of HIV with T cells. It did not account for mutations of HIV. The following is a model for evolving mutations of an HIV infection and an immune system response. This model is based on one introduced into the literature by Nowak, May, and Anderson.

Any model of an HIV infection should reflect the high mutability of the virus.

In Section 10.3 we discussed the high degree of mutability characteristic of the HIV virus, which results in a large number of viral quasi-species. The human immune system seems able to mount an effective response against only a finite number of these mutations, however. Furthermore, the activation of a latently infected helper T cell appears to stimulate viral reproduction with the result that every time a new mutant activates a T cell, vigorous viral population growth ensues. The immune system's T cell population evidently can endure this cycle only a limited number of times. The objective of this section is to modify the T cell–HIV model to reflect these facts in the model. In this, we follow Nowak, May, and Anderson [10] and [11]; see also Nowak and McMichael [4].

Key assumptions

1. The immune response to a viral infection is to create sub-populations of immune cells specific to a particular viral strain that direct immunological attack against that strain alone. The response is directed against the highly variable parts of the virus.
2. The immunological response to the virus is also characterized by a response which is specific to the virus, but which acts against all strains. In other words, it acts against parts of the virus conserved under mutations.
3. Each mutant of the initial viral infection can cause the death of all immune system cells whether those cells are directed at variable or conserved regions.

In this modified model, we keep track of three sets of populations. Let $\{v_1, v_2, \ldots, v_n\}$ designate the various sub-populations of viral mutants of the initial HIV infection. Let $\{x_1, x_2, \ldots, x_n\}$ designate the populations of specific lymphocytes created in response to these viral mutations. And let z designate the immune response that can destroy all variations of the infective agent. The variable n, for the number of viral mutations that arise, is a parameter of the model. We also include a parameter, the *diversity threshold* N_{div}, representing the number of mutations that can be accomodated before the immune system collapses.

The equation for each HIV variant, v_i, consists of a term, with parameter a, for its natural population growth rate, a term, with parameter b, for the general immune response, and a term, with parameter c, for the specific immune response to that variant,

$$\frac{dv_i}{dt} = v_i(a - bz - cx_i), \qquad i = 1, \ldots, n. \tag{10.5.1}$$

The equation for each specific immune response population, x_i, consists of a term, with parameter g, which increases the population in proportion to the amount of its target virus present, and a term, with parameter k, corresponding to the destruction of these lymphocytes by any and all viral strains,

$$\frac{dx_i}{dt} = gv_i - kx_i(v_1 + v_2 + \ldots + v_n), \qquad i = 1, \ldots, n. \qquad (10.5.2)$$

Finally, the equation for the general immune response population, z, embodies a term, with parameter h, for its increase in proportion to the sum total of virus present but also a mass action term for its annihilation upon encounter with any and all virus,

$$\frac{dz}{dt} = (h - kz)(v_1 + v_2 + \ldots + v_n). \qquad (10.5.3)$$

The fate of the immune response depends on a critical combination of parameters.

Again drawing on Reference [10], we list several results which can be derived from this modified model. The model adopts one of two asymptotic behaviors depending on a combination of parameters, denoted N_{div}, defined by

$$N_{\text{div}} = \frac{cg}{ak - bh} \qquad \text{where } a/b > h/k. \qquad (10.5.4)$$

If the number n of viral variants remains below or equal to N_{div} then the virus population eventually decreases and becomes subclinical. On the other hand, if $n > N_{\text{div}}$ then the virus population eventually grows unchecked.

Note that N_{div} depends on the immune response to the variable and conserved regions of the virus in different ways. If specific lymphocytes rapidly respond (a large g) and are very effective (a large c), then N_{div} will be large in proportion to each, meaning a large number of mutations will have to occur before the virus gains the upper hand. By contrast, the general immune response parameters, h and b, appear as a combination in the denominator. Their effect is in direct opposition to the comparable viral parameters a and k.

Naturally, the size of N_{div} is of considerable interest. Assuming that the denominator of equation (10.5.4) is positive, $ak > bh$, we make three observations; their proofs may be found in Reference [10].

Observation 1. The immune responses, the x_i's and z, in total, have only a limited response to the HIV infection. That is, letting $X = x_1 + x_2 + \ldots + x_n$, be the sum of the specific immunological responses, then

$$\lim_{t \to \infty} X(t) = g/k$$

$$\lim_{t \to \infty} z(t) = h/k$$

$$(10.5.5)$$

where the parameters g, h, and k are as defined as in equations (10.5.2)–(10.5.4). The implication is that, even though the virus continues to mutate, the immune system cannot mount an increasingly higher response.

The next observation addresses the possibility that, after some time, all the immune subspecies populations are decreasing.

Observation 2. If all mutant subspecies populations v_i are decreasing after some time τ, then the number of mutants will remain less than N_{div} and the infection will be controlled. That is, if there is a time τ such that all derivatives $v_i'(t) < 0$ are negative for $t > \tau$, $i = 1, \ldots, n$, then the number of mutations n will not exceed N_{div}.

In the next observation, we see that if the number of variations increases to some level determined by the parameters, then the viral population grows without bound.

Observation 3. If the number of mutations exceeds N_{div}, then at least one subspecies increases without bound. In fact, in this case, the sum $V(t) \equiv v_1 + v_2 + \cdots + v_n$ increases faster than a constant times e^{at} for some positive number a.

Observation 4. If $ak < bh$, the immune system will eventually control the infection.

Numerical studies illustrate the observations graphically.

In what follows we give parameters with which computations may be made to visualize the results discussed here. These parameters do not represent biological reality; likely the real parameters are not known. The ones used illustrate the features of the model. In Reference [10], the authors choose $a = c = 5$, $b = 4.5$, and $g = h = k = 1$. This choice yields the diversity threshold as 10 ($N_{div} = 10$). Thus, if there are 11 or more mutations, the virus population will increase without bound. To keep computation time small, we choose the same constants, except $b = 4$. This produces a threshold of 5. As a result, only 13 differential equations need to be integrated: six for the virus strains, six for the variable-region immune responses, and one for the conserved-region immune response. We specify the times of mutation as T_i with $T_0 = 0$. Thus, we specify that $v_i(t) = 0$ for $0 < t < T_i$ and $v(T_i)$ is some initial value. We take this initial value to be 5/100 for the initial infection and 1/100 for each succeeding mutation. Step size in the integration is often a quandary. We want enough accuracy to understand the trends, yet not so much as to cause the computations to become long and boring. Step sizes of 1/10 in the range of greatest deviation and 2/10 otherwise seem to make a compromise. We use the Runge–Kutta routine to integrate the system of thirteen differential equations.

```
> with(share): readshare(ODE,plots):
```

Observation 1 above introduces the diversity threshold. We initialize the constants and, with these constants, determine N_{div}.

```
> a:=5: b:=4: c:=5: g:=1: h:=1: k:=1: Ndiv:= c*g/(a*k-b*h);
```

This computation should give the $N_{\text{div}} = 5$.

As a result of this computation, we expect that after six mutations the level of the viral population will grow unbounded—until the virus kills the host. Before giving the syntax to run this simulation, we will show the results that should be expected in order to illustrate the nature of the results proved in this section. Suppose first that there is an initial infection and the virus runs a course as modeled with equations (10.5.1)–(10.5.3), except no mutations occur. We expect that the infection will flare up and be controlled by the immune system. This scenerio is pictured in the graph of Figure 10.5.1. The plot is time plotted against viral particles. Because $N_{\text{div}} = 5$, we expect the viral infection to be subdued.

We now allow two mutations. Because this number is still less than N_{div} for our choice of parameters, we expect the immune system to remove the virus. What we see is the growth and removal of the original infection and the growth and removal of two mutations.

Finally, we allow six mutations to occur. It is expected that the viral infections will initially be suppressed, but the repeated mutations—greater than N_{div}—eventually overwhelms the immune system.

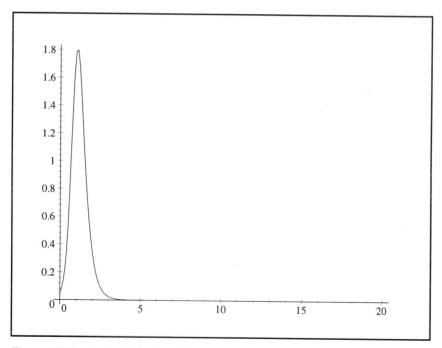

Figure 10.5.1   A viral infection with no mutation

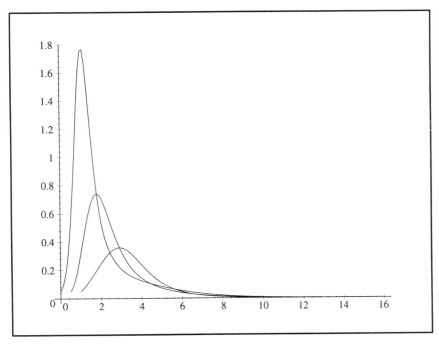

Figure 10.5.2   Infection with two mutations

```
> RSv1:= (t,v1,v2,v3,v4,v5,v6,x1,x2,x3,x4,x5,x6,z)-->(a - b*z - c*x1)*v1:
> RSv2:= (t,v1,v2,v3,v4,v5,v6,x1,x2,x3,x4,x5,x6,z)-->(a - b*z - c*x2)*v2:
> RSv3:= (t,v1,v2,v3,v4,v5,v6,x1,x2,x3,x4,x5,x6,z)-->(a - b*z - c*x3)*v3:
> RSv4:= (t,v1,v2,v3,v4,v5,v6,x1,x2,x3,x4,x5,x6,z)-->(a - b*z - c*x4)*v4:
> RSv5:= (t,v1,v2,v3,v4,v5,v6,x1,x2,x3,x4,x5,x6,z)-->(a - b*z - c*x5)*v5:
> RSv6:= (t,v1,v2,v3,v4,v5,v6,x1,x2,x3,x4,x5,x6,z)-->(a - b*z - c*x6)*v6:
> RSx1:= (t,v1,v2,v3,v4,v5,v6,x1,x2,x3,x4,x5,x6,z)-->g*v1 -
   k*(v1+v2+v3+v4+v5+v6) *x1:
> RSx2:= (t,v1,v2,v3,v4,v5,v6,x1,x2,x3,x4,x5,x6,z)-->g*v2-
   k*(v1+v2+v3+v4+v5+v6)*x2:
> RSx3:= (t,v1,v2,v3,v4,v5,v6,x1,x2,x3,x4,x5,x6,z)-->g*v3 -
   k*(v1+v2+v3+v4+v5+v6)*x3:
> RSx4:= (t,v1,v2,v3,v4,v5,v6,x1,x2,x3,x4,x5,x6,z)-->g*v4 -
   k*(v1+v2+v3+v4+v5+v6)*x4:
> RSx5:= (t,v1,v2,v3,v4,v5,v6,x1,x2,x3,x4,x5,x6,z)-->g*v5 -
   k*(v1+v2+v3+v4+v5+v6)*x5:
> RSx6:= (t,v1,v2,v3,v4,v5,v6,x1,x2,x3,x4,x5,x6,z)-->g*v6-
   k*(v1+v2+v3+v4+v5+v6)*x6:
> RSz := (t,v1,v2,v3,v4,v5,v6,x1,x2,x3,x4,x5,x6,z)-->(h-k*z) *
   (v1+v2+v3+v4+v5+v6):
```

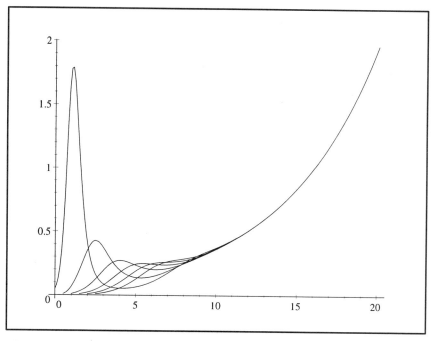

Figure 10.5.3   Six mutations, with $N_{\mathrm{div}} = 5$

What follows is the syntax for the right-hand side of the equations for each v_i in (10.5.1), each x_i in (10.5.2), and for z in (10.5.3). The first six lines are "right-hand side v_1," "right-hand side v_2," etc.

We now integrate these equations using the numerical scheme found in the ODE Runge–Kutta package. In the following syntax, a mutation is allowed to occur at each integral multiple of 1/2 until there are six mutations. Then the program is allowed to run until $t = 20$. By that time, the blow-up of the viral population will have begun. The first line starts with $t = 0$, $v_1 = 5/100$, and all other viral levels 0. This is integrated in time steps of 1/10 for 5 steps.

```
> init1:=[0,5/100,0,0,0,0,0,0,0,0,0,0,0,0];
> output1:=rungekuttahf([RSv1, RSv2, RSv3, RSv4, RSv5, RSv6, RSx1,
  RSx2, RSx3, RSx4, RSx5, RSx6, RSz],init1,.1,5):
```

Each remaining pair initializes the third, the fourth, the fifth, and finally the sixth mutation at the level of 1/100 and runs for time steps of 1/10 for five steps.

```
> init2:=subsop(3=1/100,output1[5]);
> output2:=rungekuttahf([RSv1, RSv2, RSv3, RSv4, RSv5, RSv6, RSx1,
  RSx2, RSx3, RSx4, RSx5, RSx6, RSz],init2,.1,5):
> init3:=subsop(4=1/100,output2[5]);
```

```
> output3:=rungekuttahf([RSv1, RSv2, RSv3, RSv4, RSv5, RSv6, RSx1,
  RSx2, RSx3, RSx4, RSx5, RSx6,RSz],init3,.1,5):
> init4:=subsop(5=1/100,output3[5]);
> output4:=rungekuttahf([RSv1, RSv2, RSv3, RSv4, RSv5, RSv6, RSx1,
  RSx2, RSx3, RSx4, RSx5, RSx6,RSz],init4,.1,5):
> init5:=subsop(6=1/100,output4[5]);
> output5:=rungekuttahf([RSv1, RSv2, RSv3, RSv4, RSv5, RSv6, RSx1,
  RSx2, RSx3, RSx4, RSx5, RSx6, RSz],init5,.1,5):
> init6:=subsop(7=1/100,output5[5]);
> output6:=rungekuttahf([RSv1, RSv2, RSv3, RSv4, RSv5, RSv6, RSx1,
  RSx2, RSx3, RSx4, RSx5, RSx6, RSz],init6,.2,88):
```

Observation 2 predicts that since more mutations have occurred than N_{div} the population will grow without bound. The graphs show this. Here is syntax for drawing the graphs.

```
> plot({makelist(output1,1,2), makelist(output2,1,2), makelist(output2,1,3),
  makelist(output3,1,2),makelist(output3,1,3),makelist(output3,1,4),
  makelist(output4,1,2),makelist(output4,1,3),makelist(output4,1,4),
  makelist(output4,1,5),makelist(output5,1,2),makelist(output5,1,3),
  makelist(output5,1,4),makelist(output5,1,5),makelist(output5,1,6),
  makelist(output6,1,2),makelist(output6,1,3),makelist(output6,1,4),
  makelist(output6,1,5),makelist(output6,1,6),makelist(output6,1,7)});
```

We now verify Observation 3 for these parameters. First, note that, from (10.5.2), the sum of the x_i's satisfies the differential equation

$$X' = V * (g - kX)$$

where $X = x_1 + x_2 + \cdots + x_6$ and $V = v_1 + v_2 + \cdots + v_6$. We could compute the solution of this equation and expect that

$$\lim_{t\to\infty} (x_1(t) + x_2(t) + \cdots + x_6(t)) = g/k.$$

Or, using the computations already done, add the x_i's. The level of the x_i's are kept in this syntax in the 8th through 13th positions. Where the computations are in the "time" variable is kept in the 1st position of the output. We add these x_i's in each output.

```
> y1:=seq([output1[n][1],sum(output1[n][j],j=8..13)],n=0..5):
  y2:=seq([output2[n][1],sum(output2[n][j],j=8..13)],n=1..5):
  y3:=seq([output3[n][1],sum(output3[n][j],j=8..13)],n=1..5):
  y4:=seq([output4[n][1],sum(output4[n][j],j=8..13)],n=1..5):
  y5:=seq([output5[n][1],sum(output5[n][j],j=8..13)],n=1..5):
  y6:=seq([output6[n][1],sum(output6[n][j],j=8..13)],n=1..40):
```

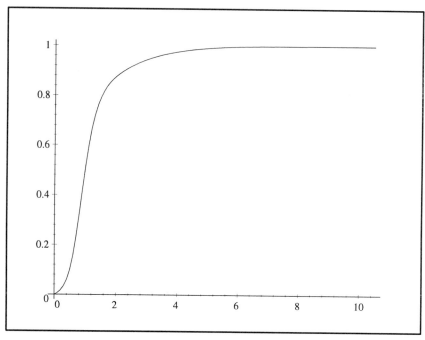

Figure 10.5.4   Graph of $x_1 + x_2 + \cdots + x_n$

Here is the plot of the sum of the x_i antigens.

```
> plot({y1,y2,y3,y4,y5,y6});
```

The plot of the response z, no matter where you start, should have asymptotic limit h/k and should look essentially the same as that for $(x_1 + x_2 + \cdots + x_6)$. Here is one way to plot the values of z. What you should see is that the z reaches a maximum.

```
> plot({makelist(output1,1,14), makelist(output2,1,14),
    makelist(output3,1,14), makelist(output4,1,14),
    makelist(output5,1,14), makelist(output6,1,14)});
```

Viral suppression is possible with some parameters.

It was stated in Observations 1 and 4 that there are two ways to achieve viral suppression. These are experiments that should be run. One could choose parameters so that $ak < bh$; then the immune system will eventually control the infection. No change need be made in the syntax, only $a = 4$ and $b = 5$. Other parameters could remain the same.

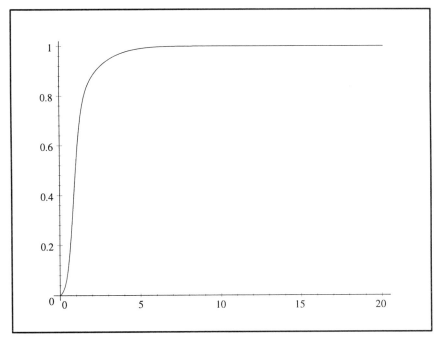

Figure 10.5.5   quad Graph of z

The simple models as presented in these two sections give a good first understanding of the progress from infection, to remission, to AIDS. Such an understanding provokes further study.

Section 10.6

Predicting the Onset of AIDS

Most diseases have a latency or incubation period between the time of infection and the onset of symptoms, AIDS is no exception. The latency period for AIDS varies greatly from individual to individual and, so far, its profile has not been accurately determined. However, assuming a given form of the incubation profile, we show the onset of symptoms occurs, statistically, as the time of infection convolved with this profile.

AIDS cases can be statistically predicted by a convolution integral.

In this chapter we have discussed the epidemiology of the HIV infection and subsequent appearance of AIDS. For most diseases, there is a period of time between infection by the causative agent and the onset of symptoms. This is referred to as the *incubation period*; an affliction is asymptomatic during this time. Research is showing that the nature of this transition for HIV is a complicated matter. Along

with trying to learn the mechanism of this process, considerable work is being
devoted to an attempt to prolong the period between HIV infection and the appearance of AIDS. This period varies greatly among different individuals and
appears to involve, besides general health, particular characteristics of the individual's immune system. See References [5]–[8] for further details.

The incubation period can be modeled as a probability density function $p(t)$,
see Section 2.6, meaning the probability that AIDS onset occurs in a Δt time
interval containing t is

$$p(t) \cdot \Delta t.$$

To discover the incubation density, records are made, when possible, of the time
between contraction of HIV and the appearance of AIDS. See Bacchetti [12] for
several comments by other researchers, and for a comprehensive bibliography. At
the present this probability density is not known, but some candidates are shown in
Figures 10.6.1(a), (b), (c), (d). Figure 10.6.1(a) is a uniform distribution over the
period of two years to 18 years. This distribution has no preferred single incubation moment but incubation is guaranteed to occur no sooner than two years after
infection and no later than 18 (18.0) years afterward. It is unlikely that this is the
operating distribution but we include it for comparison purposes. Figure 10.6.1(b)
is an exponential distribution. This distribution pertains to many "arrival time"
processes in biology such as times for prokaryotic cell division (which occurs
upon the "arrival" of cell maturation). Here again there is no preferred incubation moment but, unlike the uniform distribution, the incubation period can be
indefinitely long. (A fraction of those infected with HIV have, so far, remained
asymptomatic "indefinitely.") Figure 10.6.1(c) is a gamma distribution incorporating both a preferred incubation "window" and the possibility of an indefinitely
long incubation period. Figure 10.6.1(d) is a beta distribution. It allows for a preferred window but, like the uniform distribution, incubation must occur between
given times.

The functions we have used to draw Figure 10.6.1 are P_1, P_2, P_3, and P_4 as
defined in the following.

```
> c1:=int(t^9*exp(-t),t=0..infinity);
  c2:=evalf(Int(sqrt((t-2)/16)*(1-(t-2)/16)^4,t=2..20));
  P1:=t->1/16*(Heaviside(t-2)-Heaviside(t-18));
  P2:=t->exp(-t/6)/6;
  P3:=t->t^9*exp(-t)/c1;
  P4:=t->sqrt((t-2)/16)*(1-(t-2)/16)^4/c2;
```

Their graphs are illustrated in Figure 10.6.1.

To derive a mathematical relationship for the appearance of AIDS cases,
we will assume that the probability distribution for the incubation period can be
treated as a deterministic rate. Let $h(t)$ denote the HIV infection density, that is,

> plot({P1(t),P2(t),P3(t),P4(t), t=0..20});

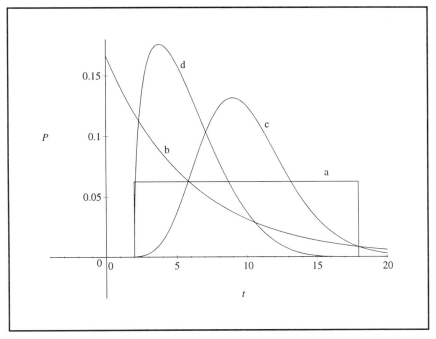

Figure 10.6.1 Some HIV incubation probability densities. Graphs of (a) uniform distribution, (b) $e^{-t/6}/6$, (c) $t^9 e^t$ normalized, (d) $\sqrt{(t-2)/16}\,(1 - (t-2)/16)^4$ normalized.

the number of new HIV infections during $[t, t + \Delta t) = h(t) \cdot \Delta t$,

and let $a(t)$ denote the AIDS density, thus

the number of new AIDS cases during $[t, t + \Delta t) = a(t) \cdot \Delta t$.

We wish to determine $a(t)$ from $h(t)$. AIDS cases at time t arise according to the probability function $p(\cdot)$ from HIV infections which occurred at some time s prior to t. Such a time interval is $[t - (s + ds), t - s)$ (see Figure 10.6.2). The number of newly infected persons during this interval is $h(t - s) \cdot ds$ and the fraction of them to become symptomatic s time later is $p(s)$. Hence the contribution to $a(t)$ here is

$$h(t - s) \cdot ds \cdot p(s).$$

Since $a(t)$ is the sum of such contributions over all previous times, we get

$$a(t) = \int_0^\infty h(t - s)p(s)\,ds. \qquad (10.6.1)$$

This sort of integral is known as a *convolution*; such integrals occur widely in science and engineering.

Convolution integrals have an alternate form under change of variable. Let $u = t - s$, then $s = t - u$ and $ds = -du$. Since $u = t$ when $s = 0$ and $u = -\infty$ when $s = \infty$, the integral of equation (10.6.1) becomes

$$a(t) = -\int_{t}^{-\infty} h(u)p(t - u)\, du.$$

Absorbing the minus sign into a reversal of the limits of integration and replacing the dummy variable of integration u by s gives

$$a(t) = \int_{-\infty}^{t} h(s)p(t - s)\, ds. \qquad (10.6.2)$$

This equation shows up a striking symmetry between the roles of h and p. Equation (10.6.2) is sometimes easier to work with than equation (10.6.1).

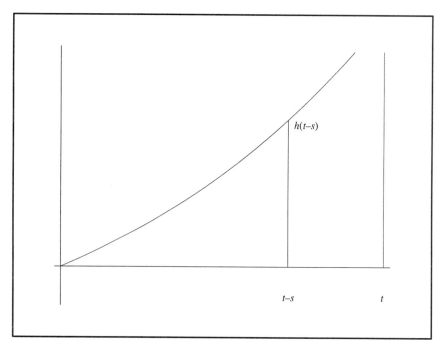

Figure 10.6.2(a) $a(t) \approx \sum_{s} h(t - s) \cdot ds \cdot p(s)$

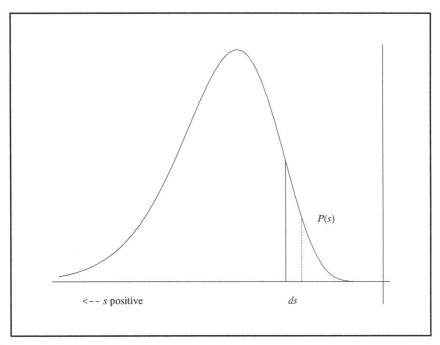

Figure 10.6.2(b) $a(t) \approx \sum_s h(t - s) \cdot ds \cdot p(s)$

The occurance of symptoms is strongly affected by the incubation distribution.

In order to determine whether a proposed incubation distribution is the correct one, we must use it in conjunction with our newly derived formula, either equation (10.6.1) or equation (10.6.2), to predict the pattern of cases. To this end we track an HIV infected *cohort*, that is, a group of people infected about the same time, through the calculation. Consider those infected over a two-year period, which we take to be $t = 0$ to $t = 2$. We will assume there are 1000 cases in each of the two years; thus the HIV density we are interested in is

$$h(t) = \begin{cases} 1000 & \text{if } 0 \le t \le 2 \\ 0 & \text{otherwise.} \end{cases} \qquad (10.6.3)$$

The total number of cases is $\int_0^2 h(s)ds = 2000$. With this choice for h, we can simplify the factor $h(t - s)$ in equation (10.6.1). Note that $0 \le t - s \le 2$ is the same as $t - 2 \le s \le t$. In other words,

$$\text{if } t - 2 \le s \le t, \text{ then } h(t - s) = 1000, \text{ otherwise } h \text{ is } 0. \qquad (10.6.4)$$

Therefore the only contribution to the integral in equation (10.6.1) comes from the part of the s axis between $t - 2$ and t (see Figure 10.6.3).

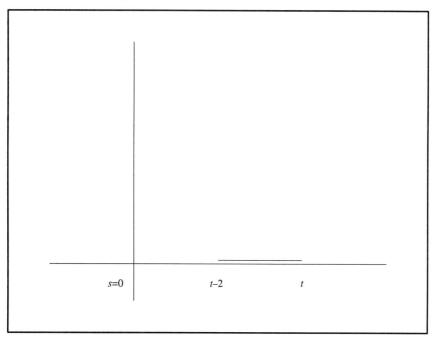

Figure 10.6.3   Contributory subinterval of the s-axis

There are three cases depending on the position of the interval $[t-2,t]$ relative to 0: to the negative side of 0, i.e., $t \leq 0$, to the right of 0, i.e., $t-2 \geq 0$, or covers 0, $0 < t < 2$. Consider each case in turn. If $t \leq 0$, then $a(t) = 0$ from equations (10.6.2) and (10.6.3). If $t \geq 2$, then $t-2 \geq 0$ and equation (10.6.1) becomes, taking into account equation (10.6.4),

$$a(t) = 1000 \int_{t-2}^{t} p(s)\,ds, \qquad t \geq 2.$$

Finally for $0 < t < 2$, the part of the interval to the left of $s = 0$ makes no contribution and in this case equation (10.6.1) becomes

$$a(t) = 1000 \int_{0}^{t} p(s)\,ds, \qquad 0 < t < 2.$$

Putting these three together we have

$$a(t) = \begin{cases} 0, & t \leq 0 \\ 1000 \int_{0}^{t} p(s)\,ds, & 0 < t < 2 \\ 1000 \int_{t-2}^{t} p(s)\,ds, & t \geq 2. \end{cases} \qquad (10.6.5)$$

Because it is inconvenient to deal with a function defined by cases, such as $a(t)$ is defined by equation (10.6.5), a standard set of "cases" type functions have been devised. One of these is the *Heaviside* function $H(t)$ and another is the *signum* function $S(t)$. The first is defined as

$$H(t) = \begin{cases} 0, & t < 0 \\ 1, & t \geq 0. \end{cases} \tag{10.6.6}$$

The signum function is just the *sign* of its argument, that is

$$S(t) = \begin{cases} -1, & t < 0 \\ 0, & t = 0 \\ 1, & t > 0. \end{cases} \tag{10.6.7}$$

Actually there is a relationship between the two, except for $t = 0$:

$$H(t) = \frac{1}{2}(S(t) + 1), \qquad S(t) = 2H(t) - 1, \qquad t \neq 0. \tag{10.6.8}$$

The Heaviside function $H(2 - t)$ cuts out at $t = 2$ while $H(t - 2)$ cuts in at $t = 2$, so in terms of Heaviside functions, equation (10.6.5) can be written as

$$a(t) = 1000H(2 - t) \int_0^t p(s)\,ds + 1000H(t - 2) \int_{t-2}^t p(s)\,ds. \tag{10.6.9}$$

For the simplest example, assume the incubation density is the uniform distribution, P_1, above, Figure 10.6.1a,

$$P_1(t) = \begin{cases} 1/16 & \text{if } 2 \leq t \leq 18 \\ 0 & \text{otherwise.} \end{cases}$$

Substituting into equation (10.6.9) and integrating gives the onset distribution $a(t)$.

```
> h:=t->1000*(Heaviside(2-t)-Heaviside(-t));
> int(h(t-s),s=2..18)/16;
> collect(collect(collect(collect(",signum(20-t)),
  signum(2-t)),signum(4-t)),signum(-t+18));
> a:=unapply(",t);
```

The output of this calculation is

$$
\begin{aligned}
a(t) = & \left(\frac{125}{4} t - \frac{1125}{2} \right) \text{signum}(18 - t) \\
& + \left(\frac{125}{4} t - 125 \right) \text{signum}(t - 4) \\
& + \left(-\frac{125}{4} t + \frac{125}{2} \right) \text{signum}(2 - t) \\
& + \left(-\frac{125}{4} t + 625 \right) \text{signum}(20 - t).
\end{aligned}
\tag{10.6.6}
$$

This provides an alternate realization for a formula for $a(t)$. Its form is different from that of equation (10.6.5). We can recover the previous one however by evaluating the signums with various choices of t. To do this, suppose that $2 < t < 4$, or $4 < t < 18$, or $18 < t < 20$, respectively, and evaluate equation (10.6.6).

```
> (125/4*t-1125/2)+(-125+125/4*t)-(-125/4*t+125/2)+(625-125/4*t);
  (125/4*t-1125/2)-(-125+125/4*t)-(-125/4*t+125/2)+(625-125/4*t);
  -(125/4*t-1125/2)-(-125+125/4*t)-(-125/4*t+125/2)+(625-125/4*t);
```

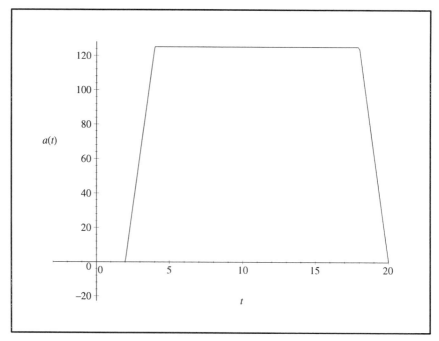

Figure 10.6.4   Graph of $a(t)$ from equation (10.6.6)

Eventually, all those infected will contract AIDS, therefore

$$\int_0^\infty a(s)ds = \int_0^2 h(s)ds,$$

but the first integral reduces to the interval $[0, 20]$ (see Figure 10.6.4). That is, the total number of people who develop AIDS during the 20-year period is the same as the total number of people in the initial two-year cohort. This computation is done as

```
> int(a(s), s=0..20);
```

which gives 2000.

Several observations should be made with the graph for each of the other distributions. There should be a gradual increase of the number of cases as the cohorts begin to develop symptoms of AIDS. Also, there should be a gradual decrease that may last past 20 years: Those in the cohorts who were infected near the end of the second year may not begin to show symptoms until the 22nd year, depending on which of P_2, P_3, or P_4 is used.

We leave the computations for the other distributions to the exercises. However, Figure 10.6.5 shows the graph for the onset of AIDS cases for a two-year cohort assuming incubation as with the gamma distribution $P_3(t)$. The function $a(t)$ defined by

$$a(t) = \int_0^\infty h(t - s)P_3(s)ds$$

is evaluated and plotted with

```
> int(1000*P3(s),s=0..t)*Heaviside(2-t)
     + int(1000*P3(s),s=t-2..t)*Heaviside(t-2):
> a:=unapply(",t);
> plot(a(t),t=0..22);
```

We verify that

$$\int_0^\infty a(s)ds = 2000.$$

```
> evalf(Int(a(s),s=0..infinity));
```

Comparing these figures we can gauge the effect of the incubation period. Note that for research purposes, it would require more than comparing figures like these with AIDS epidemiologic data to determine the incubation distribution, because one could not separate the AIDS cases into those stemming from a particular cohort—all cohort onsets are mixed together.

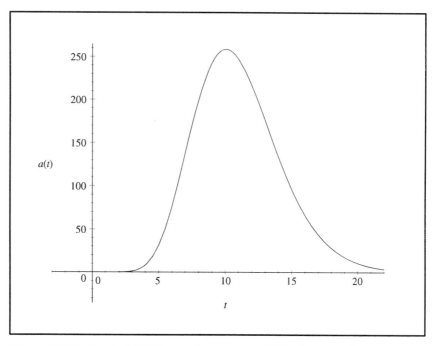

Figure 10.6.5 Onset of AIDS cases for a two-year HIV cohort assuming gamma incubation.

Exercises

1. Choose each of the four hypothetical incubation densities of Figure 10.6.1 in turn. Draw the graph of the number of AIDS cases expected to develop, $a(t)$, for the cohort of equation (10.6.2) with the assumption that the one you have chosen is correct. The following syntax draws the Figure 10.6.4 and verifies equation (10.6.6).

    ```
    > int(1000*P3(s),s=0..t)*Heaviside(2-t)
          + int(1000*P3(s),s=t-2..t)*Heaviside(t-2):
    > a:=unapply(",t);
    > plot(a(t),t=0..22);
    > evalf(Int(a(s),s=0..infinity));
    > int(P3(s),s=0..infinity);
    ```

2. We pose a *what if* exercise. Suppose that around 2010, a vaccine for HIV is developed and, while current cases cannot be cured, HIV is no longer transmitted. The number of reported new cases of HIV, $h(t)$, drops dramatically to zero by 2020. Model the reported cases of HIV with a hypothetical scenario such as

$$h(t) = \frac{(t - 1980)}{40}\left[1 - \frac{(t - 1980)^6}{40^6}\right].$$

a. Draw a graph of $h(t)$. Observe that $h(1980) = 0$ and $h(2020) = 0$.

```
> h:=t->((t-1980)/40)*(1-(t-1980)^6/40^6);
> plot(h(t),t=1980..2020,xtickmarks=0,ytickmarks=2);
```

b. Determine where the maximum value of h occurs. This represents the time when the reported new cases of HIV infected individuals peaks if this "optimistic scenario" were to happen.

```
> sol:=solve(diff(h(s),s)=0,s);
> evalf(sol[1]);
```

c. Define a "later, rather than sooner" hypothetical incubation density and draw its graph.

```
> c5:=int(1/16*(t-2)*(1-(1/16*(t-2))^2),t=2..18);
  P5:=t->1/16*(t-2)*(1-(1/16*(t-2))^2)/c5;
> plot([t,P5(t),t=2..18],t=0..20);
```

d. Find $a(t)$ as in equation (10.6.1) associated with this distribution.

```
> a15:=t->int(h(t-r)*P5(r),r=2..t-1980);
  a25:=t->int(h(t-r)*P5(r),r=2..18);
  a:=t->a15(t)*Heaviside(2000-t)+ a25(t)*Heaviside(t-2000);
```

e. Sketch the graphs for the hypothetical $h(t)$ and associated $a(t)$.

```
> Digits:=30;
  plot({[t,h(t),t=1980..2020],[t,a(t),t=1982..2028]},
    xtickmarks=2,ytickmarks=0);
```

Section 10.7

Questions for Thought and Discussion

1. What are four suspected ways that HIV kills cells?
2. Why do viral mutations lead to the development of new antibodies by the immune system?
3. Describe the life cycle of HIV.
4. Why do we continue to get colds, year after year, but seldom get mumps more than once?
5. Describe clonal selection and clonal deletion.
6. How does the clonal deletion model explain the fact that a mouse injected prenatally with a virus never will raise antibodies against the virus after the mouse is born?
7. Describe three general immunolgic mechanisms.

8. How does HIV infection result in the inactivation of both the humoral and cell-mediated immune responses?

9. Most DNA replication includes a proof-reading function that corrects mismatched DNA nucleotides during DNA replication. The reverse transcriptase of HIV seems to lack this ability, which results in high mutation rates (as much as one or more per generation). Discuss this problem in terms of antibody production by a host's immune system.

References and Suggested Further Reading

1. **Blood cells; immunity:** William T. Keeton and James L. Gould, *Biological Science*, W.W. Norton and Company, New York, 5th ed., 1993.

2. **Immunity: Special issue on the immune system:** *Scientific American*, Vol. 269, No. 3, September, 1993.

3. **HIV and AIDS:** "What Science knows about AIDS" (entire issue), *Scientific American*, Vol. 259, No. 4, October, 1988.

4. **HIV and AIDS:** Martin A. Nowak and Andrew J. McMichael, "How HIV Defeats the Immune System," in *Scientific American*, Vol. 273, No. 2, page 58, August, 1995.

5. **HIV and T cells:** A. S. Perelson, Modeling the Interaction of the Immune System with HIV, in *Mathematical and Statistical Approaches to AIDS Epidemiology*, edited by C. Castillo-Chavez, pp. 350–370. *Lecture Notes in Biomath*, Vol. 83, New York: Springer-Verlag, 1989.

6. **HIV and T cells:** A. S. Perelson, D. E. Kirschner and R. J. De Boer, The Dynamics of HIV Infection of $CD4^+$ T Cells, *Math. Biosci.* **114**, pp. 81–125, 1993.

7. **HIV and T cells:** K. E. Kirschner and A. S. Perelson, A model for the immune system response to HIV: AZT treatment studies, in *Mathematical Population Dynamics: Analysis of Heterogeneity and the Theory of Epidemics*, O. Arino, D. E. Axelrod, M. Kimmel and M. Langlais, eds., Wuerz Publishing, Winnipeg, Canada, pp. 295–310.

8. **HIV and T cells:** A. S. Perelson, Two theoretical problems in immunology: AIDS and epitopes, in *Complexity: Metaphors, Models and Reality*, G. Cowan, D. Pines and D. Meltzer, eds., Addison-Wesley, Reading, MA, pp. 185–197.

9. **The immune response:** W. C. Greene, AIDS and the Immune System, *Scientific American* **269** #3, Special Issue, September, 1993.

10. **Mutations of HIV:** Martin A. Nowak, Robert M. May and Roy M. Anderson, The evolutionary dynamics of HIV-1 quasi species and the development of immunodeficiency disease, *AIDS* 1990 **4**, pp. 1095–1103, 1990.

11. **Mutations of HIV:** Martin A. Nowak and Robert M. May, Mathematical biology of HIV infections: Antigenic variation and diversity threshold, *Mathematical Biosciences* **106**, pp. 1–21, 1991.

12. **Calculations of the time from HIV infection to AIDS symptoms:** P. Bacchetti, M. R. Segal and N. P. Jewell, "Backcalculation of HIV Infection Rates," *Statistical Science*, Vol. 8, no. 2, pp. 82–119, 1993.

Chapter 11

Genetics

Introduction to this chapter

In this chapter we will study the ways that genetic information is passed between generations and how it is expressed. Cells can make exact copies of themselves through asexual reproduction. The genes such cells carry can be turned off and on to vary the cells' behaviors, but the basic information they contain can be changed only by mutation, a process that is somewhat rare to begin with and usually kills the cell anyway.

Genetic material is mixed in sexual reproduction, but the result of such mixing is seldom expressed as a "blend" of the properties' expressions. Rather, the rules for the combination of genetic information are somewhat complex. Sexual reproduction thus results in offspring that are different from the parents. Much research shows that the ultimate genetic source of this variation is mutation, but the most immediate source is the scrambling of preexisting mutations.

The variations produced by sexual reproduction serve as a basis for evolutionary selection, preserving the most desirable properties in a particular environmental context.

Section 11.1

Asexual Cell Reproduction—Mitosis

Asexual reproduction of a cell results from the copying and equal distribution of the genetic material of a single cell. Each resultant daughter cell then possesses the same genes as the parent cell. If we are considering a single-celled organism, an environment for which the parent cell is suited should therefore also be suitable for the daughter cells. If we are considering a multicellular organism, the daughter cell may take on functions different from that of the parent cell by selectively turning genes off. This process creates the various tissues of a typical multicellular organism.

Eukaryotic mitosis gives each of two daughter cells the same genes that the parent cell had.

The actual process of eukaryotic mitosis is comparable to a movie, with some-times-complex actions flowing smoothly into one another, without breaks. For reference, however, mitosis is usually described in terms of five specific stages, named interphase, prophase, metaphase, anaphase and telophase. It is impor-tant to remember, however, that a cell does not jump from one stage to the next. Rather, these stages are like "freeze-frames," or preserved instants; they are guide-posts taken from the continuous action of mitosis (see Reference [1] for further discussion).

Most of the time a cell's nucleus appears not to be active; this period is called *interphase*. If one adds to an interphase cell a stain that is preferentially taken up by nuclei and then examines the cell through a microscope, the nucleus appears to have no internal structure over long periods of time. This appearance is actu-ally quite misleading because, in fact, the nucleus is very active at this time. Its activity, however, is not reflected in changes in its outward appearance. For ex-ample, the addition of radioactive thymine to an interphase cell often leads to the formation of radioactive DNA. Clearly DNA synthesis takes place in interphase, but it does not change the appearance of the nucleus.

Biologists further subdivide interphase into G_1, during which preparations for DNA synthesis are made, S, during which DNA is synthesized and G_2, dur-ing which preparations are made for actual cell division. (The "G" stands for "gap.") If we could see the DNA of a human skin cell during G_1 we would find 46 molecules. Each molecule, as usual, consists of two covalent polynucleotides, the two polymers being hydrogen-bonded to one another in a double helix. Genetic information is linearly encoded into the base sequence of these polynucleotides.

When we discussed DNA structure in Chapter 9 we associated a *gene* with the nucleic acid information necessary to code for one polypeptide. Thus, a gene would be a string of perhaps a few hundred to a few thousand bases within a DNA molecule. It is convenient to define a gene in another way, as a *functional unit of heredity*, a definition that has the virtue of generality. It can therefore include the DNA that codes for transfer RNA or ribosomal RNA, or it can just be a section of DNA that determines a particular observable property, such as wing shape or flower color. In this general definition, each DNA molecule is called a *chromo-some*, where each genetic region (*gene; locus*), on the chromosome determines a particular observable property.

In Figure 11.1.1 one chromosome is illustrated for a cell progressing through mitosis. (A human skin cell, for example, has 46 chromosomes, and each one behaves like the one in the figure.) The structure of the chromosome at G_2 cannot be seen in a microscope, so we must surmise its structure by its appearance in the next stage (prophase).

At *prophase* the nuclear membrane disappears and the chromosomes become visible for the first time, resembling a ball of spaghetti. If we could grab a loose end of a chromosome and separate it from the others, we would see that it looks like the one shown in the figure beside the prophase cell. It consists of two halves called sister *chromatids*, lying side by side and joined at a *centromere*. The two chromatids of each prophase chromosome are chemically and physically identical to each other because one of each pair was manufactured from the other in the preceding S phase. Each chromatid therefore contains a double stranded DNA molecule that is identical to the DNA of its sister chromatid. The two chromatids are still referred to as a single chromosome at this point.

As prophase progresses, the chromosome becomes shorter and fatter, and it moves to the center of the cell. The stage at which the chromosomes reach maximum thickness and are all oriented at the cell's center is called *metaphase*. Chromosomes at metaphase have reached their maximum visibility, and a view through a microscope often shows them all arranged neatly in the cell's equatorial plane, as shown in the photo in Figure 11.1.1.

At *anaphase* each chromosome splits into its two component chromatids, which are now referred to as individual chromosomes in their own right, and one copy moves toward each end, or *pole*, of the cell. Recall that the two sister chromatids of each chromosome are identical to each other. In summary, what happens in anaphase is that identical double-stranded DNA is delivered to each pole.

At *telophase* the chromosomes collect together at each pole and a new nuclear membrane forms around them. The cell then divides its cytoplasm in such a way that one new nucleus is contained in each half.[1] There are now two cells where there was only one, but the crucial point is that each of the daughter cells now has the same DNA code that the original cell had. Put another way, two cells have been formed, each having the same genes as the parent cell.

One way to look at asexual reproduction is to think of each chromosome as a piece of paper, with information written on it. A human skin cell has 46 pages, labeled 1 through 46. At S phase an exact copy is made of each page, and during mitosis each daughter cell gets one copy of each page. No new information is created, nor is any lost. Each daughter cell gets the same genetic information, i.e., each daughter cell ends up with 46 pages, labeled 1 through 46.

A karyotype is a picture of a cell's chromosomes.

It is not difficult to obtain a picture of most organisms' chromosomes. For example, it is a routine laboratory procedure to take a sample of a person's blood and isolate some of their white blood cells (mammalian red blood cells won't do because they lose their nuclei as they mature). These white cells are then cultured in a test tube and their nuclear material is stained as they enter metaphase, which is when chromosomes are most easily visualized. The cell, with its chromosomes, is photographed through a microscope. The chromosomes are then cut out of the photograph and arranged in a row, according to size. This picture is a *karyotype*. An example is shown in Figure 11.1.2.

---

[1] The actual splitting of the cell is called *cytokinesis*.

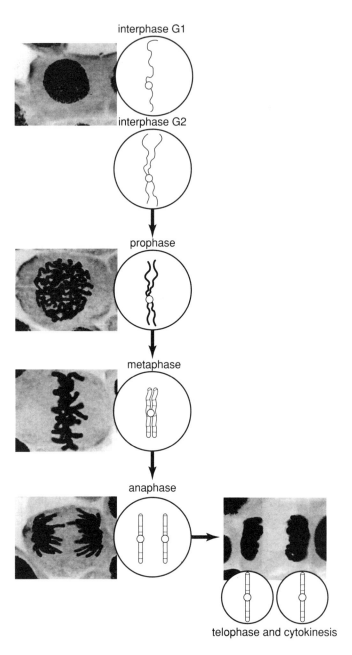

Figure 11.1.1   The stages of mitosis. The figure shows actual photographs of a dividing cell's chromosomes. The line drawings show how the individual chromosomes
behave during that stage of division. During mitosis each chromosome replicates
lengthwise and the two copies go to different daughter cells. Thus, each daughter
cell ends up with exactly the same genetic complement as the parent cell. (Photos of
mitosis taken from "Radiation and Chromosomes Biokit," Carolina Biological Supply
Company, Burlington, North Carolina; #F6-17-1148. Used with permission.)

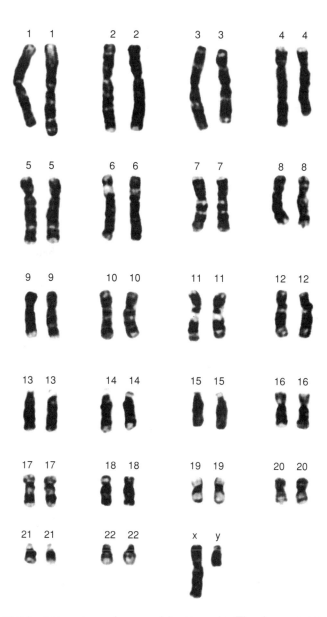

Figure 11.1.2 A karyotype of a normal human male. The chromosomes were photographed at metaphase and images of the individual chromosomes were then cut
out and arranged by size. The result is a group of 22 chromosome pairs, called homologs, each pair of which is matched by length, centromere location and staining
pattern. Because this is a male's karyotype, the 23rd pair of chromosomes (sex chromosomes; X and Y) do not match each other. Each of the chromosomes shown in the
figure consists of two identical daughter chromatids, but they are so closely associated
that they are often indistinguishable at metaphase. However, note the right hand homolog of number 18; the two chromotids can be distinguished. (Photo of karyotype
arranged from "Human Karyotypes, Normal Male," Carolina Biological Supply Company; Burlington, North Carolina; #F6-17-3832. Used with permission.)

There are several interesting features of the illustrated karyotype:

1. These are metaphase chromosomes and therefore are lengthwise doubled, joined at a centromere. Each chromosome consists of two chromatids, a feature that sometimes confuses students. The problem is that the chromosomes must be photographed at metaphase because that is when they are most easily visible and distinguishable from one another. This is also the point at which they are in a duplex form. You may want to refer to the discussion of Figure 11.1.1 to clarify the distinction between chromosome and chromatid.

2. There are 46 chromosomes in this cell. This is the number found in most of the cells of the human body, the exceptions being mature red blood cells, which lack nuclei, and certain cells of the reproductive system, called *germinal cells*, to be discussed later in this chapter. Any cells of our body that are not germinal are said to be *somatic* cells, a category that therefore includes virtually the entire bulk of our body: skin, blood, nervous system, muscles, the structural part of the reproductive system, etc. Our somatic cell chromosome number is thus 46.

3. The chromosomes in the karyotype seem to occur in identical-appearing pairs, called *homologous pairs*. Evidently our human chromosomal complement is actually two sets of 23 chromosomes. It is very important to understand the difference between a homologous pair of chromosomes and the two chromatids of a single metaphase chromosome. The karyotype shows 23 homologous pairs; each member of each pair consists of two chromatids. Each chromatid contains a double-helical DNA molecule that is identical to the DNA of its sister chromatid, but different from the DNA of any other chromatid.

Asexual reproduction can generate daughter cells that differ from each other.

We could imagine an amoeba, a common single-celled eukaryote, dividing by mitosis to yield two identical amoebas. We could just as easily imagine a skin cell of a human, a multicellular eukaryote, dividing by mitosis to give two identical human skin cells. Indeed, this is the way that our skin normally replaces those cells that die or are rubbed off. In both cases the daughter cells have the same DNA base sequence that the parent cell had, and that is reflected in the identical physiology and appearance of the daughter cells.

There is another possibility: Consider a single fertilized human egg. It divides by mitosis repeatedly to form a multicellular human, but the cells of a developed human are of many sizes, shapes and physiological behaviors. Liver cells look and behave one way, nerve cells another and muscle cells still another. Mitosis seems not to have been conservative. How could cells that have exactly replicated their DNA in mitosis and then partitioned it out equally have yielded different progeny cells?

One possibility is that cells in each unique kind of tissue of a multicellular organism have lost all their genes except those essential to the proper functioning of that particular tissue. Thus, liver cells would have retained only those genes needed for liver functioning and muscle cells would have retained only those genes needed for muscle functioning. This possibility is easy to reject by a simple experiment: In the cells of a plant stem the genes necessary for stem growth and function are obviously active, and there is no evidence of genes involving root formation. If the stem is broken off and the broken end inserted into soil, within a few weeks the plant will often start to grow roots at the broken stem end. Clearly the genes for root growth and function were in the cells of the stem all along, but were reversibly turned off. A similar experiment has been done on a vertebrate, in which a nucleus from a specialized somatic tissue, the intestinal lining of a tadpole, has been used to grow a whole tadpole and the subsequent toad. We can conclude that mitosis generates different tissues of multicellular organisms when selected genes are turned off or on in the course of, or in spite of, asexual cell division.

The process by which unspecialized cells of a multicellular organism take up specialized roles—liver, nerve, skin, etc.—is called *differentiation*. Differentiation is not restricted to embryos, but can occur all our lives, e.g., in bone marrow, where unspecialized stem cells can become specialized blood cells. Differentiation is only one part of *development*, which includes all the changes in an organism in its life, from conception to death. Other aspects of development would include tissue growth and deterioration as described in Section 7.2.

Some cell types rarely divide.

Certain cells of multicellular organisms seem to have a very limited, even nonexistent, capacity for division. For example, muscle cells don't divide; the muscle enlargement associated with exercise comes from cellular enlargement. Fat cells get larger or smaller, but their numbers stay the same (which is why cosmetic liposuction works—the lost fat cells can't be replaced). Cells of the central nervous system don't divide, which explains the seriousness of spinal injuries. Liver cells rarely divide unless part of the liver is cut away—in which case the liver cells undergo division to replace those removed. Note the implication here: Genes controlling liver cell division haven't been lost. They were shut off, and can be reactivated.

Section 11.2

Sexual Reproduction—Meiosis and Fertilization

Sexual reproduction involves the creation of an offspring that contains genetic contributions from two parents. A type of cell division called meiosis halves the chromosome number of germinal cells to produce sperm or eggs. A sperm and an egg then combine in fertilization to restore the double chromosome number. The new offspring now has genetic information from two sources for every characteristic. The ways that these two sets of information combine to produce a single property is complex, and is the subject of the study of classical genetics.

Sexual reproduction provides variation upon which evolutionary selection can act.

Recall the Darwinian model: More organisms are born than can survive, and they exhibit variability. Those with favored characteristics survive and may pass the favored properties to their offspring. It is tempting to credit genetic mutation with this variability and let it go at that. The fact is that all of the ten (non-twin) children in a hypothetical large family look different and virtually *none* of the variations among them are the result of mutations in their, or their parents', generation. This fact, surprising at first, seems more reasonable when we consider the accuracy of DNA base pairing, the "proof-reading" capability of some kinds of DNA polymerase and the existence of repair mechanisms to correct DNA damaged by such mutagens as radiation. Thus, DNA sequences tend to be conserved over many generations. We can therefore conclude that most of the variations among the ten children of the same family are the result of scrambling of existing genes, not the result of recent mutation. The cause of this shuffling of the genetic cards is sexual reproduction. Of course, the variant genes *originated* through mutation, but virtually all of them originated many generations earlier (see Reference [2] for further discussion).

Sexual reproduction involves the combination of genetic material from two parents into one offspring.

Refer to the karyotype in Figure 11.1.2. The human chromosome complement consists of 23 homologous pairs or, put another way, of two sets of 23 each. The sources of the two sets of 23 can be stated simply: We get one set from each of our parents when a sperm fertilizes an egg. What is not so simple is how the genetic material in those 46 chromosomes combines to make each of us what we are. The rules for combination will be the subject of Section 11.3. Our more immediate concern, however, is the means by which we generate cells with 23 chromosomes from cells having 46.

Meiosis halves the chromosome number of cells.

A special kind of reductional cell division, called *meiosis*, creates *gametes* having half the number of chromosomes found in somatic cells.[2]

The chromosomes are not partitioned at random however; rather, every gamete winds up *with exactly one random representative of each homologous pair*, giving it one basic set of 23 chromosomes. Such a cell is said to be *haploid*. A cell that has two basic sets of chromosomes is said to be *diploid*. We see that somatic cells are diploid and germinal cells are haploid. Thus, meiosis in humans converts diploid cells, with a chromosome number of 46, to haploid cells with a chromosome number of 23.

Meiosis is diagrammed in Figure 11.2.1 for a hypothetical organism having two homologous pairs; its diploid number is 4. Each chromosome is replicated in

---

[2]Gametes are often called *germinal cells* to distinguish them from somatic cells.

Meiosis

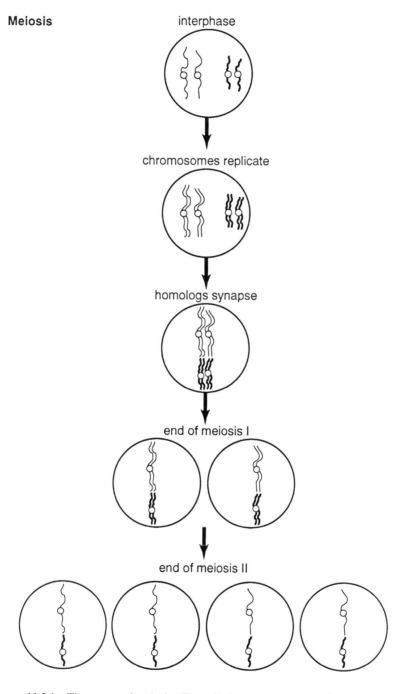

Figure 11.2.1 The stages of meiosis. The cell shown has two homologous pairs. Each chromosome replicates lengthwise to form two chromotids, synapses to its homolog and then two cell divisions ensue. The daughter cells each end up with exactly one representative of each homologous pair. Thus, a diploid cell at the start of meiosis results in four haploid cells at the end of meiosis.

interphase and thus contains two identical chromatids joined at a centromere. In a departure from meiosis, homologs bind together, side by side, in a process called *synapsis*, to form *tetrads* consisting of two chromosomes (four chromatids). The homologs then separate to end the first meiotic division. Next, the chromatids separate to complete the second meiotic division. The result is four cells, each containing two chromosomes, the haploid number for this hypothetical organism. Note that the *gametes' chromosomes include exactly one representative of each homologous pair.*

The process of meiosis (perhaps followed by developmental maturation of the haploid cell) is called *gametogenesis*. Specifically in animals, the formation of male gametes is called *spermatogenesis* and it yields four sperm, all similar in appearance. The formation of female gametes is called *oogenesis* and it yields four cells, but three of them contain almost no cytoplasm. The latter three are called *polar bodies* and they die. Thus, oogenesis actually produces only one living egg, and that one contains all the cytoplasm of the diploid precursor. The reason for this asymmetry is that, once the egg is fertilized, the first several cell divisions of the fertilized egg (called a *zygote*) remain under the control of cytoplasmic factors from the mother. Evidently all the cytoplasm from the egg precursor is needed in a single egg for this process.

The concept of sexual reproduction can be incorporated into The Alternation of Generations.

We can diagram the alternation of the diploid and haploid generations:

$$\cdots \longrightarrow \text{diploid} \xrightarrow{\text{meiosis}} \text{haploid} \xrightarrow{\text{fertilization}} \text{diploid} \longrightarrow \cdots .$$

Note that the diploid and haploid generations are equally important, because they form a continuous string of generations. On the other hand, the two generations are not equally *conspicuous*. In humans, for instance, the haploid generation (egg or sperm) is microscopic and has a lifetime of hours to days. In other organisms, mainly primitive ones like mushrooms and certain algae, the haploid generation is the conspicuous one, and the diploid generation is very tiny and short-lived.

Another way to show the alternation of generations is in the following diagram.

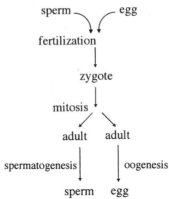

Section 11.3

Classical Genetics

Classical genetics describes the many ways that the genetic material of two parents combines to produce a single observable property. For instance, a red-flowered plant and a white-flowered plant usually produce an offspring with a single color in its flower. What that color will be is not predictable unless a geneticist has already studied flower-colors in that plant—because there are about a dozen ways that parental genes can combine. We describe many of those ways in this section.

Classical genetics describes the result of interactions in genetic information.

A diploid human cell carries 23 homologous pairs of chromosomes: One member of each pair comes from a sperm cell of the male parent and the other member comes from an egg cell of the female parent. Other diploid organisms may have chromosome numbers ranging from a few up to hundreds, but the same principle about the origin of homologous pairs holds. What we will consider now is how the genetic information from the two parents combines to produce the characteristics that appear in the offspring and why the latter are so variable. Let us first examine a chromosome at G_1 phase, because that is the usual condition in a cell.

Genes, defined generally as functional units of heredity, are arranged linearly along the chromosome (Figure 11.3.1). Each gene locus affects some property, say flower color or leaf shape in a plant. The order in which these loci appear is the same on each member of the homologous pair. Thus, it is common to refer to the "flower color" locus, meaning the section of either member of a homologous pair that is the gene that determines flower color. Clearly, each property is determined by two such sections, one on each homolog. *Each parent, then, contributes to each genetic property in the offspring.*

The behavior of chromosomes provides a basis for the study of genetics.

The pioneering geneticist was Gregor Mendel, who studied the genetics of sweet peas, a common flowering plant. Sweet peas, like many flowering plants, have male and female reproductive structures in the same flower. The male part makes pollen that is carried to the female part of that or another plant; the pollen then produces a sperm cell and fertilizes an egg. It a straightforward matter to dissect out the male part of a flower to prevent the plant from self-pollinating. Further, it is simple to use pollen from the male part of one plant to fertilize an egg of another plant and thus to make controlled matings. The seed that results from fertilizing an egg can be planted and the appearance of the offspring studied. The principles of chromosomal behavior and gene interaction in sweet peas are the same as for humans.

Mendel had two groups, or populations, of plants that were *true breeding*. A population is true-breeding if its freely interbreeding members always give rise to progeny that are identical to the parents, generation after generation. Members

homologous pair of chromosomes

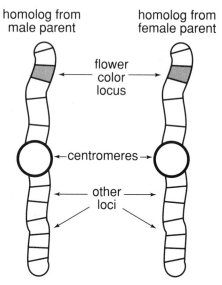

homolog from
male parent

homolog from
female parent

flower
← color →
locus

←centromeres→

other
← loci →

Figure 11.3.1 This shows a simple model of the chromosome. The genes are lined up along the length of the chromosome, like beads on a string. A hypothetical flower color locus is labelled.

of a population of true-breeding red-flowered sweet peas fertilize themselves or other members of the population for many generations, but only red-flowered plants ever appear. Mendel made a cross between a plant from a true-breeding red-flowered population and one from a true-breeding white-flowered population.

Mendel did not know about chromosomes, but we do and we will make use of that knowledge, which will simplify our learning task in the discussion to follow. We will therefore represent the cross in the following way: The gene for flower color is indicated by the labelled arrow in Figure 11.3.1. Note that each of the two homologs has such a gene locus.[3] The genetic information for red-flower color is symbolized by the letter R and the plant has two copies, one from each parent. (The reason for the copies being alike will become clear shortly.) Using the same convention, the genetic information at the flower color locus of the two homologs in the other (white-flowered) parent is symbolized by w.

Meiosis produces gametes containing one, and only one, representative of each homologous pair, as shown in Figure 11.3.2. A gamete from each parent combines at fertilization to reestablish the diploid condition. The offspring has flower color genetic information Rw. It turns out that this pea plant produces only

[3]For learning purposes we will ignore all other chromosomes, as if they do not have loci that affect flower color. In actual fact, this may not be true.

a) parental generation

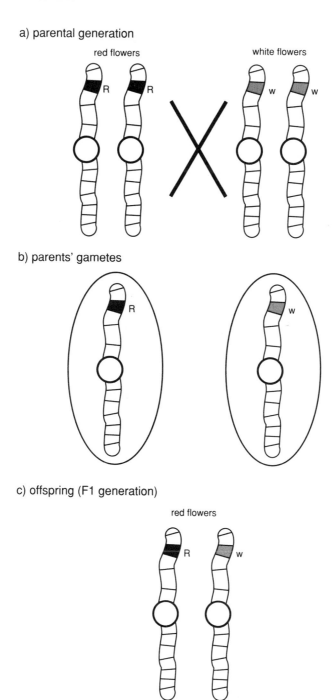

b) parents' gametes

c) offspring (F1 generation)

Figure 11.3.2 The behavior of chromosomes and their individual loci during a cross between two homozygous parents. The parents (*RR* and *ww*) each contribute one chromosome from the homologous pair to form gametes. The gametes combine in fertilization to restore the diploid number of two. The offspring's flowers will be red.

red flowers, indistinguishable from the red parent. Evidently red somehow masks white; we say that red information is *dominant* to white, and white is *recessive* to red.

At this point we need to define several terms. The variant forms of information for one property, symbolized by R and w, are *alleles*, in this case flower color alleles. The allelic composition is the organism's *genotype*; RR and ww are *homozygous* genotypes and Rw is the *heterozygous* genotype. What the organism actually looks like, red or white, is its *phenotype*. Thus, the initial, or parental, cross, was between a homozygous red plant and a homozygous white plant. The result in the first filial, or F1, generation was all heterozygous, red-flowered plants.

To obtain the F2 generation, we self-cross the F1, which is equivalent to crossing it with one just like itself. Figure 11.3.3 shows the gametes obtained from each parent in the F1 generation. They combine in all possible ways at fertilization. The result is a ratio of 1 RR, 2Rw and 1ww, which gives a 3:1 ratio of red-to-white phenotypes.

An experiment of the sort just described, involving a single property like flower color, is called a monohybrid cross. We used the chromosome model, whereas Mendel actually ran the experiment; satisfyingly, both give the same results. Let us now make a *dihybrid* cross, involving the two properties of flower color and stem length, which we specify to be *unlinked*, which means that their genetic loci are on different homologous pairs. The cross is diagrammed in Figure 11.3.4. Note that we have quit drawing in the chromosomes—we understand that the genes are on chromosomes and that drawing the latter is redundant. The F1 self-cross now can be represented as RwLs × RwLs. Note the phenomenon of *independent assortment*: Each gamete gets one, and only, one representative of each homologous pair, and the behavior of one pair in meiosis is independent of the behavior of the other pair. Thus, meiosis in the F1 generation results in equal numbers of gametes containing RL, Rs, wL and ws. The outcome of the cross is shown in the array, called a *Punnett square*, at the bottom of the figure.

The dihybrid cross yields a 9:3:3:1 phenotypic ratio of offspring. We should ask whether the inclusion of stem length in any way interferes with the 3:1 ratio of flower color. Among the 16 offspring in the Punnett square we see 12 red and 3 white, which gives the 3:1 ratio. We might have anticipated this—that the two properties would not affect their separate ratios—after all, they are unlinked and the two homologous pairs assort independently.

We must obtain large numbers of progeny in order to get the expected ratios of offspring.

Suppose we make a cross like Rw × Rw in sweet peas (red × red), and get only four progeny. We should not expect an exact 3:1 ratio of phenotypes in this experiment. After all, if we flipped a coin two times we would not be certain to get one head and one tail. Rather, we expect to get the 1:1 ratio only if we flip the coin many times, say 2000. The same reasoning holds in genetics—we must make enough Rw × Rw crosses to get many offspring, say 4000, and *then* we would obtain very close to 3000 red and 1000 white offspring.

a) F1 (self-crossed)

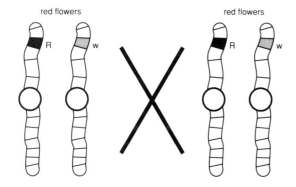

b) F1 gametes

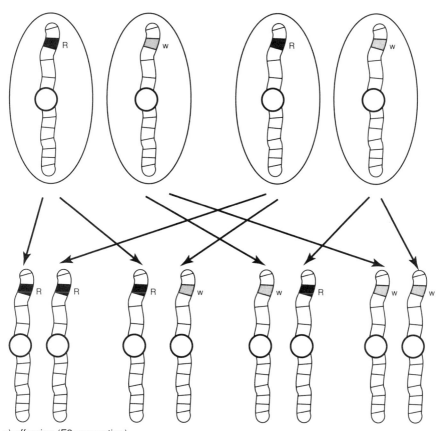

c) offspring (F2 generation)

Figure 11.3.3 A cross between two heterozygotes. Each F1 from Figure 11.3.2 makes gametes having the genes R and w with equal probability. When the gametes combine to make the F2 generation, there results offspring of genotypes RR, Rw and ww in the ratio 1:2:1.

R = red flowers
w = white flowers
L = long stems
s = short stems

a) parental generation

RRLL x wwss

b) parental gametes

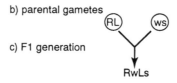

c) F1 generation

RwLs

d) self-cross F1

RwLs x RwLs

e) F1 gametes

f) Punnett square to give F2 generation

	RL	wL	Rs	ws
RL	RRLL	RwLL	RRLs	RwLs
wL	RwLL	wwLL	RwLs	wwLs
Rs	RRLs	RwLs	RRss	Rwss
ws	RwLs	wwLs	Rwss	wwss

9:3:3:1 ratio of phenotypes

Figure 11.3.4   A complete dihybrid cross between plants whose flower color locus and stem length locus are on different homologous pairs, i.e., the two properties are not linked. The result is a 9:3:3:1 ratio of phenotypes in the F2. In this figure only the allelic symbols are shown; the chromosomes are not drawn.

The ratios 3:1 and 9:3:3:1 are often called *Mendelian ratios*, because they are what Mendel reported. There is a bit of a problem here: Statisticians have examined Mendel's data and some have concluded that the experimental data is too good, i.e., consistently too close to the 3:1 and 9:3:3:1 expected ratios. For the sample sizes Mendel reported, it would be expected that he would have gotten somewhat larger deviations from Mendelian ratios.

Sexual reproduction leads to variation in several ways.

We shall concern ourselves with organisms in which the diploid generation is the most conspicuous, e.g., humans, and we will examine the variations introduced into the diploid organism by sexual reproduction. It should always be borne in mind, however, that haploid organisms are under genetic control also.

Earlier it was pointed out that, while mutation is the ultimate cause of genetic variation, there is only a very small chance that a given locus will mutate between two generations, will be unrepaired and subsequently not kill the cell. In spite of this, there are great variations among even the offspring of a single mating pair. We are now in a position to understand the sources of this immediate variation. First, look at the Punnett square of the dihybrid cross in Figure 11.3.4. Note that the F1 (RwLs) yields the gametes RL, Rs, wL, and ws, and yet the gametes of the parental generation were RL and ws. Thus, two new combinations have turned up in the gametes of the F1. The reason is that the flower-color locus and the stem-length locus are unlinked—they are on different homologous pairs—and every homologous pair assorts independently of every other pair. Thus, in the gametes of the F1, R paired up with L as often as R paired up with s. There were therefore $2^2 = 4$ combinations of chromosomes in the gamete. A human has 23 homologous pairs, all of which assort independently; thus, a person can produce 2^{23} different combinations of chromosomes in their gametes, using independent assortment alone!

Second, when homologous chromosomes synapse they can exchange pieces in a process called *crossing over*. Let us cross two true-breeding parents, AABB × aabb, as shown in Figure 11.3.5. Notice that the two gene loci are *linked*, i.e., on the same chromosome. The F1 genotype is AaBb and we *test cross* it.[4] Some of the gametes of the F1 are the expected ones, AB and ab, but, as the figure shows, crossing over, in which the homologs break and rejoin in a new way, produces gametes with two new allelic combinations, Ab and aB. These two new kinds of gametes, called *recombinant* gametes, are different from the gametes of either members of the parental generation. When the various gametes are paired up with the ab gametes in the test cross, the following *phenotypes* appear in the F2 generation: Ab, aB, AB and ab. The last two of these are the same phenotypes as the parental generation and the first two are recombinant offspring, having phenotypes not seen in the previous crosses. We see that crossing over rearranges genetic material and presents novel phenotypes upon which selection can act.

How often does such crossing over occur? Actually, it is not unusual to find at least one example in every tetrad. Furthermore, crossing over is predictable: The farther apart two loci are, the more likely crossing over is to occur between them. The frequency of crossing over, measured by the frequency of recombinant offspring, is used by geneticists as a measure of the distance between two loci.

Note that we could account for an immense number of allelic combinations just by using independent assortment and crossing over, without a mention of

---

[4]A *test cross* is a cross with a homozygous recessive individual.

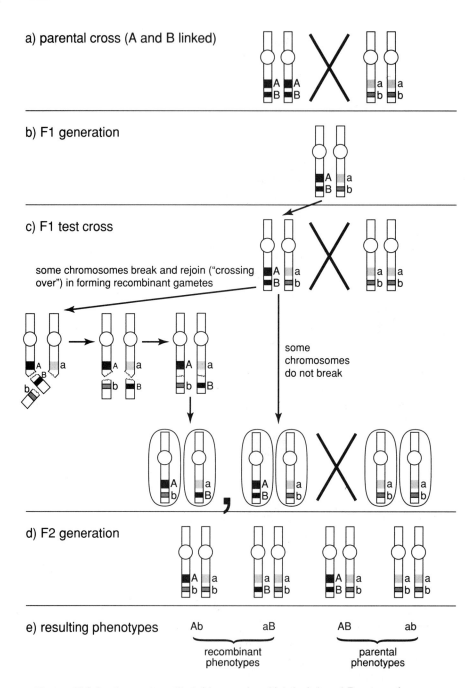

a) parental cross (A and B linked)

b) F1 generation

c) F1 test cross

some chromosomes break and rejoin ("crossing over") in forming recombinant gametes

some chromosomes do not break

d) F2 generation

e) resulting phenotypes

Ab   aB   AB   ab

recombinant phenotypes

parental phenotypes

Figure 11.3.5 A complete dihybrid cross, in which loci A and B are on the same chromosome, i.e., the two properties are linked. The results are predictable until the F1 test-cross at (c), when the chromosomes may break, yielding new combinations of the two loci. Notice that the resulting phenotypes at (e) include two (Ab and aB) that are unlike either of the two original parents.

mutation. Independent assortment and crossing over account for virtually all the phenotypic variation seen in members of a single family generation. This variation, in the main, is what Darwinian selection works on.

A final point is worth mentioning here: Self-fertilization might be considered to be a limiting form of sexual reproduction.[5] Suppose that allele A is completely dominant to allele a: If we self-cross an individual of genotype Aa, variant offspring appear in the ratio of 3:1, a mark of sexual reproduction. Asexual reproduction in the same organism yields only one kind of offspring—Aa. Where self-fertilizing organisms might run into evolutionary problems is in *continued* self-fertilization, which minimizes variation. This is shown by the following example: Take a population that is 100% heterozygotes (Aa) and self-cross all individuals. Note that the result is 50% heterozygotes and 50% homozygotes. Now self-cross all of that generation and note that 75% of the next generation will be homozygotes. After a few more generations of self-fertilization virtually the entire population will be homozygous, either AA or aa. This can create problems for the population in two ways; First, suppose that the recessive allele is an unfavorable one that is usually masked by the dominant allele. As shown above, self-fertilization increases homozygosity, and homozygous recessive individuals would be selected out. Second, when homozygotes fertilize themselves, independent assortment and crossing over can occur, but they cannot generate variation. (You should verify this statement by schematically working out the cross.)

Here is an idea to think about: We sometimes hear about the "rescue" of a species that is near extinction. The last few members of the species are brought together to be bred in a controlled environment, free from whatever forces were causing the extinction in the first place. Suppose now that a particular species has been depleted until only one male and one female are left. This mating pair must serve to reestablish the species. It is to be expected that each number of this pair would be heterozygous for at least a few unpleasant recessive genes. In light of the information in the preceding paragraph, what unique problems will the reconstituted species face?

A group of questions for practice and for extending Mendelian genetics

1. Refer to the definition of "true-breeding" two sections back. In the discussion of the monohybrid cross and Figures 11.3.2 and 11.3.3 true-breeding was asserted to mean "homozygous." Suppose for a moment that a member of a supposedly true-breeding population were a heterozygote. Show that being heterozygous is inconsistent with the definition of true breeding.

2. Suppose you are given a red-flowered sweet pea. A *test cross* will enable you to determine whether this dominant phenotype is a heterozygote (Rw) or a homozygote (RR). Cross it with a homozygous recessive individual (ww); the cross is therefore either RR × ww or Rw × ww. Note the different results

---

[5]Think of it this way: A self-cross is just like a cross between two separate, but genetically identical, parents.

obtained, depending on the genotype of the dominant phenotype. How do we know that a white-flowered plant is homozygous?

3. The red flower allele in sweet peas completely masks the white flower allele, i.e., red is *completely dominant* to white. If we cross a true-breeding red snapdragon with a true-breeding white one, the F1 offspring are all pink. We say that dominance is *incomplete*, or *partial*, for snapdragon flower color; partial dominance is a very common phenomenon. Cross two pink snapdragons to get offspring with a phenotypic ratio of 1 red:2 pink:1 white.

4. Foxes with platinum fur have the genotype Pp and silver foxes have the genotype pp. The genotype PP kills the fetus right after conception, i.e., it is *lethal*. Evidently the gene locus for fur color controls other properties as well, among them at least one very basic metabolic process. Show that a cross of two platinum foxes gives a 2:1 phenotypic ratio of offspring.

5. There is a notable exception to the statement that every chromosome in a mammalian diploid cell has an exact homolog. Mammalian males have one chromosome called an X chromosome and one called a Y chromosome. Females have two Xs and no Ys. These *sex chromosomes* carry a number of genes having to do with gender and many others that do not. Despite the fact that they are not homologous the X and Y chromosomes in a male can synapse over a portion of their length to facilitate meiosis. A well-known recessive gene on the X chromosome is for hemophilia, a blood-clotting disorder. Let us represent a heterozygous ("carrier") female as X^hX^+, where "X" indicates X-linkage, "h" indicates the hemophilia allele and "+" represents the normal allele. Note that a male of genotype X^hY will show the disorder because there is no possibility of a dominant allele on his Y chromosome to mask the hemophilia allele on his X chromosome. Cross a carrier female with a hemophilic male to show that a female can get hemophilia. Cross a carrier female with a normal male to show that no daughters and half the sons would be affected.

6. Often there are more than two choices for the alleles for a property, a phenomenon called *multiple alleles*. The presence of certain molecules on red blood cells is determined by the alleles A, B and O. For example, the genotypes AA and AO yield the A molecule, the genotypes BB and BO yield the B molecule, the genotype OO yields neither molecule and the genotype AB yields both the A and B molecules. The latter case, expression of both alleles, is called *codominance*. Cross an AB parent with an O parent; what ratio of offspring is obtained? Could an O-type man be the parent of an AB child? Can you conclude that a particular A-type man is the father of an A-type child by an A-type mother?

7. The expression of some genes is determined by the environment. The gene for dark pigmentation in Siamese cats is expressed only in cool parts of the cat's body—nose, ears and tail tip. The expression of the gene for diabetes mellitus, a deficiency in sugar metabolism, is affected by diet and the person's age. As an example, environmental effects might cause a dominant allele not to be expressed under certain conditions, and an individual with genotype AA or Aa might show the recessive phenotype. How might you determine that such an individual is actually of the dominant genotype?

Section 11.4

A Final Look at Darwinian Evolution

We close out our discussion of biology with a last look at the Darwinian model of evolution, which we introduced in Section 3.1. Fitness is measured by the persistence of a property in subsequent generations. If a property cannot be inherited, it cannot be selected. Thus, acquired properties like facelifts cannot be selected, nor can genetic properties of sterile individuals, like a mule's hardiness.

Populations evolve; individuals do not. An individual is born with a fixed set of genes; mutations in somatic cells are not transmitted to offspring and mutations in germinal cells can only be seen in the offspring.

Some organisms do not exhibit sexual reproduction but, rather, reproduce only asexually. Their only source of variation is therefore mutation. Nevertheless, such organisms have long evolutionary histories.

Fitness is measured by the ability to project genes into subsequent generations.

Common phrases like "struggle for survival" and "survival of the fittest" can be very misleading because they bring to mind vicious battles to the death between two contestants. The fact is that, except arguably among humans, violence is rarely the route by which Darwinian fitness is achieved in the biological world. Even the noisy, aggressive encounters between male animals seen on television nature programs seldom result in serious injury to participants. We must look to much more subtle interactions as a source of fitness.

One group of organisms may be slightly more able than another to tolerate heat, to thrive on available food or to elude predators. Subtle pressure is the norm in evolution; it works slowly, but there is no hurry. *Drosophila*, a common fruit fly, is used in many genetic experiments because it is easy to raise, has a short life span and has many simple physical properties, like eye color, whose modes of genetic transmission are easy to follow. If a large number of red-eyed and white-eyed *Drosophila* are put together in an enclosure and left to their own devices, the fraction of flies with white eyes will decrease steadily for several tens of generations and finally reach zero. Close observation reveals the reason: A female *Drosophila*, either red-eyed or white-eyed, will generally choose not to mate with a white-eyed male if a red-eyed male is available. Thus, there is a definite selection for the red-eye genetic trait.

Humans are not excluded from such subtle pressures: Personals ads in newspapers contain wish lists of traits people prefer in a mate. Height and affluence (control of territory?) are prized male traits, and hour-glass figures and youth (ability to bear children?) are valued female traits.

Regardless of the strength of the selective pressure or the nature of the properties being selected, there is really only one way to measure the evolutionary value of a trait, and that is the degree to which it is propagated into future generations. A shy, ugly person who has lots of fertile children has a high degree of fitness. We see that one generation of propagation is not enough; the trait must be persistent. For example, mules are known for their hardiness, but they are sterile

offspring of horses and donkeys. As a result, the hardiness of a mule cannot con-
fer any evolutionary advantage.[6] A discussion of human sexual selection can be
found in Reference [3].

Populations evolve, individuals do not.

A *population* is a group of organisms of the same species, living in the same area.
As before, we will restrict our discussion here to populations of organisms for
which the diploid generation is most conspicuous, e.g., humans.

If we observe a population over many generations, the "average" phenotypic
property will change, in keeping with our earlier discussion of species formation.
Thus, the average height may increase, or the typical eye color may darken. We
now ask and answer two questions: At which points in the alternation of generations do the changes occur, and what kinds of changes are relevant to evolution?

The Darwinian model stipulates that favored properties may be transmitted
to offspring; in any case, they certainly must be *capable* of transmission for the
model to apply. A diploid individual is conceived with a set of genes that are
relatively fixed for that individual's lifetime. Exceptions to this statement might
involve mutations in somatic cells and infection by lysogenic viruses (See Chapter
10). As long as these changes do not occur in germinal cells or germinal cell
precursors they cannot be transmitted to the next generation, and thus they have
no evolutionary effect. In addition, there are many phenotypic properties that
favor reproduction but that cannot be transmitted to offspring because they are
not of a genetic nature. Examples are suntans, exercise-strengthened bodies and
straightened teeth.

Genetically transmissible variations must originate via one of at least three
routes, all of which require sexual reproduction (in other words, an *intervening
haploid generation*) for their expression:

1. independent assortment;
2. crossing over;
3. mutation in a sperm, or an egg, or in their precursors in a parent prior to
    conception, or in a zygote at a very early stage of development. The altered
    genetic material in any one of these cases should turn up in those cells of
    the reproductive system that undergo meiosis to form the next generation of
    gametes.

We can conclude that, because the Darwinian model requires changes that
are inheritable, and because the observation of inheritable changes requires
the observation of more than one generation, *it is the population that evolves.*
Changes restricted to the somatic cells of individuals are not genetically transmit-

---

[6]There is a peculiar example of a non-inheritable trait—a desire for a large family—that might be
passed from one generation to another by teaching, and which could have a strong positive selective
value. This was discussed in Section 4.1.

ted to offspring; thus, in terms of evolution, an individual is fixed. Over a period of time, however, the average, or typical, characteristics of the population evolve.

Some organisms do not exhibit sexual reproduction.

Sexual reproduction is unknown (and probably nonexistent) in several kinds of organisms, for example, most bacteria, blue-green algae and some fungi. In those cases, all reproduction is asexual, which would seem to limit severely the possibilities of variation. Nonetheless, these organisms seem to have gotten along fine over long periods of history. We must conclude that some combination of three things applies: Either these organisms have not been exposed to large fluctuations in their environments, or they possess an innate physiological flexibility that permits them to get along in different environments, or their spontaneous mutation rates are sufficiently high to generate the variation necessary for adapting to new environmental situations.[7]

Section 11.5

The Hardy–Weinberg Principle

Diploidism and sexual reproduction complicate the calculation of inheritance probabilities. But, remarkably, the results are the same as if alleles were balls selected for combination from an urn. This is the Hardy–Weinberg principle. Although its veracity depends on random mating, among other properties, it continues to provide good approximations in many other situations as well.

Mendelian inheritance follows the laws of probability.

We will be concerned with probabilities associated with Mendelian inheritance for a diploid organism. As explained in Section 11.2, meiosis produces 4 haploid cells of 2 different kinds each equally likely to participate in fertilization. Then the probability is $1/2$ that a given kind of gamete will do so.

Consider first a single locus for which there are only two alleles, say A and a. Hence there are three distinct genotypes, the homozygotes AA and aa, and the heterozygote Aa (or aA). If one parent is AA and the other Aa, then the possible zygote genotypes resulting from a mating may be represented by an *event tree* as follows.

---

[7]There is now good evidence that bacteria, including asexual bacteria, can pass small pieces of DNA, called plasmids, to other bacteria.

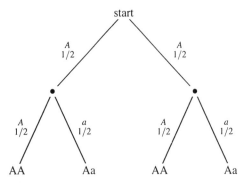

Figure 11.5.1 Probabilities for the offspring of an AA with Aa mating

Let the first branch point in Figure 11.5.1 correspond to the allele donated by the first parent, AA. There are two possible alleles and so here the diagram will have two branches. But for the parent AA both branches lead to the same result, namely the contribution of allele A to the offspring. Let the second branch point correspond to the allele donated by the second parent. Again there are two possibilities, but this time the outcomes are different as indicated.

Now the resulting probabilities may be calculated in several ways, (see Section 2.7). Since all the legs, or *edges*, of the diagram are equally likely, so are the resulting outcomes, each having probability $1/4$. Hence

$$\text{Pr(AA)} = 1/2 \qquad \text{and} \qquad \text{Pr(Aa)} = 1/2.$$

Alternatively, starting at the top node, the *root node*, and traversing the two edges to the left leading to AA, gives a probability of $1/4$ for this outcome by multiplying the probabilities along each edge of the path $(1/2 \cdot 1/2)$. This way of calculating the probabilities is the method of *conditional probabilities* since the probabilities along the branches leading away from any node are conditioned on the sequence of events leading to the node. Altogether the probability of an AA zygote by this method is $\frac{1}{2} \cdot \frac{1}{2} + \frac{1}{2} \cdot \frac{1}{2} = \frac{1}{2}$ since AA can occur in two different ways according to the tree.

Finally, the probabilities can be calculated by the principle of independence (see Section 2.7). The selection of a gamete from the AA parent will result in an A with probability 1. The selection of a gamete from the Aa parent is independent and will result in an A with probability $1/2$. Therefore, the probability of an AA zygote is $1 \cdot 1/2$.

The complete list of Mendelian inheritance probabilities are given in Table 11.5.1. The calculation above represents the second line of the table.

Random allelic combination preserves allelic fractions.

Let n_{AA} denote the number of AA genotypes in a population and likewise let n_{aa} denote the number of aa genotypes. For reasons that will shortly become clear,

Table 11.5.1 Mendelian Inheritance Probabilities

Parent Genotypes	Zygote Genotypes		
	AA	Aa	aa
AA×AA	1		
AA×Aa	1/2	1/2	
AA×aa		1	
Aa×Aa	1/4	1/2	1/4
Aa×aa		1/2	1/2
aa×aa			1

let n_{Aa} denote one-half the number of Aa genotypes. Then the size of the entire population N is the sum $N = n_{AA} + 2n_{Aa} + n_{aa}$. Let n_A and n_a denote the number of A alleles and a alleles, respectively, carried by the population. Thus, $n_A + n_a = 2N$ since the population is diploid.

Similarly let p_{AA}, p_{Aa}, p_{aa}, p_A, and p_a denote their corresponding fractions of the population. Then $p_A + p_a = 1$ and $p_{AA} + 2p_{Aa} + p_{aa} = 1$. Moreover

$$p_A = \frac{n_A}{2N} = \frac{2n_{AA} + 2n_{Aa}}{2N} = p_{AA} + p_{Aa}$$

and similarly

$$p_a = p_{Aa} + p_{aa}.$$

Now imagine all the alleles of the population are pooled and two are selected at random from the pool to form a pair. The selection of an A happens with probability p_A while the selection of an a happens with probability p_a. (We assume the pool is so large that the removal of any one allele does not appreciably change the subsequent selection probability.) Then, for example, the probability of forming an AA pair is p_A^2 since we assume the selections are made independently. In the same way, the other possible pair selections are calculated with the results shown in Table 11.5.2. As always, these probabilities are also the (approximate) fractions of the various outcomes in a large number of such pairings.

Table 11.5.2 Mendelian Inheritance Probabilities

Female Gametes (Frequencies)	Male Gametes (Frequencies)	
	A (p_A)	a (p_a)
A (p_A)	AA (p_A^2)	Aa ($p_A p_a$)
a (p_a)	aA ($p_A p_a$)	aa (p_a^2)

From the table we can calculate the fraction, p_A', of A alleles among the resultant pairs. Each pair of type AA contributes two A alleles and, while each Aa pair only contributes one, there are twice as many such pairs. Hence

$$p_A' = \frac{2p_A^2 + 2p_A p_a}{2} = p_A(p_A + p_a) = p_A.$$

In this it is necessary to divide by 2 because each pair has two alleles. Thus the fraction of A alleles among a large number of pairings is the same as their fraction in the original gene pool, p_A. The same is (consequently) true for the a allele, $p_a' = p_a$.

Of course the process of gene maintenance for bisexual diploid organisms is much more complicated than the simple random pairing of alleles selected from a common pool that we explored here (see Section 11.6). Nevertheless we will see in the next subsection that the results are the same if mating is random.

Random mating preserves allelic fractions.

Again consider a one-locus two-allele system and suppose mating is completely random. Then the probability of an AA×Aa mating, for example, is $2p_{AA}(2p_{Aa})$ since the first parent could be AA and the second Aa or the other way around. Altogether there are six different kinds of matings; their probabilities are listed in Table 11.5.3.

Table 11.5.3 Mendelian Inheritance Probabilities

Genotype Mating	Probability
AA×AA	$(p_{AA})^2$
AA×Aa	$2p_{AA}(2p_{Aa})$
AA×aa	$2p_{AA}p_{aa}$
Aa×Aa	$(2p_{Aa})^2$
Aa×aa	$2(2p_{Aa})p_{aa}$
aa×aa	$(p_{aa})^2$

Now apply the Mendelian inheritance laws to calculate the probability of the various possible zygotes, for example, an AA zygote. First, an AA results from an AA×AA parentage with probability 1. Next an AA results from an AA×Aa parentage with probability $1/2$ (see Figure 11.5.1), and finally an AA results from an Aa×Aa cross with probability $1/4$. Now, by the method of conditional probabilities as discussed at the begining of this section, we have

$$\Pr(\text{AA}) = p_{AA}^2 \cdot 1 + 2p_{AA}(2p_{Aa}) \cdot \frac{1}{2} + (2p_{Aa})^2 \cdot \frac{1}{4}$$
$$= p_{AA}^2 + 2p_{AA}p_{Aa} + p_{Aa}^2$$
$$= (p_{AA} + p_{Aa})^2 = p_A^2.$$

Similarly we leave it to the reader to show that

$$\Pr(\text{aa}) = (p_{aa} + p_{Aa})^2 = p_a^2$$

and

$$\Pr(Aa) = 2(p_{AA} + p_{Aa})(p_{Aa} + p_{aa}) = 2p_A p_a.$$

But this shows that the fractions of alleles A and a are again p_A and p_a respectively among the offspring just as among their parents, assuming the various genotypes are equally likely to survive. This is the same result we calculated in the last section. In other words, the effect of random genotype mating is indistinguishable from that of random gamete recombination. This is the Hardy–Weinberg Principle (see Reference [4]).

Hardy–Weinberg Principle.   Under the condition that mating is random and all genotypes are equally fit, the fractions of alleles will stay the same from generation to generation.
    A consequence of the Hardy-Weinberg Principle is that after at most one generation, the fractions of genotypes also stabilize and at the values

$$p'_{AA} = p_A^2$$
$$2p'_{Aa} = 2p_A p_a$$
$$p'_{aa} = p_a^2.$$

For example, suppose initially 70% of a population is AA and the remaining 30% is aa. Then the fractions of alleles in subsequent generations are also 70% and 30% for A and a respectively. Therefore, after one generation, the fractions of genotypes will be

$$\begin{aligned} \text{AA}: &\quad (.7)^2 = .49 \\ \text{Aa}: &\quad 2(.7)(.3) = .42 \\ \text{aa}: &\quad (.3)^2 = .09. \end{aligned}$$

In some cases the Hardy–Weinberg Principle is applicable even when mating is not random. Mating would fail to be random, for example, if the homozygote for a recessive gene is impaired or unviable. But, in fact, the homozygote in these cases are so rare that the induced error is very small. Keep in mind that, for a recessive gene a, the homozygote AA and heterozygote Aa are indistinguishable, so that random mating among them is a reasonable assumption.
    The Hardy–Weinberg Principle breaks down when there is migration, inbreeding, or non-random mating, that is, phenotypes are selected for some attribute.

Sex-linked loci give rise to different rates of expression between males and females.

In the event that males (or females) have one or more non-homologous chromosomes, the foregoing derivations must be modified. One consequence of non-homologous chromosomes is that there can be a large difference in expression of a sex-linked character between males and females. Suppose the male has the non-homologous pair XY while the female has the homologous pair XX.[8] Then fractions of alleles for genes on either the X or the Y chromosome are identical to genotype fractions for the male. For example, suppose a recessive sex-linked allele occurs with frequency p among a population. Then p is also the rate at which the allele will occur in males. However the rate at which the homozygous condition will occur in females is p^2.

An example of such an allele is color blindness in humans. Through various studies it is believed that the frequency of the recessive allele is 8% as derived from the incidence rate in males. Therefore the incidence rate in females ought to be $(.08)^2 = .0064$ or 0.6%. Actually, the female incidence of the disease is about .4%. The discrepancy is an interesting story in its own right and stems from the fact that there are four different kinds of color blindness, two of which are red blindness and green blindness. The bearer of defective genes for different types, such as these two, can still see normally.

Another possibility that can arise relative to sex-linked genes is that the allelic fractions are different between males and females. This can happen, for instance, when males and females of different geographical backgrounds are bought together. Let F be the fraction of allele A in the females and M be its fraction in males. Then $f = 1 - F$ is the fraction of a in females and $m = 1 - M$ is its fraction in males. Assuming an equal number of males and females, the population frequencies work out to be

$$p_A = \frac{M+F}{2} \qquad \text{and} \qquad p_a = \frac{m+f}{2} = 1 - p_A$$

and these will remain constant by the Hardy–Weinberg Principle. However, the values of M and F will change from generation to generation.

To follow these fractions through several generations we need only keep track of F and M since f and m can always be found from them. Let F_n and M_n refer to generation n with $n = 0$ corresponding to the initial fractions.

Since a male gets his X chromosome from his mother, the allelic frequencies in males will always be what it was in females a generation earlier, thus

$$M_{n+1} = F_n.$$

On the other hand the frequency in females will be the average of the two sexes in the preceeding generation since each sex contributes one X chromosome, hence

$$F_{n+1} = \frac{1}{2}M_n + \frac{1}{2}F_n.$$

---

[8]This is a mammalian property. In birds the situation is reversed.

In matrix form this can be written

$$\begin{bmatrix} M_{n+1} \\ F_{n+1} \end{bmatrix} = \begin{bmatrix} 0 & 1 \\ \frac{1}{2} & \frac{1}{2} \end{bmatrix} \begin{bmatrix} M_n \\ F_n \end{bmatrix}.$$

In the exercises we will investigate where this leads.

Before leaving this example, there is another observation to be made. We used the the matrix T above,

$$T = \begin{bmatrix} 0 & 1 \\ \frac{1}{2} & \frac{1}{2} \end{bmatrix},$$

in conjunction with multiplication on its right to update the column of male/female fractions M_n and F_n. But in this example there is a biological meaning to left multiplication on the matrix T. In each generation there will be a certain fraction of the alleles on the X chromosome in males which originally came from the females. It is possible to track that distribution.

To fix ideas, suppose a ship of males of European origin runs aground on a South Seas island of Polynesian females. Further suppose (hypothetically) that the alleles for a gene on the X chromosomes of the Europeans, the E-variant, are slightly different from those of the Polynesians, the P-variant, in, say, two base pairs. So the distribution of E-variant and P-variant chromosomal alleles of the emigrating males can be described by the (row) pair

$$(1 \quad 0)$$

where the first element is the fraction originating with the males and the second is the fraction originating with the females. The distribution of these fractions can be traced through the generations by a matrix calculation similar to that above, only this time using matrix multiplication on the left. In the first generation we have

$$(1 \quad 0) \begin{bmatrix} 0 & 1 \\ \frac{1}{2} & \frac{1}{2} \end{bmatrix} = (0 \quad 1),$$

showing that all the alleles in the males in this generation come from the females. In the second generation the fraction works out to

$$(0 \quad 1) \begin{bmatrix} 0 & 1 \\ \frac{1}{2} & \frac{1}{2} \end{bmatrix} = (\tfrac{1}{2} \quad \tfrac{1}{2}),$$

or 50–50. Of course the calculation can be continued to obtain the fractions for any generation.

The same calculation can give the female ratios, by starting with the initial female ratio of $(0 \quad 1)$.

Section 11.6

The Fixation of a Beneficial Mutation

A beneficial mutation does not necessarily become a permanent change in the gene pool of its host species. Its original host individual may die before leaving progeny for example. Under the assumption that such a mutation is dominant (rather than recessive) and that individuals with the mutation behave independently, it is possible to derive the governing equations for calculating the fixation probability. One way of measuring the value of a vital factor is the expected number of surviving offspring, beyond self-replacement, an adult will leave. For an r-strategist (see Section 4.1) the chance a beneficial mutation will become permanent is about twice the over-replacement value of the mutation to its holder.

Probability of fixation of a beneficial mutation is the complement of the fixed point of its probability generating function.

Let p_k be the probability that a chance mutation appearing in a zygote will subsequently be passed on to k of its offspring. A convenient method of organizing a sequence of probabilities, such as p_k, $k = 0, 1, \ldots$, is by means of the polynomial

$$f(x) = p_0 + p_1 x + p_2 x^2 + \ldots$$

in which the coefficient of x^k is the kth probability. This polynomial is called the *probability generating function* for the sequence p_k. The probability generating function is purely formal, that is, it implies nothing more than a bookkeeping device for keeping track of its coefficients. Note that $f(1) = 1$. And $f(0) = p_0$ is the probability the mutation disappears in one generation. Also note that the expected number of offspring to have the mutation is given (formally) by

$$\sum_{k=1}^{\infty} k p_k = f'(1)$$

(see Section 2.7). To say the mutation is beneficial is to say this expectation is greater than 1, that is,

$$\sum_{k=1}^{\infty} k p_k = 1 + a > 1,$$

for some value a, which is a measure of the benefit in terms of over replacement in fecundity.

    Now, if two such individuals with this mutation live and reproduce independently of each other (as in a large population), then the probability generating function for their combined offspring having the mutation is

$$p_0^2 + 2p_0 p_1 x + (2p_0 p_2 + p_1^2)x^2 + (2p_0 p_3 + 2p_1 p_2)x^3 + \ldots \qquad (11.6.1)$$

which is proved by considering each possibility in turn. There will be no mutant offspring only if both parents leave none; this happens with probability p_0^2 by independence. There will be one mutant offspring between the two parents if one leaves none and the other leaves exactly one; this can happen in two ways. There will be two mutant offspring if the first leaves none while the second leaves two, or they both leave one, or the first leaves two while the second leaves none; this is $(2p_0p_2 + p_1^2)$. Similarly the other terms of equation (11.6.1) may be checked.

But note that equation (11.6.1) is exactly the polynomial product $f^2(x)$,

$$(p_0 + p_1x + p_2x^2 + \ldots)(p_0 + p_1x + p_2x^2 + \ldots)$$
$$= p_0^2 + 2p_0p_1x + (2p_0p_2 + p_1^2)x^2 + \ldots.$$

More generally, m independent individuals with the mutation as zygotes will pass on the mutation to their combined offspring with probability generating function given by the mth power $f^m(x)$.

Now start again with one mutant zygote and consider the probability generating function, f_2, for Generation 2. Of course the outcome of Generation 2 depends on the outcome of Generation 1. If there are no mutants in Generation 1, and this occurs with probability p_0, then there are none for certain in Generation 2. Hence this possibility contributes

$$p_0 \cdot 1$$

to f_2. On the other hand, if the outcome of Generation 1 is 1, then the probability generating function for Generation 2 is $f(x)$; so this possibility contributes

$$p_1 f(x).$$

If the outcome of Generation 1 is two mutant individuals (and they behave independently), then the probability generating function for Generation 2 is, from above, $f^2(x)$; so this possibility contributes

$$p_2 f^2(x).$$

Continuing this line of reasoning yields the result that the probability generating function for Generation 2 is the composition of the function f with itself, $f_2(x) = f(f(x))$,

$$f_2(x) = p_0 + p_1 f(x) + p_2 f^2(x) + p_3 f^3(x) + \ldots = f(f(x)).$$

More generally, the probability generating function for generation n, $f_n(x)$, is given as the composition $f \circ f \circ \ldots \circ f$ of f with itself n times or

$$f_n(x) = \underbrace{f(f(\ldots f(x)\ldots))}_{n \text{ times}}.$$

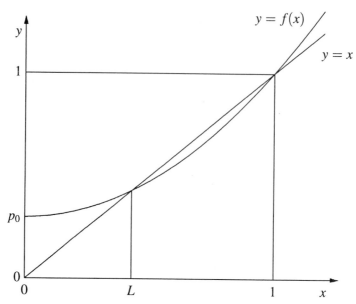

Figure 11.6.1 L is the fixed point of $f(x)$

Now the probability the mutation dies out by the nth generation is the constant term of $f_n(x)$ or $f_n(0)$. Hence the probability the mutation dies out or vanishes some time is the limit

$$V = \lim_{n \to \infty} f_n(0) = \lim_{n \to \infty} \underbrace{f(f(\dots f(0)\dots))}_{n \text{ times}}.$$

Applying f to both sides of this equality shows that V is a fixed point of f,

$$f(V) = V.$$

The fixed point of $f(x)$ is where the graphs $y = f(x)$ and $y = x$ intersect; see Figure 11.6.1. Since $f'(x)$ and $f''(x)$ are non-negative for $x > 0$ (having all positive or zero coefficients), and since $f(1) = 1$, we see there can be either 0 or 1 fixed point less that $x = 1$. If there is a fixed point less than 1, then V is that value; otherwise $V = 1$.

For example, suppose a mutation arose on the Y-chromosome of a human female about the time that Lucy walked the earth (3 million years ago). Further, suppose the following probabilities of producing surviving (female) offspring pertained to the holder of such a mutation,

probability of leaving no female offspring, $p_0 = .35$,
probability of leaving 1 female offspring, $p_1 = .25$,
probability of leaving 2 female offspring, $p_2 = .20$,
probability of leaving 3 female offspring, $p_3 = .1$,

probability of leaving 4 female offspring, $p_4 = .1$,
and 0 probability of leaving more than four female offspring.
Then the probability generating function is

$$f(x) = .35 + .25x + .2x^2 + .1x^3 + .1x^4.$$

Its fixed points can be found by solving the roots of the fourth degree polynomial

$$.1x^4 + .1x^3 + .2x^2 + (.25 - 1)x + .35 = 0.$$

With the following code, the appropriate root is found to be 0.62.

```
> f:=.1*x^4+.1*x^3+.2*x^2+ (.25-1)*x+.35;
> fsolve(f,x,0..1);
```

Hence the probability of fixation is the complementary probability 0.38.

*The chance a mutation will become permanent for an r-strategist is about twice
its over-replacement benefit.*

Under certain conditions, the probability that an individual will have k offspring
over its life is $b^k e^{-b}/k!$ for some constant b;[9] this is known as the Poisson distribution. The conditions are approximately satisfied by many r-strategists. In this
case, the probability generating function is

$$f(x) = e^{-b}\left(1 + \frac{b}{1!}x + \cdots\right) = e^{-b}e^{bx} = e^{b(x-1)}.$$

Let the benefit of the mutation be a, then

$$1 + a = f'(1) = be^{b(1-1)} = b,$$

so $b = 1 + a$. Now let F be the fixation probability of the beneficial mutation, that
is the probability the mutation will become permanent; then $F = 1 - V$. Since
$V = f(V)$ (from the previous section), we have

$$1 - F = e^{-(1+a)F}.$$

Taking logarithms,

$$(1 + a)F = -\ln(1 - F) = F + \frac{F^2}{2} + \frac{F^3}{3} + \cdots.$$

[9]The conditions are: (a) the probability of an offspring over a short period of time Δt is proportional to Δt, (b) the probability of 2 or more offspring over a short period of time is essentially 0,
and (c) offspring occur independently of one another. The distribution would also apply if offspring
occurred in batches; then k counts batches.

The infinite series is the Taylor series for the middle term. Divide by F and subtract 1 to get

$$a = \frac{F}{2} + \frac{F^2}{3} + \ldots.$$

If a is small, then approximately

$$a \approx \frac{F}{2},$$

so the fixation probability is about $2a$.

Exercises

1. In this problem assume a diploid organism having 3 loci per homologous chromosomal pair and two alleles per loci.
   a. If the organism has only one such chromosomal pair, how many different genotypes are possible?
   b. Same question if there are two chromosomal pairs?
   c. Suppose there are two chromosomal pairs with genes α, β, and γ on one of them while genes δ, ε, and ϕ lie on the other. How many different haploid forms are there?
   d. For a given genotype as in (c), how many different gametes are possible? That is, suppose a particular individual has the homologous chromosomes: (1) (A, b, C) and (A, B, C) and (2) (d, e, F) and (D, e, F). How many haploid forms are there?
   e. What is the maximum number of different offspring possible from a mating pair of organisms as in (d)? What is the minimum number? How could the minimum number be achieved?
   f. Work out a graph showing how the number of haploid forms varies with (i) number of chromosomal pairs, or (ii) number of genes per chromosomal pairs. Which effect leads to more possibilities?
2. For a given diploid two allele locus the initial fractions of genotypes are $AA{:}p$, $Aa{:}q$, and $aa{:}r$. Recall that the frequencies in the next generation will be $p_{AA} = x$, $p_{Aa} = y$, and $p_{aa} = z$ where

$$x = \left(p + \frac{1}{2}q\right)^2, \qquad y = 2\left(p + \frac{1}{2}q\right)\left(r + \frac{1}{2}q\right), \qquad z = \left(r + \frac{1}{2}q\right)^2.$$

Under the assumption that the various genotypes are neither selected for or against, show that these ratios will be maintained in all future generations, i.e., show that

$$x = (x + \frac{1}{2}y)^2$$

$$y = 2(x + \frac{1}{2}y)(z + \frac{1}{2}y)$$

$$z = (z + \frac{1}{2}y)^2.$$

Hence, when the Hardy–Weinberg Principle holds, genotype frequencies stablize in one generation.

```
> x:=(p+q/2)^2;
> y:=2*(p+1/2)*(r+q/2);
> z:=1-x-y;
> X:=(x+y/2)^2;
> simplify(X);
> simplify(x);
```

etc.

3. In this problem we want to see how many homozygous recessives for a trait result from homozygous parents and how many result from heterozygous parents.

parents/frequency

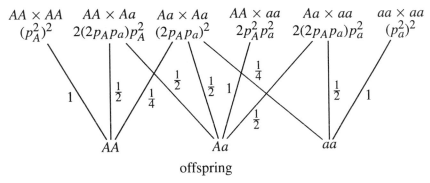

offspring

The question is, given an *aa* progeny, what is the probability the parents were *aa* × *aa*?

Since the progeny is known to be *aa*, the universe for this problem are the paths of the tree leading to *aa* and its frequency is given by

$$u = (2p_A p_a)^2 * \frac{1}{4} + 2(2p_A p_a)p_a^2 * \frac{1}{2} + (p_a^2)^2 * 1.$$

So the relative frequency this occurs via $aa \times aa$ parents is

$$\frac{(p_a^2)^2 \cdot 1}{u} .$$

a. Calculate the probable parentage of an aa progeny via $Aa \times Aa$ genotypes and $Aa \times aa$ genotypes.
b. Make three graphs of these probable parentages over the range of frequencies of allele a from .25 to .001, say.
c. If $p_a = 0.01$, then what is the chance that an aa individual had heterozygous parents? Same question for at least one heterozygous parent?

Hints for part (a)

```
> u:= (2*pA*pa)^2*(1/4)+2*(2*pA*pa)*pa^2*(1/2)+(pa^2)^2;
> pA:=1-pa;
> aaxaa:=simplify((pa^2)^2/u);
```

Hints for part (b)

```
> plot(aaxaa,pa=0.001..0.25);
```

Hints for part (c)

```
> pa:=0.01; eval(aaxaa);
```

4. This problem refers to the Sex-linked loci subsection of Section 11.5.
   a. Let the starting fraction of allele a in males be $M_0 = 0.1$ and in females be $F_0 = 0.3$. By performing the matrix calculation

$$\begin{bmatrix} M_{t+1} \\ F_{t+1} \end{bmatrix} = \begin{bmatrix} 0 & 1 \\ \frac{1}{2} & \frac{1}{2} \end{bmatrix} \begin{bmatrix} M_t \\ F_t \end{bmatrix}$$

   repeatedly, find the limiting fractions M_∞ and F_∞. What is the ratio M_∞/F_∞?
   b. Do the same for the starting ratios $M_0 = 1$ and $F_0 = 0$. What is the limiting ratio M_∞/F_∞?
   c. Let T be the matrix in part (a)

$$T = \begin{bmatrix} 0 & 1 \\ \frac{1}{2} & \frac{1}{2} \end{bmatrix} .$$

   Show that T satisfies

$$T \begin{pmatrix} 1 \\ 1 \end{pmatrix} = \begin{pmatrix} 1 \\ 1 \end{pmatrix} .$$

We say this column vector, with both components 1, is a right eigenvector for T with eigenvalue 1.

d. As in part (a), iterate the calculation

$$(0 \quad 1) = (1 \quad 0) \begin{bmatrix} 0 & 1 \\ \frac{1}{2} & \frac{1}{2} \end{bmatrix},$$

this time multiplying the matrix on the left by the vector, to obtain the limit. This will represent the ultimate distribution of the original male versus female alleles. Show that $(\frac{1}{3} \quad \frac{2}{3})$ is a left eigenvector for T. What is the eigenvalue?

```
> with(linalg):
> T:=matrix(2,2,[0,1,1/2,1/2]);
> v:= vector([0.1,0.3]);
> w:=evalm(T &* v);
> v:=evalm(T &* w);
```

etc. Do this a few times to see the trend.

```
> # do ten multiplies at once as follows
> doten:=proc(v)
> local i,x,y;
   y:=v;
   for i from 1 to 10 do
      x:= y;
      y:= evalm(T &* x);
   od
   RETURN( eval(y))
> end:
> v:= vector([0.1,0.3]);
> w:= doten(v);
> v:= doten(w);
```

etc. as desired.

```
> # the ratio
> v[1]/v[2];
> # show (1  1) and eigenvector with eigenvalue 1.
> v[1]:= 1;
> v[2]:= 1;
> evalm(T &* v);
> # also use built-in routine
> eigenvals(T);
> eigenvects(T);
```

In the reply, the first item is the eigenvalue (as above), the second is its multiplicity (how many times repeated, should be 1 here) and the third is the eigenvector. Eigenvectors may be multiplied by any constant so if (1   1) is an eigenvector, so is (3   3).

5.   Two hypothesis that explain the greater incidence of early baldness in males than in females are (1) an autosomal dominance that is normally expressed only in males and (2) an X-linked recessive. If the first is correct and Q is the frequency of the gene for baldness, what proportion of the sons of bald fathers are expected to be bald? What proportion are from nonbald fathers? What are the corresponding expectations for the X-linked recessive hypothesis.

Data gathered by Harris (*Ann. Eugen.* **13**, 172–181, 1946) found that 13.3% of males in the sample were prematurely bald. Of 100 bald men, 56 had bald fathers. Show that this is consistent with the sex-limited dominance hypothesis but not the sex-linked recessive. (Note that it is easier to get data about the fathers of bald sons than it is to wait for the sons of bald fathers to grow up to get data about bald sons.)

6.   Suppose an organism, which is capable of it, reproduces c fraction of the time asexually (by cloning) and $1 - c$ fraction of the time sexually with random mating. Let P_t be the fraction of the genotype AA in generation t and let p be the frequency of allele A. Assume c is independent of genotype and consequently p will remain constant from generation to generation. However the frequency of genotype AA can change. Using the Hardy-Weinberg Principle, show that the change in this fraction is given by

$$P_{t+1} = cP_t + (1 - c)p^2.$$

Find the limiting fraction P_∞.

```
> # First try
> F:= proc(x);
   RETURN( c*x+(1-c)*p^2);
> end;
> x:=0;
> y:=F(x);
> x:=F(y);
```

and so on to see the trend.

```
> # Now try a slight redefine of F.
> F:=proc(x)
> local y;
   y:= c*x+(1-c)*p^2;
   RETURN( simplify(y));
> end;
> # and repeat
```

7. Suppose the frequency of a recessive allele is p (equal to $1/1000$ say) therefore the frequency of homozygotes under the hypothesis of random mating
will be p^2. But what if mating is not random? In this problem we want to
investigate this somewhat.

First suppose the species is capable of self-fertilization. Then clearly the
offspring of a homozygous adult will again be homozygous. On the other
hand, the heterozygous Aa will produce A and a haploid cells in 50–50 mix as
before. Hence, as before an offspring will be AA with 1/4 chance, aa with 1/4
chance and Aa with 1/2 chance. We record these observations in the following
3×3 matrix

$$T = \begin{bmatrix} 1 & 0 & 0 \\ \frac{1}{4} & \frac{1}{2} & \frac{1}{4} \\ 0 & 0 & 1 \end{bmatrix}.$$

In this, the rows correspond to the genotypes AA, Aa, and aa in that order and
so do the columns.

Next suppose we start out with a mix of genotypes, say their fractions are
p, q, and r respectively, $p + q + r = 1$. Then after one generation the new
fractions p', q', and r' will be given by the matrix product

$$(p' \quad q' \quad r') = (p \quad q \quad r) T.$$

a. Using specific values for the starting fractions, find the limiting fractions
after many generations.

Next consider parent/child matings and calculate the probability that a
homozygous recessive aa will be the result. First condition on the parent
(free-hand a tree diagram), from the root node, there will be three edges
corresponding to the possibilities that the parent is AA, Aa, or aa. The
AA branch cannot lead to a aa grandoffspring so no need to follow that
edge further. The Aa parent occurs with frequency $2p(1-p)$ as we have
seen, and the aa parent with frequency p^2.

Next condition on the genotype of the child. Use the Hardy–Weinberg
Principle for probabilities of allele A and a. Starting from the Aa node,
the possibilities are AA with probability $\frac{1}{2}(1-p)$, Aa with probability
$\frac{1}{2}p + \frac{1}{2}(1-p) = \frac{1}{2}$, and finally aa with probability $\frac{1}{2}p$. You do the
possibilities from the aa node.

Now assign the offspring probabilities using Mendelian genetics.
From the Aa node along the path from root through the Aa parent the
probability of an aa offspring is $\frac{1}{4}$. From the aa node through the Aa
parent the probability is $\frac{1}{2}$ and so on.

b. Altogether the result should be

$$P(aa \text{ offspring}) = \frac{1}{2}p(\frac{3}{4} + \frac{3}{2}p - p^2).$$

Finally consider sibling matings. As in part (a) above we want to investigate the trend of the population toward homozygosity. Starting with the parents, there are 6 possible matings by genotype, $AA \times AA$, $AA \times Aa$, and so on through $aa \times aa$. Consider the $AA \times Aa$ parents. Their offspring are AA and Aa both with frequency $\frac{1}{2}$. Therefore the sibling mating possibilities are $AA \times AA$ with frequency $\frac{1}{4}$, $AA \times Aa$ with frequency $\frac{1}{2}$, and $Aa \times Aa$ with frequency $\frac{1}{4}$.

Justify the rest of the following table.

	Sibling Mating Frequencies					
Parent Genotypes	AA×AA	AA×Aa	Aa×Aa	AA×aa	Aa×aa	aa×aa
AA×AA	1	0	0	0	0	0
AA×Aa	1/4	1/2	0	1/4	0	0
Aa×Aa	1/16	1/4	1/8	1/4	1/4	1/16
AA×aa	0	0	1	0	0	0
Aa×aa	0	0	0	1/4	1/2	1/4
aa×aa	0	0	0	0	0	1

The corresponding transition matrix T is

$$T = \begin{bmatrix} 1 & 0 & 0 & 0 & 0 & 0 \\ 1/4 & 1/2 & 0 & 1/4 & 0 & 0 \\ 1/16 & 1/4 & 1/8 & 1/4 & 1/4 & 1/16 \\ 0 & 0 & 1 & 0 & 0 & 0 \\ 0 & 0 & 0 & 1/4 & 1/2 & 1/4 \\ 0 & 0 & 0 & 0 & 0 & 1 \end{bmatrix}.$$

c. Make up an initial distribution of genotypes $(p \quad q \quad r \quad s \quad t \quad u)$, track the change in distribution over a few generations, and find the limiting distribution.

Section 11.7

Questions for Thought and Discussion

1. Discuss the concept of fitness as it is used in the Darwinian model. What kinds of selection factors might be involved in the case of humans?
2. A woman with type A blood has a child with type O blood. The woman alleges that a certain man with type B blood is the father. Discuss her allegation and reach a conclusion, if possible.
3. In Drosophila, females are XX and males are XY. On the X chromosome there is an eye-color gene such that red is dominant to eosin and to white, and eosin is dominant to white. What is the result of crossing an eosin-eyed male with a red-eyed female whose mother had white eyes?

4. Mitosis is a conservative form of cell replication, because each daughter cell gets an exact copy of the genetic material that the parent cell had. How can we explain the fact that most of our tissues were formed by mitosis and yet are different?

5. Suppose there is an organism that reproduces only by self-fertilization, which is the highest degree of inbreeding. Start with a heterozygote for a single property and let it and its descendants reproduce by self-fertilization for three generations. Note how the fraction of homozygotes increases with each generation. What implication does this have if the recessive allele is harmful? Or, suppose it is not harmful?

6. Combining the concepts of the Central Dogma of Genetics with that of meiosis, trace the path of hereditary control of cellular chemisty from one generation to another.

7. In a hypothetical laboratory animal, solid color allele is dominant to striped, and long hair is dominant to short hair. What is the maximum number of phenotypes that could result from the mating of a long, solid animal with a short, striped animal?

References and Suggested Further Reading

1. **Cell division and reproduction:** William S. Beck, Karel F. Liem and George Gaylord Simpson, *Life – An Introduction To Biology*, 3rd ed., Harper-Collins Publishers, New York, 1991.

2. **Genetics:** David T. Suzuki, Anthony J. F. Griffiths, Jeffrey H. Miller and Richard C. Lewontin, *An Introduction to Genetic Analysis*, W.H. Freeman and Co. New York, 3rd ed., 1986.

3. **Sexual selection in humans:** David M. Buss, "The Strategies of Human Mating," *American Scientist*, vol. 82, May–June, pp. 238–249, 1994.

4. **Mathematical genetics:** J. F. Crow, M. Kimura, "An introduction to population genetics theory," Harper & Row, New York, 1970.

5. **Mathematical genetics:** J. B. S. Haldane, "A Mathematical Theory of Natural and Artificial Selection, Part V, Selection and Mutation," *Camb. Philos. Soc. Proc.* **23**, Pt VII, 838–844, 1927.

Index